Composites Science and Technology

Series Editor

Mohammad Jawaid, Lab of Biocomposite Technology, Universiti Putra Malaysia, INTROP, Serdang, Malaysia

Composites Science and Technology (CST) book series publishes the latest developments in the field of composite science and technology. It aims to publish cutting edge research monographs (both edited and authored volumes) comprehensively covering topics shown below:

- Composites from agricultural biomass/natural fibres include conventional composites-Plywood/MDF/Fiberboard
- Fabrication of Composites/conventional composites from biomass and natural fibers
- Utilization of biomass in polymer composites
- Wood, and Wood based materials
- Chemistry and biology of Composites and Biocomposites
- Modelling of damage of Composites and Biocomposites
- Failure Analysis of Composites and Biocomposites
- Structural Health Monitoring of Composites and Biocomposites
- Durability of Composites and Biocomposites
- Biodegradability of Composites and Biocomposites
- Thermal properties of Composites and Biocomposites
- Flammability of Composites and Biocomposites
- Tribology of Composites and Biocomposites
- Bionanocomposites and Nanocomposites
- Applications of Composites, and Biocomposites

To submit a proposal for a research monograph or have further inquries, please contact springer editor, Ramesh Premnath (ramesh.premnath@springer.com).

More information about this series at http://www.springer.com/series/16333

Mohammad Jawaid · Tanveer Ahmed Khan ·
Mohammed Nasir · Mohammad Asim
Editors

Eco-Friendly Adhesives for Wood and Natural Fiber Composites

Characterization, Fabrication and Applications

Springer

Editors
Mohammad Jawaid
Laboratory of Biocomposite Technology
Institute of Tropical Forestry and Forest Products (INTROP)
Universiti Putra Malaysia
Serdang, Malaysia

Mohammed Nasir
Collage of Forestry
Banda University of Agriculture and Tech
Banda, Uttar Pradesh, India

Tanveer Ahmed Khan
Laboratory of Adhesion and Bio-Composites
Seoul National University
Seoul, Korea (Republic of)

Mohammad Asim
Laboratory of Biocomposite Technology
Institute of Tropical Forestry and Forest Products (INTROP)
Universiti Putra Malaysia
Serdang, Selangor, Malaysia

Composites Science and Technology
ISBN 978-981-33-4748-9 ISBN 978-981-33-4749-6 (eBook)
https://doi.org/10.1007/978-981-33-4749-6

This Springer imprint is published by the registered company Springer Nature Singapore Pte Ltd.
The registered company address is: 152 Beach Road, #21-01/04 Gateway East, Singapore 189721, Singapore

Preface

Bioadhesives are the substitute of petroleum-based adhesives that have received great attention from environmental concerns, academic researchers and industries. They are available in both forms: thermosets and thermoplastics. Several researchers have explored the easy way to synthesize more bioadhesives having acceptable properties for industrial applications. Bioadhesives have some peculiar advantages such as biodegradability, biocompostibility, involvement of non-hazardous chemicals, good mechanical properties, handling, toughness, excellent flame retardance, good heat resistance and good interfacial bonding. Several bioadhesive-based products have been produced and are being used daily usage.

This book will explore the potentiality of the chemical structure, synthesis process and derivatives of thermoset and thermoplastic bioadhesives. Several synthesis processes of bioadhesives from biodegradable sources have been studied, and their mechanical and thermal properties, flammability and rheological properties are tested. The obtained data helped in comparative studies with synthetic polymers. Bioadhesives are very much equal to synthetic polymers in their properties but in some requirements, they performed better, e.g. Biodegradability. Due to the versatile properties, they can be used for both outdoor and indoor applications. With this approach, Bioadhesives can easily enter eco-friendly market segments and can be a promising material for the automotive, marine, aerospace, construction and building, wind energy and consumer goods, etc.

This book has clearly shown the syntheses, derivations and utilization of bioadhesive which helps to understand this polymer and its application in the real polymer world as per the need of markets. This book also coveres challenges, advantages and future perspectives of bioadhesives as compared to the polymers/composites industries, waste material-based bioadhesives, thermoplastic- and thermoset-based bioadhesives, physical, mechanical, thermal and fire-retardant properties of bioadhesives, natural fiber-based biocomposites, wood-based biocomposites and nanocellulose biocomposites. This versatile version of bioadhesives and their composites help to develop lightweight, durable and biodegradable components which can be used for heavy duty.

We are highly thankful to all the authors who contributed to and provided their valuable ideas and knowledge in this edited book. We attempted to gather all the scattered information of authors from diverse fields around the world (Malaysia, India, Libya, Canada, Thailand, Italy, China, Algeria, France, Iran, etc.) in the areas of polymers and biocomposites and finally completed this venture in a fruitful way. We greatly appreciate the contributor's commitment for their support to compile our ideas in reality. We are highly thankful to Springer Nature, Singapore team, for their generous cooperation at every stage of the book production.

Serdang, Malaysia — Mohammad Jawaid
Pune, India — Tanveer Ahmed Khan
Banda, India — Mohammed Nasir
Serdang, Malaysia — Mohammad Asim

Contents

About the Editors

Dr. Mohammad Jawaid is currently working as Senior Fellow (Professor) at Biocomposite Technology Laboratory, Institute of Tropical Forestry and Forest Products (INTROP), Universiti Putra Malaysia (UPM), Serdang, Selangor, Malaysia, and also has been Visiting Professor at the Department of Chemical Engineering, College of Engineering, King Saud University, Riyadh, Saudi Arabia since June 2013. He has more than 16 years of experience in teaching, research, and industries. His area of research interests includes hybrid composites, lignocellulosic reinforced/filled polymer composites, advance materials: graphene/nanoclay/fire retardant, modification and treatment of lignocellulosic fibers and solid wood, biopolymers and biopolymers for packaging applications, nanocomposites and nanocellulose fibers, and polymer blends. So far, he has published 38 books, 65 book chapters, more than 350 peer-reviewed international journal papers, and several published review papers under top 25 hot articles in science direct during 2013–2018. He also obtained 2 Patents and 5 Copyrights. H-index and citation in Scopus are 53 and 12632 and in Google scholar, H-index and citation are 61 and 17568; H-index = 51 (Web of Science). He is founding Series Editor of Composite Science and Technology Book Series from Springer-Nature, and also Series Editor of Springer Proceedings in Materials, Springer-Nature. Beside that he is also a member of International Advisory Board of Springer Series on Polymer and Composite Materials. He worked as guest editor of special issues of SN Applied Science, Current Organic Synthesis and Current Analytical Chemistry, International Journal of Polymer Science, IOP Conference Proceeding. He is also an Editorial Board Member of Journal of Polymers and The Environment, Journal of Plastics Technology, Applied Science and Engineering Progress Journal, Journal of Asian Science, Technology and Innovation and the Recent Innovations in Chemical Engineering. Recently he get Fellow and Charted Scientist from Institute of Materials, Minerals and Mining (IOM), UK. He is a life member of Asian Polymer Association, and Malaysian Society for Engineering and Technology. He is also a member of Society of Plastics Engineers, American Chemical Society, Athens Institute for Education and Research, and International Association of Advanced Materials. Besides that, he is also a reviewer of several high-impact international peer-reviewed journals of

Elsevier, Springer, Wiley, Saga, ACS, RSC, Frontiers, etc. Presently, he is supervising 12 Ph.D. students (4 Ph.D. as Chairman, and 8 Ph.D. as Member) and 7 Master's students (1 Master as Chairman, and 6 Master as Member) in the fields of hybrid composites, green composites, nanocomposites, natural fiber-reinforced composites, nanocellulose, etc. 24 Ph.D. and 12 Master's students graduated under his supervision in 2014–2020. He has several research grants at university, national, and international levels on polymer composites of around 3 million Malaysian ringgits (USD 700,000). He also delivered plenary and invited talks in international conferences related to composites in India, Turkey, Malaysia, Thailand, the United Kingdom, France, Saudi Arabia, Egypt, and China. Besides that, he is also a member of technical committees of several national and international conferences on composites and material science. Recently Dr. Mohammad Jawaid received Excellent Academic Award in Category of International Grant-Universiti Putra Malaysia-2018 and also Excellent Academic Staff Award in industry High Impact Network (ICAN 2019) Award. Beside that Gold Medal-Community and Industry Network (JINM Showcase) at Universiti Putra Malaysia. He also Received Publons Peer Review Awards 2017, and 2018 (Materials Science), Certified Sentinel of science Award Receipient-2016 (Materials Science) and 2019 (Materials Science and Cross field). He is also Winner of Newton-Ungku Omar Coordination Fund: UK-Malaysia Research and Innovation Bridges Competition 2015.

Tanveer Ahmed Khan is a Planner, implementer & achiever with multi-cultural experience of over 8 years in spearheading the Development functions across Biomass, Bio-Adhesive, and Bio-Composites. He is currently an Innovation Center Manager at Modern Mill Inc. & Gigantic Bag Co. llc. McComb, Mississippi, USA. Prior to this, He was working as a post-doctoral researcher in the Laboratory of Adhesion & Bio-Composites, Seoul National University. He has completed his Ph.D. in October' 2015 in the Faculty of Chemical and Natural Resources Engineering at University Malaysia Pahang. His project "Woody Biomass Advanced Materialization Technology" bagged Brilliant Korea (BK21) Fellowship for his postdoc. He secured many national and international awards at various international conferences and scientific exhibitions. He has over 8 years of experience in the areas of Research and Development, Teaching & Training, Project Management, Laboratory Operations, besides others across Synthesis of carbon Fibers, carbon nanomaterials, Bio-Adhesives, Wood Composites, Wood Plastic Composites, Natural Fiber composites, Biomass, Nano-Cellulose, Lignin Chemistry, Polymer Synthetic Chemistry, Polymer Nano-composites, Advance Materials, Superabsorbent composites, Polymer Materials, and Engineering. He has published 18 papers in high-end journals, 2 book chapters, 14 conference proceedings, and several other publications are currently under review. He has successfully formulated nano-based polymer composites for 3D printing. He has considerable experience in securing external funding and managing large-scale research projects. He has been a key researcher in various grants related to the synthesis of nano/micro-structured carbon materials, Woody Biomass Advanced Materialization Technology, Development of Oil Palm Empty Fruit Bunch/Sugarcane Bagasse Fiber-Based Hybrid Composites Insulation Board for Construction Applications, and Development of UV Curable Polymer Nano Composites Based on Inorganic Nano Fillers for 3D Printing. He received a Gold medal and Out Standing Achievement Award for his Ph.D. project at Seoul International Youth Invention Contest (IYIC), South Korea & at the International Innovation Festival (INNOFEST), Malaysia in 2014.

Dr. Mohammed Nasir is currently working as assistant professor at forest Products Utilization, College of Forestry, Banda University of Agriculture and technology, Banda (UP) India. He was born on 17th June 1982, in Mau Uttar Pradesh, India. He did his B.Sc. Degree in Forestry (Hons.) from Chandra Shekhar Azad University of Agriculture & Technology Kanpur, M.Sc. in Wood Science and Technology from Forest Research Institute Dehradun and Ph.D. in Chemical Engineering (wood composites) from Universiti Malaysia Pahang, Malaysia. During his Ph.D. he worked on "Bio-composite fabrication from enzyme treated fiber from lignin based bio adhesive". He extended his research work further as postdoc fellow at School of Industrial Technology, Universiti Sains Malaysia for two years, where he focused on developing an "Eco-friendly method of nano-cellulose synthesis from oil palm trunk waste". He has also worked as a National postdoc fellow at Forest Research Institute Dehradun for one year before joing the current service. His focused research interests are: 1) Lignin based Bio adhesive preparation, 2) Researching the scope of wood alternative, 3) Bamboo utilization in composite products manufacturing. He has published 39 research paper and 3 book chapters in various national and international journals. Furthermore, his two Malaysia patents for adopting a new method and preparing a composite from lignin based bio-adhesive are under consideration.

Dr. Mohammad Asim is currently a post-doctoral fellow in the Laboratory of Biocomposite Technology at the Institute of Tropical Forestry and Forest Products (INTROP), Universiti Putra Malaysia. He was born on the 2nd August 1988 in Mau, Uttar Pradesh, India. He completed his bachelor's degree in forestry from C. S. Azad University of agriculture and technology, Kanpur, India in 2011. Afterward, he continued his study and obtained his MSc in wood science and technology from Forest Research institute, Dehradun, India in 2013 and finally Ph. D degree in the field of biocomposite technology from Universiti Putra Malaysia in 2017. His main research areas are: Treatment and modification of natural fibres, Hybrid Reinforced/Filled Polymer Composites, Advance Materials: Nanoclay/Fire Retardant, Lignocellulosic Reinforced/Filled Polymer Composites, Nano Composites and Nanocellulose fibres, thermosets and thermoplastics. Dr. Asim has published more than 35 international journal papers, 5 review papers, 6 conference proceedings, 7 book chapters and edited 2 book in springer Nature. Dr. Asim also have copyrights the design of Meat Tray Prototype by Using 3D Printing and Vacuum Bagging Process for Natural Fibre Based Composite, Dr. Asim is also a regular reviewer of different international journals published by Elsevier, Wiley, Springer, etc. H-index and citation in Scopus: 13 and 650 respectively, in Google scholar: H-index and citation are 15 and 900 respectively.

Introduction of Eco-Friendly Adhesives: Source, Types, Chemistry and Characterization

Anisah Sajidah Saud, Gaanty Pragas Maniam, and Mohd Hasbi Ab. Rahim

Abstract Natural and wood-based fiber composites and adhesive industry have shown remarkable progress during the last decades. Adhesives play a central role in wood-based panel production. Knowledge about wood bonding is quite limited although it is one of the ancient techniques. Wood structure having so many variability including; cell structures, complex morphology at all spatial scales that make it hard to model the process. The main driving forces today are to develop cheaper but more complex structured adhesives. Despite these problems, many adhesives have been commercialized that are stronger, more durable than the wood and allowing the wood pieces to be glued together in economical manner. Adhesives play a central role in wood-based panel production. The quality of bonding, and hence the properties of the wood-based panels, are determined mainly by the type and quality of the adhesive. The development of the physical properties during the setting process and the interaction of the adhesive with the wood need to be further investigated to develop cost effective new adhesives. Development in wood-based panels, therefore, is always linked to development in adhesives and resins.

Keywords Eco-friendly adhesives · Bio-based adhesives · Synthetic adhesives · Natural adhesives · Wood composite adhesives

---

A. S. Saud · G. P. Maniam · M. H. Ab. Rahim (✉)
Faculty of Industrial Sciences and Technology, Universiti Malaysia Pahang, Lebuhraya Tun Razak, Gambang, 26300 Kuantan, Pahang, Malaysia
e-mail: mohdhasbi@ump.edu.my

G. P. Maniam · M. H. Ab. Rahim
Earth Resources & Sustainability Centre, Universiti Malaysia Pahang, Lebuhraya Tun Razak, Gambang, 26300 Kuantan, Pahang, Malaysia

G. P. Maniam
Central Laboratory, Universiti Malaysia Pahang, Lebuhraya Tun Razak, Gambang, 26300 Kuantan, Pahang, Malaysia

M. Jawaid et al. (eds.), *Eco-Friendly Adhesives for Wood and Natural Fiber Composites*, Composites Science and Technology,
https://doi.org/10.1007/978-981-33-4749-6_1

1 Introduction

Manufactured wood composite panels commercially used to produce furniture such as particleboard, plywood, fibreboard and other composites (Li et al. 2018; Valarmathi et al. 2020). The process typically involved adhesives with the aim to create require bonding between the wooden elements or different materials. For centuries, they were bonded using various fossil-derived polymers such as formaldehyde, isocyanate, vinyl acetate-based adhesives in which, many wood-based materials are solely prepared from formaldehyde-based adhesives (DeVito 2016; Ling et al. 2018; Catto et al. 2017). These are likely to be cost-effective and a benign to bonding performance, mechanical properties, thermal stability and water resistance (Jin et al. 2010; Zhang et al. 2013; Jivkov et al. 2013a, b; Yang et al. 2015). Adhesives comprise formaldehyde have already been used industrially for many large-scale applications due to their very highly reactive compound, making it well suited for use as a binder.

However, the US Environmental Protection Agency (EPA) has identified formaldehyde as a "probable human carcinogen" (Marshall 1987) and in fact numerous committee's review addressing all aspects of formaldehyde toxicity and potential for adverse health effects profile in the regulatory and other authoritative bodies, including the National Toxicology Program (NTP), US Department of Health and Human Services (Panel 2009) and the World Health Organization (WHO) (Chemicals and Organization 2010). There is sufficient evidence that long-term formaldehyde exposure can cause cancer of the nasopharynx, leukaemia and sinonasal cancer (Cancer 2006). Thus, to reduce the free formaldehyde emissions from wood composites, especially in indoor-related applications, along with the more environmentally conscious, focus on the sustainability of raw materials and final products, as well as more stringent environmental legislation, are the main driving factors for shifting the scientific and industrial interest from the traditional formaldehyde-based synthetic adhesives to the new bio-based adhesives for production of eco-friendly wood composites and related materials (Dunky 2004; Frihart 2005; Kües 2007; Pizzi 2006; Navarrete et al. 2013; Valyova et al. 2017).

Increasing environmental reports and legislation about emission of formaldehyde putting an adverse effect on production of these adhesives. Moreover, it is a move forward to reduce dependency on fossil-derived resources that has spurred the wood industry to develop the renewable eco-friendly adhesives. Nowadays there are only a few bio-based adhesives for production of wood composites and they are highly likely that these adhesives are not economically feasible through both niche and mainstream production of wood composites. For these adhesives and for the ones still in development, due to their low cost, a synthetic cross-linker are still more commonly required (Hemmilä et al. 2017). Thus, on-going research in eco-friendly and renewable based adhesives could partially-substitute and supplement synthetic adhesives subsequently that make the wood-based adhesives are more sustainable, green, and biodegradable (Pizzi 1991; Pang et al. 2020).

Tannin and lignin polymeric-based adhesives are found to be among eco-friendly types of adhesives (Pizzi 1980; Nimz 1983). Both adhesives are extensively studied due to their applications as wood panel adhesives (Trosa and Pizzi 2001; Pizzi 2006; Ballerini et al. 2005; Pichelin et al. 2006; Pizzi and Salvadó 2007; Mansouri et al. 2011). In addition to polymeric-based adhesives, several types of adhesives such as plant protein adhesives (Khosravi et al. 2010; Xu et al. 2011; Jang et al. 2011), crop-based, carbohydrate adhesives (Tondi et al. 2012; Sulaiman et al. 2013), wheat gluten (Khosravi et al. 2011) and animal protein glue (Konnerth et al. 2009) have also been developed. The adhesive properties provided by these materials are comparable to conventional formaldehyde-based adhesive. As they are derived from renewable raw materials, they can have favourable sustainability characteristics, like reduced the rate of accumulation of CO_2 in the atmosphere and reduced or no toxicity. Table 1 shows the lignocellulosic materials and adhesives reported in prior-art.

1.1 Source, Types, Chemistry, and Characterization of Wood Composites Adhesives

The utilization of renewable raw materials for the production of wood adhesives have taken much attention since the world's oil crisis of the 1970s (Pizzi 2006). Bio-based adhesives are indeed derived from natural materials requiring or using novel technologies, formulations, and methods. Examples for adhesives include starch, tannin, lignin, proteins, carbohydrates, and oil (Ferdosian et al. 2017b; Pizzi 2006). They were found to be less petroleum dependent, sustainable sourcing and meet the requirements for indoor applications compared to formaldehyde adhesives (He and Wan 2017).

Economical and fast production of adhesives to fulfil industrial standards and requirement initiate the early development of wood composite industry. For example, early plywood were made by conventional hand method, with very slow adhesive setting speed (Muller 1992). To increase the rate of production, protein adhesives manufacture could be used. The curing time, final use of panel and the speed at which the panel can made has become a generic point for manufacturing of medium density fiberboard. As an example, in 1930s, for the development of wood-based panel (particle board and medium density fibreboard, MDF), urea–formaldehyde, was found to be a cheap commodity adhesive. However, following advanced performance criteria, melamine substituted the urea–formaldehyde due to its better moisture resistance for manufacturing of more moisture resistant boards.

Moreover, in nineteenth century, phenol formaldehyde (PF) adhesive was originally developed while their commercialization begins in 1930s for the manufacturing of plywood and engineered wood products. PF adhesives got more attention due to high moisture resistance in comparison to urea–formaldehyde adhesives. In recent years, due to be overwhelmed environmental concerns changed the driving force for adhesive development. For example, in the 1980s, the European Union started to strict the limits on formaldehyde emission in panel products. In 1985,

Table 1 Reported lignocellulosic materials and adhesives

Lignocellulosic materials	Adhesives	References
Pine, Spruce	Wheat gluten	(Khosravi et al. 2010)
Industrial wood	Crop-based starch and sugar	(Tondi et al. 2012)
Wood specimen	Polyvinyl acetate/ cellulose nanocrystals	(Mabrouk et al. 2020)
Poplar (*Populus deltoides*)	Oxidized demethylated corn lignin	(Chen et al. 2020)
Maple wood veneer	Cottonseed protein-based	(Cheng et al. 2019)
Poplar veneers, Pine wood pencil sandwich slats	Cottonseed meal-based	(He et al. 2019)
Sweet sorghum bagasse	Citric acid	(Kusumah et al. 2017)
Bagasse, wood	Urea–formaldehyde	(Hazrati-Behnagh et al. 2015)
Recycled softwood particles	Citric acid and sucrose	(Umemura et al. 2014)
Oil Palm trunk	Binderless	(Baskaran et al. 2015)
Rubberwood	Urea–formaldehyde	(Hua et al. 2015)
Apple tree pruning, sunflower stalk	Urea–formaldehyde	(Ghofrani et al. 2015)
Rhizophora spp.	Gum Arabic	(Abuarra et al. 2014)
Date Palm	Urea–formaldehyde and Phenol formaldehyde adhesive	(Amirou et al. 2013)
Macadamia nut shells	Castor oil derived adhesives	(Wechsler et al. 2013)
Bagasse	Binderless	(Nonaka et al. 2013)
Tobacco stalks, *Paraserianthes falcataria*	Urea–formaldehyde	(Acda and Cabangon 2013)
Willow, Pine	Urea–formaldehyde	(Warmbier et al. 2013)
Pine	Pine tannin	(Valenzuela et al. 2012)
Reed, wood	Urea–formaldehyde	(Ghalehno et al. 2011)
Sugar cane bagasse, Pine	Phenol–formaldehyde	(Hein et al. 2011)
Rice husk	Soybean protein concentrate-based adhesive	(Ciannamea et al. 2010)
White birch	Urea–formaldehyde adhesive	(Pedieu et al. 2009)
Eucalyptus	Urea–formaldehyde adhesive	(Colak et al. 2009)

Austria, Denmark, Germany and Sweden started to regulate formaldehyde emissions E1 (0.1 ppm boards) became obligatory. Likewise, in 2002, European standard for wood-based panels (EN 13,986) became obligatory across the Europe. The EN 13,986 was classified into two classes E1 and E2. The E1 referred to panels with no formaldehyde additions while E2 referred to rest of formaldehyde adhesives-based panels. The existing E1 standard was further modified in 2006, by the European panel federation (EPF7) and 'Blue Angel' (an additional environmental label) was

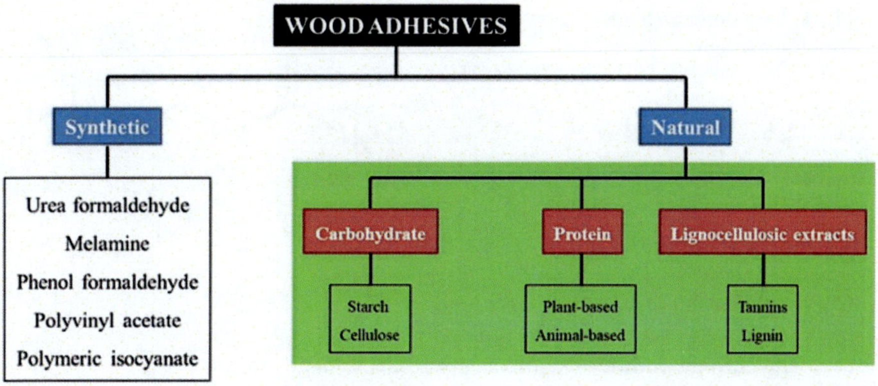

Fig. 1 Classification of wood adhesives

specified a steady-state concentration less than 0.05 ppm formaldehyde. Other testing method such as Japanese- F limits and the legislation of the state of California in the USA are among the important methods and limit around the globe.

However, the only possible move away from commodity adhesives is to manufacture and commercialization of a niche premium product. In contrast to above scenario, any product and process adaptation to new adhesive system is forbidden and insignificant in manufacturing of panel prodcuts.

Figure 1 shows the classification of wood adhesives.

1.2 Synthetic Adhesives

The urea–formaldehyde (UF), melamine urea formaldehyde (MF), and phenolic (PF) adhesives are common synthetic adhesives that have been using for current composite technologies. The other currently used synthetic adhesives are including poly vinyl acetate and isocyanate adhesives. Meanwhile, the future natural adhesives for composites applications are carbohydrate, protein, and lignocellulosic extracts adhesives.

1.2.1 Urea–formaldehyde Adhesives

In early nineteenth century, urea–formaldehyde (UF) adhesive was developed (Dinwoodie 1979) and widely used in the composites industry (Fig. 2). Around 90% of the particleboard is produced using UF adhesive (Dinwoodie 1979; Pizzi 1994a; Pizzi 1994b; Xian-qing et al. 2020; Khan et al. 2020). The main features of UF adhesives were summarised as follows:

Fig. 2 Urea–formaldehyde chemical structure

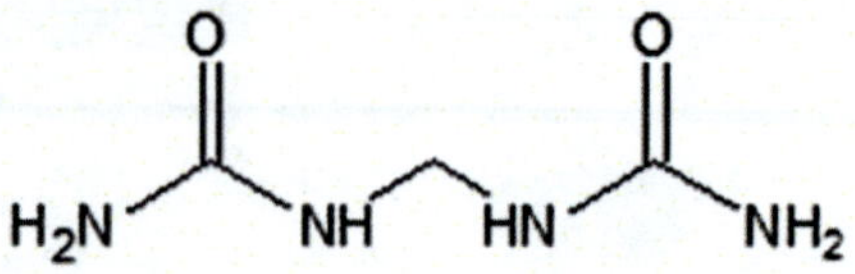

(a) Adhesive hardness.
(b) Low flammability.
(c) High thermal properties.
(d) The colour absence in the crude polymer.
(e) The curing conditions with variety adjustability.

Urea–formaldehyde adhesives are considered to be suitable for largeness and low-priced production due to the initial water solubility. However, the presence of moisture or acid opens up the hydrolytic degradation of adhesives. This is caused by the hydrolysis process towards amino plastic and the methylene bridges.

The Production of Urea–formaldehyde (UF) Adhesive

The most common method for UF adhesive formation involved the addition of urea to formaldehyde while keeping the ratio of both reactants in between 1:2 and 1:2.2 along with methylation at temperatures between 90 and 95 °C under reflux. The completion of adhesive formation is indicated when the exotherm has subsided followed by the addition of acid to maintain the lower level of 5.0–5.3 which let the polymer to build.

To avoid the increase in polymer size, pH of the mixture is increased soon after the correct viscosity is attained. Another urea is added to complete the repeating chain of free formaldehyde until a ratio range from 1:1.1 to 1:1.7.

Finally, different structured polymers such as linear, branched and 3D matrix can be found in cured adhesive due to different functionalities of urea and formaldehyde. Typically, formaldehyde has a functionality of two while urea has four due to two and four replaceable hydrogen respectively (Fig. 3).

The size and viscosity of the adhesive molecule has crucial importance towards the commercial production of UF adhesives. The size of the molecules increases during the change in viscosity, and the change of the properties of the adhesive can be observed (Goulding et al. 1994). The condensation process propagated the random splitting off the adhesive molecules due to water causing an increase in molecular weight. However, aqueous conditions are suitable for the condensation reaction. The quality of the adhesive based on the viscosity, pH, concentration and solubility of the adhesive while preparation.

Therefore, it can be summarized that there are three main factors that affect the properties of the finished urea–formaldehyde product;

(a) The relative molarities of the reactants.
(b) The reaction temperature.

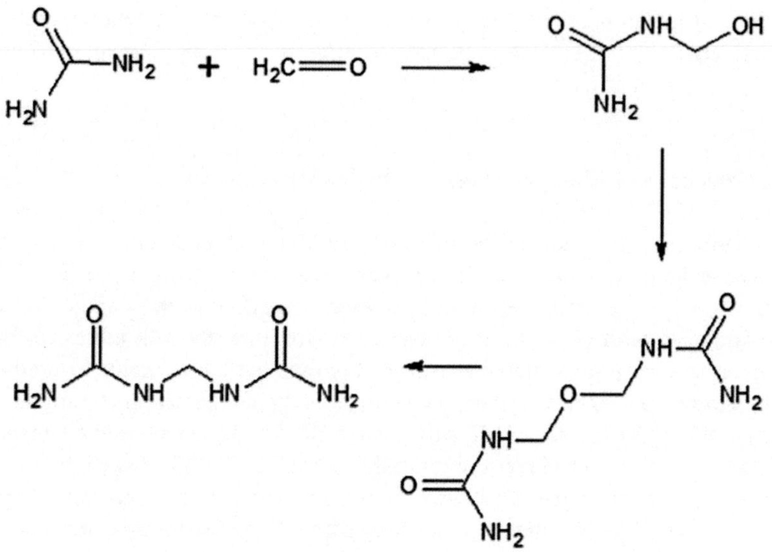

Fig. 3 Basic reactions of urea and formaldehyde

(c) The pH of the condensation reaction.

Urea–formaldehyde Curing Mechanism

Citric acid and formic acid that act as acid catalyst in the UF curing process is performed under ambient conditions. However, the process is a slow process. To boost up the speed or production rate at industrial level, heat is needed to the curing reaction. Another possible of two condensation reactions of hot UF curing that resulted the ridged three-dimensional structure, which are methylene and ether bridge. Methylene bridges are formed during first step of reaction through condensation between adjacent nitrogen from amide group and adjacent polymers. While ether bridges are formed through condensation reaction between methylol groups. As a result of UF curing process, it increases the viscosity of adhesives to form a gel until the process is complete.

1.2.2 Melamine Urea–formaldehyde (MF) Adhesives

Due to high water solubility and less resistance of water attack to UF adhesives, there is a need to find an alternatives material possessing high water resistance. Among them, melamine formaldehyde adhesives taken over the market for domestic use especially for outdoor project (exterior rated) and furniture panel. However, the cost of melamine formaldehyde is 2.5 times more expensive than that of urea–formaldehyde and therefore a number of urea must be supplemented to adhesives offering the best cost and performance.

The Production of Melamine Formaldehyde (MF) Adhesive

The formation MF adhesives initiated by the condensation reaction between melamine and formaldehyde. In a typical reaction, formaldehyde attacks on the amino group of melamine resulting formation of methylol compounds. This reaction has similarity with UF adhesives synthesis. However, the MF adhesive formation occurs relatively more freely and completely than the UF adhesive formation.

Goulding and Pizzi with their own co-researchers reported that complete methylolation take place in case of MF adhesives while it is practically not occurred during UF adhesive formation (Pizzi 1983; Goulding et al. 1994). The hydrophobic intermediaries appeared during early stage of MF adhesive formation reaction as compared to UF formation mainly due to rapid proceeding of the hydrophilic stage of the reaction. Goulding et al. (Goulding et al. 1994) investigated that the condensation and curing of MF adhesive continued not only in acid conditions but also in neutral and alkaline conditions.

The mechanism of reaction involved the similar steps involved in UF adhesive formation including the urea–formaldehyde, and the bridges (methylene and ether) formation along with a rapid increase in molecular weight of the adhesive. The final desired insoluble, infusible adhesives are formed through amino and methylol groups reaction during transformation step of final curing process. Several researchers have investigated the ether bridges besides un-reacted methylol groups and methylene bridges (Koehler 1943; Frey 1935). It was also found that no considerable amounts of formaldehyde are liberated for MF adhesive curing up to 100 °C temperature whereas urea–formaldehyde is considerable.

In term of commercial production, the current process is almost equally approach to the UF adhesive while the development of MF adhesive system relies mostly of the intended applications. In certain cases, the compounds such as acetoguanamine and E-caprolactam have been added at 3–5% (w/w) for desired application, which is for the impregnation of paper and fibres (Goulding et al. 1994). The function is to reduce the brittleness of MF adhesives when the degree of cross-linking is decrease in the cured adhesive by adding the such modifying compounds acetoguanamine and E-caprolactam.

Another important parameter in tuning the MF property is through viscosity control. In order to prevent the adhesive penetration into wood substrate higher viscosity of adhesives is use on the wood panel. The penetration may cause the cell wall swells or components of the cell wall may be chemically modified. High viscosity adhesives could produce by increase the level of condensation degree.

In case of low viscosity MF adhesives which theoretically having good penetration power, it can be produce through few approaches including by increase the content of methylol group and lowering the level of condensation. Another method is applied by adjusting the ratio of (melamine:formaldehyde) to (1:1.8 to 1:2) along with lower level of condensation.

In term of economic perspectives with the aims to lessen the cost of the adhesive, sugars have been used as modifiers. However, at the same time, it adversely affects the life-time of adhesive typically indicated by colour change and cracking which degrade to the adhesive's properties for long periods of time.

Melamine Formaldehyde-Based Adhesives Curing Process

Typically, the curing methods for MF-derived adhesives are quite similar with UF-derived adhesives. The only difference is the process is carried out without presence of acid catalyst and perform curing under hot conditions (Marra 1992). The curing process completed via formation of the methylene bridges.

The pure MF-derived adhesives have advantage over UF (and MUF) adhesives due to that can be cured under 100 °C avoiding the release of formaldehyde. Moreover, a small amount of formaldehyde release between 100 and 150 °C while curing MF-derived adhesives-based panels which is lower than UF –derived adhesives under similar conditions.

1.2.3 Phenolic Adhesives

The development of synthetic polymer of Phenolic adhesive (PF) is started at the end of nineteenth century. However, these adhesives were not fully commercialised until 1930s. Typically, PF adhesives are used in the wafer board and the plywood.

The Production of Phenolic Adhesives

There are two main manufacturing routes for Phenolic adhesives which is Novolac and Resol process. The later process is commonly used throughout the panel industry. The Resol process involve the reaction of phenol with formaldehyde in excess form catalysed by an acid catalyst and produced the quinine methide molecule. The ratio of phenol and formaldehyde is crucial and it's kept within the range of 1.6:1 to 2.5:1. Linear structures could be produced by selecting the lower range while cross-linked structures produced following the higher ratio for both the reactants (Ormondroyd 2015).

The condensation of quinone methide is occurred under alkaline conditions intended to form the methylol bridges due to C–C bonds which ultimately strengthens the phenolic adhesives. Marra et al. reported that the bridges are having the bonds which are strong and durable (Marra 1992).

On heating, reactive methylol groups in Resol combine firstly to make larger molecules, afterwards gel formation and finally transformed to solid state adhesives.

During the whole manufacturing process an alkaline pH is maintained which ensures the stability and shelf life of the adhesives.

The Curing of Phenolic Adhesives

The curing of PF adhesives undergoes three phases which are phase A, B, and C. In initial phase (A) results in the formation of adhesives having low molecular weight (<200) (Sellers 1985; Bliem et al. 2020). In this stage, the viscosity of adhesives may be a low to high. The products would soluble in sodium hydroxide solution and several polar solvents. The stage also makes the solid melts when being heated.

Phase B is the polymerization and methylol bridges are occurred in the presence of heat. This indicated the formation of second stage adhesive called as Resitol. Due to partial polymerization, Resitol adhesive is having the variance molecular weight of molecules. The Resitol adhesive is insoluble in most of the solvents; however, it become soft and rubbery on heating and can be in swollen state due to existing crosslinking.

In third curing stage, the Resitol become soft and again start polymerisation on heating where upon the process completed, Resitol is converted to Resite. After this stage no further re-melting or dissolution of adhesive into any solvent occurs.

1.2.4 Polyvinyl Acetate Adhesives

Poly vinyl acetate (PVA) is a thermoforming adhesive. The applications are based on wood DIY (Do it yourself) industries. The history of PVA begins from June 1920 when a German State patent office registered a patent for manufacturing of PVA (Pizzi 1983). While research on free radical vinyl polymerisation is in between 1915 to 1930 highlighted the practical applications of PVA and commercial production of PVA was begins since 1930 and onwards. The production of PVA has expanded enormously since the Second World War due to widely application in DIY wood work glue and also replacement of animal glue.

In comparison to conventional UF adhesives, PVA is much different in a sense, on curing PVA become flexible and converted to linear polymer with aliphatic backbone. PVA is characterized by its strong hydrogen bonding between acetates groups. This leads to an adhesive interaction within wood cell wall with OH^{-1} group. According to Rowell et al., a good adhesive should have the ability to maintain strength even after there is a wood expand. The internal and applied energies may be dissipated at the flexing in the bonded polymer (Rowell 2012). Their work further elaborated that creep and moisture resistance is almost negligible for adhesives. However, by adding cross linker can give them better improvement prior to the application of adhesive which essentially modified the thermoforming adhesive into a thermosetting system.

1.2.5 Isocyanate Adhesives

Isocyanate adhesives were discovered around 1884, but they were not fully developed as adhesives until Second World War (Pizzi 1983). Isocyanate adhesives found their early applications in tyre manufacturing industry followed by wide applications in wood industry (Ormondroyd 2015).

Wood panel industry utilised numerous types of Isocyanate adhesives, the most commonly form is MDI (4, 4-diphenylmethane diisocyanate). MDI has certain advantages over other adhesive belonging to same category. Some of the feature of MDI resins are summarised as below;

(a) Lower vapour pressure at ambient temperatures.
(b) Lower viscosity in comparison to other isocyanates.
(c) Lower cost in comparison to other isocyanates.
(d) High moisture tolerance (i.e. required less energy for fiber drying).
(e) High OH^{-1} bond making ability in wood comparative to other isocyanate adhesives.

However, MDIs have their disadvantages as listed;

(a) Highly toxic when uncured especially in gas or droplet form.
(b) High risk of lung infections (for unprotected operators at plant due bonding of gas with reactive sites in lungs).
(c) MDI composites exhibit low tack and fragile.

Based on aforementioned disadvantages, the usage of MDI is limited in wood industry especially in Europe. However, MDI exhibit certain applications including MDI adhesive without formaldehyde and outdoor applications of MDI adhesive composites. Since MDI adhesives in oriented strand board (OSB) and strand board; though, another modified form of MDI known as MDF is also successfully manufactured.

The Manufacture of MDI Adhesives

The basic manufacturing of MDI adhesives consisted of a chemical reaction between 1° amines and phosgene. In 1884, Hentschel, performed this reaction for the first time while further improvement to this synthesis method continued over the years.

In a typical reaction, a non-aqueous (xylene) solution of 1° amine is mixed with non-aqueous (xylene) solution of phosgene at a temperature > 60 °C. The mixture is digested in multiple steps with an increment of 20 °C along with addition of additional phosgene. Finally, to recover the excess phosgene and any impurities, distillation process is performed.

MDI Curing Process

MDI curing mechanism involved two main steps. The first step involves chemical reaction between MDI and OH^{-1} groups containing compounds. While second step involved direct reaction between MDI and water. A polyurethane formation take place through reaction between MDI and water molecules which then proceeded to setting of adhesives in a way similar to UF adhesive. The whole process of polyurethane formation involved several reactions such as the production of carbamic acid (unstable) in first step, which on decomposition produced an amine and CO_2. A substituted urea compound formed via direct reaction between released amines and isocyanate molecules. Substituted urea molecule and carbon dioxide are the final prodcuts produced by decomposition of carbamic acid aldehyde. In addition, to aforementioned process, a very significant reaction takes place between MDI and wood. The reaction involved the covalent bond formation between wood and MDI whilst the MDI undergoing curing process (Ormondroyd 2015).

1.3 Natural Adhesives

1.3.1 Bio-Adhesives

Due to increasing industrialization and strict legislation for environment, protection around the globe triggers the research on wood adhesives and adhesives from natural and sustainable resources. However, after the world's first oil crisis in the 1970s the awareness and commercialization dramatically increased. It is worth mentioning that the key factor for extensive research work on various types of adhesives in late twenty first century was not only due to public perception, but also due to an rise in national and international regulations (Pizzi 2006).

Bio-based adhesives include the adhesives derived from natural, non-mineral and organic sources. A variety of renewable materials such as tannins, carbohydrates, oils and liquefied cellulosic can be utilized for production of adhesive products. A brief discussion about synthesis, curing and application of important bio-based adhesives is carried out in this book chapter.

Tannin-Based Adhesives

Hydrolysable tannins and condensed tannins are two main types of phenolic nature tannins (Kües 2007; Péguy et al. 2020). The bark, wood, leaves and fruits of different plant species are the main sources of tannins while a fewer number of plants can provide enough concentration of extractable tannins. The plants, which can be used to extract tannins, includes pine, oak, chestnut, wattle, eucalyptus, myrtle, maple, birch, willow, etc. The adhesive properties of tannins are closely related to their method of extraction such as powdered tannins are produced by plant extraction, followed by purification and finally subjected to spray drying process (Pizzi 2003). Other components of the extraction include sugars, pectins and other polymeric carbohydrates, amino acids, as well as other substances (Antov et al. 2020).

Phenol–formaldehyde resins used upto 50% hydrolysable tannins as partial substitutes (Kulvik 1976, 1977). However, lower macromolecular structure, the lower degree of phenol substitution, limited the global production and being expensive hinders their commercialization in comparison to condensed tannins (Pizzi 2003, 2006). Condensed tannins with an annual production of 200 000 tons covers around 90% of the global production (Pizzi 2003, 2006).

The adhesives application of tannins for wood-based panels depends mainly on two factors: the content of reactive polyphenols, and the reactivity of these components towards formaldehyde. Tannins can be used as adhesives alone (with a formaldehyde component as cross linker) or in combination with aminoplastic or phenolic resins. MDF produced with tannins replacing parts of phenol in phenol-urea formaldehyde resins or even with 100% tannin resin can meet interior grade requirements but still unable to fulfil exterior grade specifications (Roffael et al. 2000; López-Suevos and Riedl 2003). Characteristics of resulting boards remarkably effected by the tannin source and tannin addition time during production of MDI chain (Müller et al. 2007). Adhesive formulations for wood applications, prepared using different hardeners and tannin powder from Turkish red pine bark, have been developed and tested by (Gonultas 2018; Ucar et al. 2013). A thermosetting tannin-based wood adhesive system from formaldehyde reaction with both condensed and hydrolysable tannin has been studied (Özacar et al. 2006).

On the other hand, there are studies focused on formaldehyde free resins by combination of bio-based materials e.g. protein and tannins (Li et al. 2004; Brossard et al. 2020). The development of tannins based particleboard adhesives have been studied by Santos et al. (Santos et al. 2017) following the complete removal of formaldehyde from adhesive formulations. They typically extracted from raw industrial lignocellulosic wastes, namely chestnut shell, chestnut bur and eucalyptus bark. Nath et al. have studied the properties and possibilities for production of particleboard with tannin-based adhesive from mangrove species (Nath et al. 2018). Cui et al. added cellulose nanofibers into tannin-based adhesives for particleboard production and reported significant increase of the mechanical properties of the produced panels (Cui et al. 2015).

Countries such as Australia, Argentina, Brazil, Chile, New Zealand, South Africa, and Zimbabwe used tannins as adhesives for production of wood-based panels (Li and Maplesden 1998; Dunky and Pizzi 2002). The application in Europe is rather limited, only for special products with specific properties.

Lignin-Based Adhesives

Lignin is the second most important component of plant biomass after cellulose. Lignin has an estimated production of 300 billion total tons in the biosphere and about 20 billion tons as an annual resynthesis (Mandel 1988). Lignocellulosic present in plants including agricultural residues, wood, grass, and other plants (Fu et al. 2010; Yousuf et al. 2020). Annually more than 50 million tons of lignin produced as by-product of pulp production worldwide (Kües 2007). About 10% of the technical lignin is commercially utilised. While the remaining amount is either combusted or not utilized at all (Kharazipour et al. 1991; Gargulak and Lebo 2000; Chakar and Ragauskas 2004; Gosselink et al. 2004). The possibilities for use of lignin in adhesive applications have been extensively studied by many authors (Hemmilä et al. 2017; Ferdosian et al. 2017a; Ghaffar and Fan 2014; Pizzi 2016; Nasir et al. 2019). The main interest in lignin is due to its phenolic structure with several favourable properties for formulation of wood adhesives such as high hydrophobicity and low polydispersity. However, the reactivity of the resin lowered by chemical structure of the lignin, which acts as disadvantage towards its applications, needed fast curing time.

Adhesives based on lignin can be classified into two clusters - formaldehyde-free lignin-based adhesives and lignin-based phenol–formaldehyde adhesives. In different studies, lignin is often combined with synthetic resins such as phenol–formaldehyde (Cetin and Özmen 2002; Çetin and Özmen 2003; Ghaffar and Fan 2014; Olivares et al. 1995; Guo et al. 2015; Pizzi 2016) or urea–formaldehyde resins (Podschun et al. 2016) to decrease the cost (Shimatani et al. 1994) or free formaldehyde emissions (Yang et al. 2015). Certain interest in this field represent the laboratory studies for manufacturing MDF by adding lignin as a binder (Zhou et al. 2011; Nasir et al. 2014) which allow the production of ecological low-toxic panels. Another studies in the MDF field considered addition of lignin activated by laccase enzyme to the fibers or lignin activated by enzyme treatment to the fibers (Kharazipour et al. 1991, 1998; Nasir et al. 2015). This method required the addition of 1% isocyanate to the panel to press at acceptably short press times or significantly extending the pressing time (Felby et al. 1997). There are also a number of successful attempts to produce MDF in laboratory conditions on the basis of lignosulfonates (Yotov et al. 2017; SAVOV and MIHAILOVA 2017a, 2017b; Savov et al. 2019; Antov et al. 2019).

The use of lignosulfonate as a replacement of phenol for phenolformaldehyde resin synthesis (up to 50 wt. % replacement level) were studied by Akhtar et al. (Akhtar et al. 2011). Phenol–formaldehyde resin contains 50% of lignosulfonate showed improved strength than commercial adhesives. The highest values of shear strength in both wet and dry conditions were attained at 20% substitution of phenol by lignosulfonate.

Another study was focused on the application of Kraft lignin-based phenol–formaldehyde resin (50 wt. % Kraft lignin) as an adhesive for production of oriented strand board panels (Cavdar et al. 2008). Addition of 3 - 8% Kraft lignin in hard-boards facilitates in reduction of swelling, (Westin et al. 2001). In another study, water stability, inner bond strength, and mechanical properties were stated to be enhanced in panels made from fibers of softwood residues defibrated in presence of extra lignin (Anglès et al. 2001).

Jin et al. studied the application of enzymatic hydrolysis lignin for phenol–formaldehyde resin synthesis (Jin et al. 2010; Qiao et al. 2015). The research data reported on the production of formaldehyde-free lignin-based adhesives is relatively low. Geng and Li (2006) prepared a formaldehyde-free wood adhesive using Kraft lignin and polyethylenimine. While the physical and mechanical properties of a modified ammonium lignosulfonate/ polyethylenimine mixture as a green adhesive for MDF production was studied by Yuan et al. (2014). In addition, a novel process to substitute formaldehyde in wood adhesive for particleboard production developed in 2007 by El Mansouri et al. (Pizzi and Salvadó 2007). The as prepared formulation met the EN 312 standards 312 for exterior grade panels and proved equivalent reactivity to formaldehyde-based adhesives.

Furthermore, a novel synthetic resin free bio-based adhesive prepared by Navarrete et al. (Navarrete et al. 2010) mainly derived from a low molecular mass lignin and tannin. The as prepared adhesive proved to be zero formaldehyde emission based on the desiccator method. A novel bio-adhesive for wood using kenaf core lignin and glyoxal was developed and tested (Hussin et al. 2019).

Starch-Based Adhesives

Starch being a natural polymer found applications including food, papermaking, additives, and adhesives, mainly because of its renewability, abundance, good adhesion, and low price (Imam et al. 1999; Norström et al. 2018; Zhao et al. 2018; Gu et al. 2019a; Lamaming et al. 2020). Starch is the combination of amylose and amylopectin, two distinct polysaccharide composed of glucose with different sizes and shapes (Tester et al. 2004). The proportion of amylose to amylopectin varies in accordance with the botanical origin of the starch and affects the properties of the wood adhesive (Norström et al. 2018).

The strength of Starch-based adhesives depends on degree of hydrogen bonding. These adhesives can easily develop hydrogen bonds with water molecules, leading to poor water resistance. Therefore, it is essential to modify starch in order to increase its performance for adhesive applications. Higher bonding strength and better water resistance can be achieved by combining starch with another component, for example polyvinyl alcohol, formaldehyde, isocyanates, and tannins (Qiao et al. 2016). There are several studies about starch-based adhesives prepared by grafting vinyl acetate onto starch using ammonium persulfate as the initiator (Wang et al. 2011, 2012).

Wang et al. demonstrated graft efficiency as an important parameter, which directly related to bonding performance of the starch adhesive (Wang et al. 2015). Esterification is a typical method for chemical modification of starch by converting OH^{-1} groups into esters in order to improve the hydrophobic properties of starch is esterification. Qiao et al. obtained esterified corn starch by reacting with maleic anhydride and then crosslinking with a polyisocyanate pre-polymer (Qiao et al. 2016). The optimal amount of pre-polymer was determined to be 10 wt %, resulting in the dry and wet shear strengths of 12 and 4 MPa, respectively. In another study, Tan et al. modified the starch-based adhesive by addition of blocked isocyanate and auxiliary agent (Tan et al. 2011). Gu et al. synthesized an environment friendly starch-based adhesive for wood-based panels by grafting polymerization of vinyl acetate monomer onto corn starch and crosslinking polymerization with N-methylol acrylamide (Gu et al.

2019b). The water resistance of the obtained adhesive was significantly improved to more than one MPa. The enhancement in overall performance achieved mainly due to increase in degree of crosslinking density and complex network structure formation.

Epoxy resins are another class of bio-based adhesives used as cross linker of starch. They are tested in combination with polyvinyl acetate grafted starch adhesives for veneer gluing. Epoxy resin provides good shear strength in both dry and humid conditions due to three-dimensional networks (Nie et al. 2013).

Protein-starch composite (Anderson et al. 2015; Gadhave et al. 2017b) and tannin-starch composite (Moubarik et al. 2010) could be used for wood and wood composites adhesives being eco-friendly (zero formaldehyde emissions). A partial substitution of phenol–formaldehyde with corn starch-quebarcho tannin-based resin was reported for the production of plywood (Moubarik et al. 2010). The optimal replacement value was determined to be 20% (15% cornstarch and 5% quebracho tannin). The addition of this resin enhanced the water-resistant properties and lowered the formaldehyde emissions. The same authors developed a non-volatile and non-toxic corn starch-tannin adhesive for interior plywsood (Moubarik et al. 2010). The mechanical properties of the produced plywood were greater in comparison with the conventional phenol–formaldehyde resin along with less toxicity, more environment-friendly, and having advantage of short reaction time.

The urea–formaldehyde resin was reactively blended with various concentration of starch (Osemeahon et al. 2013), esterified starch (Gadhave et al. 2017a; Zhu 2014), and oxidized starch (Ni 2014; Sun et al. 2018) as wood and wood composite adhesive. The modified system with urea starch blending has lower brittleness and formaldehyde emission. The as prepared adhesive has good chemical stability, insulating properties, temperature resistance, aging resistance, and oil resistance making it suitable for application towards wood adhesion (Dunky 1998; Bloembergen et al. 2005).

The performance of starch-based adhesives can be improved by incorporating the additives or fillers into the formulated adhesives (Wang et al. 2011). The addition of silica nanoparticles was reported to enhance the adhesive properties such as thermal stability, bonding strength, and water resistance. The shear strength improved by a factor of 5.12 MPa and 2.98 MPa respectively on addition of 10% silica both in dry and wet conditions.

Soy-Based Adhesives

A new type of adhesives using Soy as biomass raw materials commonly known as Soy-based adhesives. These adhesives are synthesized by partial hydrolysis of soy protein and their properties in accordance with panel standards (Xi et al. 2019; Xu et al. 2020).

Soy-derived adhesives based on soy protein which could be available in three different forms: At first, a non-effective adhesive based crude extract; at second a protein rich concentrate; and at third; an effective adhesive base protein isolate.

A number of factors are responsible for the efficient performance of soy-based adhesive such as morphology, viscosity, particle size, and nature of bonding with substrate. However, proteins in the adhesive could be altered to enhance the water resistance and bond strength however, the applications of soy-based adhesives composites are still limited in manufacturing industry. One of the method commonly used to improve the bonding strength and water resistance includes the soy protein modification with alkali, urea and guanidine hydrochloride (Graham and Krinski 1983).

Carbohydrate Adhesives

Carbohydrate adhesives are the class of adhesive where these sugars molecules are utilised as modifiers for formaldehyde adhesives. Typically, the modification process involved the formation of degradation compounds, which ultimately used as source for adhesives formation. Certain carbohydrates compound is reduced under acid environment, led to the formation of furfuraldehyde and furfuryl alcohol which finally used for manufacturing of panel adhesives (Brown 1952).

Extensive studies using variety of biomass have been done on direct liquefaction of carbohydrate materials in adhesive industry. Among these, a large variety of lignocellulosic biomass such as wood (Alma et al. 2003), wheat straw (Chen and Lu 2009), corn bran and cornstalks gain a lot of interest for production of biopolyols (Alma et al. 2003; Chen and Lu 2009). However, consumption of high volumes of petroleum-derived solvents as liquefaction agents considerably increase the production cost and ultimately acts as barrier towards commercialization of these biopolyols. As a rough estimate; approximately 100–125 g solvent is required for 20 g of lignocellulosic biomass to obtain high-quality biopolyols (Hu et al. 2012).

Research efforts have been carried out to evaluate the usage of liquefied biomass as a suitable adhesive panel product. Among them, Medved et al. (Medved et al. 2011) investigated that liquefied wood addition to a MUF adhesive surge in properties upto 30% substitution rate and vice versa fall down.

Oil-Based Adhesives

Due to acceptable prices of unsaturated oils, their conversion into adhesives has been the subject of research over the years. For example, linseed oil on epoxidation and cross linking with cyclic polycarboxilic acid anhydride converted to oil-based adhesive having appropriate molecular weight. In another examples, cardanol-based adhesives utilised the cashew liquor as a precursor for the manufacture of (Tomkinson 2003; Wool et al. 1998). Cashew nut-based adhesive (CNSL) manufacturing process in first step involved the ozonolysis of the cardanol followed by generation of an alkenyl chain. In subsequent step, these groups undergo self-condensation to forma-strengthened network. These CNSL adhesives showed good internal bond (IB) strength at laboratory level under wet and dry condition to a level of 1.05 and

0.58 MPa, respectively. Although a lot of research work has been completed on these oil-based adhesives but still efforts are undertaken to advance the fabrication of these adhesives and commercial utilization for panel production. Transformation of commodity oils could be fruitful for production of oil-based adhesives while their application is yet limited in the panel industry based on wood.

2 Future Aspects of Bio-Adhesive

Bio-based materials especially adhesives are getting more attention due to their renewability, eco-friendly nature and due to their independence from petroleum resources. This trend could be visualized via extended research work reported on renewable materials. Research efforts are being carried out to develop the novel molecular structures having advantages of vegetable oils such as reduced toxicity (human and environmental), hydrophobicity, and better efficiency compared to their petroleum counter parts.

2.1 Polysaccharides and Proteins

Constrains: Reduced vulnerability to hydrolytic degradation.

Future opportunities: Gu et al. developed an interesting approach for wood adhesives based on Konjac glucomannan, polyvinyl alcohol (PVC) and chitosan (Gu et al. 2016). A Box-Behnken statistical design was developed to optimize the interactions between aforementioned polymers based on their bonding strength. Adhesive with a composition of 2.3% chitosan, 5% of a 10% PVC solution, and 2.3% Konjac glucomannan showed an optimal bond strength of 3.4 MPa. Ruediger et al. developed another approach having application for protein adhesives (Ruediger et al. 2016; Rüdiger 2017). This approach involved the adsorption of an aspartic protease and enzyme chymosin on a glass surface mainly to destabilize the casein micelles causing the deposition of casein layer at the surface. The two surfaces are placed at a distance of 375 μm apart from each other, where casein adhesive layer developed between them.

Marine plants are reported as a new source of polysaccharides, which can effectively utilized for production of polysaccharide-based adhesives. In a same row, algae could also serve as non-food competitive and abundant source for renewable materials. For example, *Chhatbar* et al. used marine-based polysaccharides for adhesive applications. They grafted polyvinylpyrrolidone onto a sulphated seaweed polysaccharide under microwave irradiation in weight ratio of 1–2.5 to 1(Chhatbar and Siddhanta 2015). In another study by Lacoste et al. alginates were used as a binder between recycled cotton fibres and wood fibres for a composite material (Lacoste et al. 2018). The resulting composite showed good insulating properties having thermal conductivities in a range between 0.078 and 0.089 W m^{-1} K^{-1}. It was further investigated that the bending Young modulus could be increased from 10 to 17 MPa upon crosslinking of alginate binder with 8% glutaraldehyde. In addition, development of bio-based adhesives for special application such as batteries, fuel cells and capacitors. Among them, lithium ion batteries containing silicon and carbon conductive anodes (e.g. graphite) are one of the common example with respect to aforementioned applications (Foyer et al. 2016; Li et al. 2016).

2.2 Lignin

Constrains: Condensation reaction increased reactivity, reproducibility and structure.

Future opportunities: Zhang et al. developed an interesting functionality to lignin adhesives to explain the repartition and recyclability (Zhang et al. 2018). For example, crosslinking between ozonated lignin compared to Kraft lignin using an epoxy compound produced from sebacic acid and epochlorohydrin. Resulting adhesive revealed a lap shear strength of 6.5 MPa for aluminium sheets bonding which is comparable to epoxy adhesives revealing 4–8 MPa strength. The separated aluminium could be rebounded by further treatment at 190 °C over a period of one hour resulting a lap shear strength of 5 MPa.

2.3 Vegetable Oils and Small Molecule Renewable Monomers

Constrains: Non-food sources.

Future opportunities: Li et al. reported a bio-based pressure sensitive adhesive by combining soyabean oil and lactic acid oligomers followed by copolymerization of aforementioned oligomers with epoxidised soybean oil, coating on PET film and finally curing under UV radiation (Li et al. 2015). A peel strength of 3.8 N cm^{-1} and tack strength of 8 N cm^{-1} was reported by Li et al.

In a similar way, tensile strength of soyabean oil based adhesives was improved by reduction of brittleness due to addition of castor oil nylon oligomer (Jian et al. 2017). They resulting modified samples showed a tensile strength of 26 MPa as compared to samples containing to nylon oligomer (0.4 MPa).

Another promising approach to enhance the functionality and value is to introduce the renewable monomers into adhesives. For example, development of flame-retardant adhesive tapes by Wang et al. which mainly based on soyabean oil and vanillin making use of incorporation of phosphorous compounds into polymer based on renewable monomer (Wang et al. 2017; Ci et al. 2017). In another recent example, Ghosh et al. developed a self-healing polyurethane by combining dimer acid (bio-based) and glycerol (Ghosh et al. 2013). An intrinsic self-healing of 100% was reported under microwave irradiation.

2.4 *Non-Adhesive Bonding*

In order to enhance the sustainability of adhesive for lignocellulosic prodcuts is research in area of bonding without adhesives. This approach involved the surface modification of wood in order to activate adhesive forces without involving any polymeric resins. In a recent study, Nakaya et al. depolymerise some of wood surface components (polysaccharides, lignin) using ionic liquid (imidazole hydrochloride: 83 g m^{-2}), water & glucose in a ratio of 9/3/2 respectively and repolymerised them to generate adhesion for applications in plywood (Nakaya et al. 2018). The process involved the hot processing of both surfaces of plywood at 2.9 MPa for one hour. A bond strength of 0.6 MPa was achieved so far which is quite low.

3 Conclusions

This chapter thoroughly emphasized on various types of synthetic and natural based adhesives. The highlighted adhesives showed wide applications in many specific areas of interest. There are three main factors controlling the overall production, consumption and commercialization of these adhesives. These include legislation, emission rates and cost. Day by day, legislative process being tighten due to increasing pollution and greenhouse gas (GHG) effect. Secondly, emission of toxic chemicals such as formaldehyde and volatile organic compounds (VOCs) should be lowered to make adhesive manufacturing a sustainable green process. Next, an initial cost and market price should have a right balance for the adhesive's commercialization. To address aforementioned concerns and commercialization targets, the develop adhesives would have to be eco-friendly, cheap, efficient, and easy accessibility.

References

Abuarra A, Hashim R, Bauk S, Kandaiya S, Tousi ET (2014) Fabrication and characterization of gum Arabic bonded Rhizophora spp. particleboards. Mater Des 60:108–115

Acda MN, Cabangon RJ (2013) Termite resistance and physico-mechanical properties of particleboard using waste tobacco stalk and wood particles. Int Biodeterior Biodegrad 85:354–358

Akhtar T, Lutfullah G, Ullah Z (2011) Lignonsulfonate-phenolformaldehyrde adhesive: a potential binder for wood panel industries. J Chem Soc Pak 33(4):535–538

Alma M, Basturk MA, Digrak M (2003) New polyurethane-type rigid foams from liquified wood powders. J Mater Sci Lett 22(17):1225–1228

Amirou S, Zerizer A, Pizzi A, Haddadou I, Zhou X (2013) Particleboards production from date palm biomass. Eur J Wood Wood Prod 71(6):717–723

Anderson KR, Porter MA, Satyavolu JV (2015) Protein and starch compositions, methods for making and uses thereof. Google Patents

Anglès M, Ferrando F, Farriol X, Salvadó J (2001) Suitability of steam exploded residual softwood for the production of binderless panels. Effect of the pre-treatment severity and lignin addition. Biomass and Bioenergy 21(3):211–224

Antov P, Savov V, Neykov N (2020) Sustainable bio-based adhesives for eco-friendly wood composites. A Review. Wood Research 65(1):51–62

Antov P, Valchev I, Savov V (2019) Experimental and statistical modeling of the exploitation properties of eco-friendly MDF through variation of lignosulfonate concentration and hot pressing temperature. In: Proceedings of the 2nd International Congress of Biorefinery of Lignocellulosic Materials (IWBLCM 2019), pp 104–109

Ballerini A, Despres A, Pizzi A (2005) Non-toxic, zero emission tannin-glyoxal adhesives for wood panels. Holz Als Roh-Und Werkstoff 63(6):477–478

Baskaran M, Hashim R, Sulaiman O, Hiziroglu S, Sato M, Sugimoto T (2015) Optimization of press temperature and time for binderless particleboard manufactured from oil palm trunk biomass at different thickness levels. Mater Today Commun 3:87–95

Bliem P, Konnerth J, Frömel-Frybort S, Gartner C, Mauritz R, van Herwijnen HW (2020) Influence of drying and curing parameters on phenol-formaldehyde impregnated wood veneers. J Adhesion 96(1–4):253–271

Bloembergen S, Kappen F, Beelen B (2005) Environmentally friendly biopolymer adhesives and applications based thereon. Google Patents

Brossard N, Bordeu E, Ibáñez RA, Chen J, Osorio F (2020) Rheological study of tannin and protein interactions based on model systems. J Text Stud

Brown LH (1952) Furfuryl alcohol resins. Google Patents

Cancer IAfRo (2006) Formaldehyde, 2-Butoxyethanol and 1-tert-Butoxypropan-2-ol. In: Formaldehyde, 2-Butoxyethanol and 1-tert-Butoxypropan-2-ol

Catto AL, Montagna LS, Santana RM (2017) Abiotic and biotic degradation of post-consumer polypropylene/ethylene vinyl acetate: wood flour composites exposed to natural weathering. Polym Compos 38(3):571–582

Cavdar AD, Kalaycioglu H, Hiziroglu S (2008) Some of the properties of oriented strandboard manufactured using kraft lignin phenolic resin. J Mater Process Technol 202(1–3):559–563

Cetin NS, Özmen N (2002) Use of organosolv lignin in phenol–formaldehyde resins for particleboard production: I. Organosolv lignin modified resins. Int J Adhesion Adhesives 22(6):477–480

Çetin NS, Özmen N (2003) Studies on lignin-based adhesives for particleboard panels. Turkish J Agricul Forest 27(3):183–189

Chakar FS, Ragauskas AJ (2004) Review of current and future softwood kraft lignin process chemistry. Ind Crops Prod 20(2):131–141

Chemicals I-OPftSMo, Organization WH (2010) WHO Recommended Classification of Pesticides by Hazard and Guidelines to Classification 2009. World Health Organization

Chen F, Lu Z (2009) Liquefaction of wheat straw and preparation of rigid polyurethane foam from the liquefaction products. J Appl Polym Sci 111(1):508–516

Chen X, Xi X, Pizzi A, Fredon E, Du G, Gerardin C, Amirou S (2020) Oxidized demethylated lignin as a bio-based adhesive for wood bonding. J Adhesion, 1–18. doi:https://doi.org/10.1080/00218464.2019.1710830

Cheng H, Kilgore K, Ford C, Fortier C, Dowd MK, He Z (2019) Cottonseed protein-based wood adhesive reinforced with nanocellulose. J Adhes Sci Technol 33(12):1357–1368

Chhatbar MU, Siddhanta AK (2015) Synthesis of seaweed polysaccharide-based polyvinylpyrrolidone copolymer with adhesive properties. Journal of Applied Polymer Science 132 (33)

Ci J, Cao C, Kuga S, Shen J, Wu M, Huang Y (2017) Improved performance of microbial fuel cell using esterified corncob cellulose nanofibers to fabricate air-cathode gas diffusion layer. ACS Sustain Chem Eng 5(11):9614–9618

Ciannamea EM, Stefani PM, Ruseckaite RA (2010) Medium-density particleboards from modified rice husks and soybean protein concentrate-based adhesives. Biores Technol 101(2):818–825

Colak S, Colakoglu G, Kalaycioglu H, Aydin I (2009) Effects of log storage conditions and steaming process on the formaldehyde emissions of particleboard manufactured from eucalyptus (E. camaldulensis) wood. Eur J Wood Wood Prod 67(4):379–382

Cui J, Lu X, Zhou X, Chrusciel L, Deng Y, Zhou H, Zhu S, Brosse N (2015) Enhancement of mechanical strength of particleboard using environmentally friendly pine (Pinus pinaster L.) tannin adhesives with cellulose nanofibers. Annals Forest Sci 72(1):27–32

DeVito SC (2016) On the design of safer chemicals: a path forward. Green Chem 18(16):4332–4347

Dinwoodie J (1979) The properties and performance of particle board adhesives. J Ins Wood Sci

Dunky M (1998) Urea–formaldehyde (UF) adhesive resins for wood. Int J Adhes Adhes 18(2):95–107

Dunky M (2004) Adhesives based on formaldehyde condensation resins. In: Macromolecular symposia, vol 1. Wiley Online Library, pp 417–430

Dunky M, Pizzi A (2002) Wood adhesives, adhesive science and engineering-2: surfaces. Chemistry and Applications, Elsevier, Amsterdam

Felby C, Pedersen LS, Nielsen BR (1997) Enhanced auto adhesion of wood fibers using phenol oxidases. Holzforschung-Int J Biol Chem Phy Technol Wood 51(3):281–286

Ferdosian F, Pan Z, Gao G, Zhao B (2017) Bio-based adhesives and evaluation for wood composites application. Polymers (Basel) 9(2):70. https://doi.org/10.3390/polym9020070

Ferdosian F, Pan Z, Gao G, Zhao B (2017) Bio-based adhesives and evaluation for wood composites application. Polymers 9(2):70

Foyer G, Chanfi B-H, Boutevin B, Caillol S, David G (2016) New method for the synthesis of formaldehyde-free phenolic resins from lignin-based aldehyde precursors. Eur Polymer J 74:296–309

Frey K (1935) Kondensationsreaktionen von Anilin mit Formaldehyd und deren Bedeutung für die Herstellung plastischer Massen. Helv Chim Acta 18(1):491–513

Frihart CR (2005) Wood adhesion and adhesives. CRC Press, Boca Raton, FL

Fu D, Mazza G, Tamaki Y (2010) Lignin extraction from straw by ionic liquids and enzymatic hydrolysis of thsse cellulosic residues. J Agric Food Chem 58(5):2915–2922. https://doi.org/10.1021/jf903616y

Gadhave RV, Mahanwar PA, Gadekar PT (2017) Factor affecting gel time/process-ability of urea formaldehyde resin based wood adhesives. Open J Poly Chem 7(02):33

Gadhave RV, Mahanwar PA, Gadekar PT (2017b) Starch-based adhesives for wood/wood composite bonding

Gargulak J, Lebo S (2000) Commercial use of lignin-based materials. In. ACS Publications

Geng X, Li K (2006) Investigation of wood adhesives from kraft lignin and polyethylenimine. J Adhes Sci Technol 20(8):847–858

Ghaffar SH, Fan M (2014) Lignin in straw and its applications as an adhesive. Int J Adhes Adhes 48:92–101

Ghalehno MD, Madhoushi M, Tabarsa T, Nazerian M (2011) The manufacture of particleboards using mixture of reed (surface layer) and commercial species (middle layer). Eur J Wood Wood Prod 69(3):341–344

Ghofrani M, Haghdan S, NicKhah V, Ahmadi K (2015) Improvement of physical and mechanical properties of particleboard made of apple tree pruning and sunflower stalk using titanium oxide nanoparticles. Eur J Wood Wood Prod 73(5):661–666

Ghosh D, Giri S, Das C (2013) ACS Sustain Chem Eng 1:1135–1142

Gonultas O (2018) Properties of Pine Bark Tannin-based adhesive produced with various hardeners. BioResources 13(4):9066–9078

Gosselink R, Troosters T, Decramer M (2004) Exercise testing: why, which and how to interpret. Breathe 1(2):120–129

Goulding T, Pizzi A, Mattal K (1994) Handbook of Adhesive Technology. Marcel Dekker, New York, 531

Graham PM, Krinski TL (1983) Heat coagulable paper coating composition with a soy protein adhesive binder. Google Patents

Gu R, Mu B, Yang Y (2016) Bond performance and structural characterization of polysaccharide wood adhesive made from Konjac glucomannan/chitosan/polyvinyl alcohol. BioResources 11(4):8166–8177

Gu Y, Cheng L, Gu Z, Hong Y, Li Z, Li C (2019) Preparation, characterization and properties of starch-based adhesive for wood-based panels. Int J Biol Macromol 134:247–254. https://doi.org/10.1016/j.ijbiomac.2019.04.088

Guo Z, Liu Z, Ye L, Ge K, Zhao T (2015) The production of lignin-phenol-formaldehyde resin derived carbon fibers stabilized by BN preceramic polymer. Mater Lett 142:49–51

Hazrati-Behnagh M, Zarea-Hosseinabadi H, Daliri-Sosefi M, Abginehchi Z, Hemmati A (2015) Mechanical and insulating performances of ultralight thick particleboard from sugarcane residues and woods planer shaving. Eur J Wood Wood Prod 74(2):161–168. https://doi.org/10.1007/s00107-015-0989-7

He Z, Cheng HN, Klasson KT, Ford C, Barroso VAB (2019) Optimization and practical application of cottonseed meal-based wood adhesive formulations for small wood item bonding. Int J Adhes Adhes 95:102448. https://doi.org/10.1016/j.ijadhadh.2019.102448

He Z, Wan H (2017) Bio-based wood adhesives research advances and outlooks. In: Bio-based Wood Adhesives. CRC Press, pp 340–353

Hein PRG, Campos ACM, Mendes RF, Mendes LM, Chaix G (2011) Estimation of physical and mechanical properties of agro-based particleboards by near infrared spectroscopy. Eur J Wood Wood Prod 69(3):431–442

Hemmilä V, Adamopoulos S, Karlsson O, Kumar A (2017) Development of sustainable bio-adhesives for engineered wood panels–a review. Rsc Advan 7(61):38604–38630

Hu S, Wan C, Li Y (2012) Production and characterization of biopolyols and polyurethane foams from crude glycerol based liquefaction of soybean straw. Bioresour Technol 103(1):227–233. https://doi.org/10.1016/j.biortech.2011.09.125

Hua LS, Ashaari Z, Chen LW, San H, Peng T, MAY C, (2015) Properties of particleboard with oil palm trunk as core layer in comparison to three-layer rubberwood particleboard. J Oil Palm Res 27(1):67–74

Hussin MH, Aziz AA, Iqbal A, Ibrahim MNM, Latif NHA (2019) Development and characterization novel bio-adhesive for wood using kenaf core (Hibiscus cannabinus) lignin and glyoxal. Int J Biol Macromol 122:713–722

Imam SH, Mao L, Chen L, Greene RV (1999) Wood adhesive from crosslinked poly (vinyl alcohol) and partially gelatinized starch: preparation and properties. Starch-Stärke 51(6):225–229

Jang Y, Huang J, Li K (2011) A new formaldehyde-free wood adhesive from renewable materials. Int J Adhes Adhes 31(7):754–759

Jian X-Y, An X-P, Li Y-D, Chen J-H, Wang M, Zeng J-B (2017) All plant oil derived epoxy thermosets with excellent comprehensive properties. Macromolecules 50(15):5729–5738

Jin Y, Cheng X, Zheng Z (2010) Preparation and characterization of phenol–formaldehyde adhesives modified with enzymatic hydrolysis lignin. Biores Technol 101(6):2046–2048

Jivkov V, Simeonova R, Marinova A (2013) Influence of the veneer quality and load direction on the strength properties of beech plywood as structural material for furniture. Innov Woodwork Ind Eng Des 3:86–92

Jivkov V, Simeonova R, Marinova A, Gradesva G (2013b) Study on the gluing abilities of solid surface composites with different wood based materials and foam PVC. In: Wood is good-user oriented material, technology and design. Proceedings of the 24th International Scientific Conference, Zagreb, Croatia, 18th October 2013. Faculty of Forestry, University of Zagreb, pp 49–55

Khan TA, Gupta A, Jamari SS, Nasir M, Jang S, Kim H-J, Asim M (2020) Synthesis of micro carbonaceous material by pyrolysis of rubber wood and its effect on properties of urea-formaldehyde (UF) resin. Int J Adhes Adhes 99:102589

Kharazipour A, Haars A, Shekholeslami M, Hüttermann A (1991) Enzymgebundene holzwerkstoffe auf der basis von lignin und phenoloxidasen. Adhäsion 35(5):30–36

Kharazipour A, Mai C, Hüttermann A (1998) Polyphenoles for compounded materials. Polym Degrad Stab 59(1–3):237–243

Khosravi S, Khabbaz F, Nordqvist P, Johansson M (2010) Protein-based adhesives for particleboards. Ind Crops Prod 32(3):275–283

Khosravi S, Nordqvist P, Khabbaz F, Johansson M (2011) Protein-based adhesives for particleboards—effect of application process. Ind Crops Prod 34(3):1509–1515

Koehler R (1943) Kunststoffe Tech. 11:1 (1941). Kolloid Z 103:138

Konnerth J, Hahn G, Gindl W (2009) Feasibility of particle board production using bone glue. Eur J Wood Wood Prod 67(2):243–245

Kües U (2007) Wood production, wood technology, and biotechnological impacts. Universitätsverlag Göttingen

Kulvik E (1976) Chesnut wood tannin extract in plywood adhesives

Kulvik E (1977) Tannin extract as cure accelerator for phenol-formaldehyde wood adhesives. Adhes Age 19:33–34

Kusumah SS, Umemura K, Guswenrivo I, Yoshimura T, Kanayama K (2017) Utilization of sweet sorghum bagasse and citric acid for manufacturing of particleboard II: influences of pressing temperature and time on particleboard properties. J Wood Sci 63(2):161–172. https://doi.org/10.1007/s10086-016-1605-0

Lacoste C, El Hage R, Bergeret A, Corn S, Lacroix P (2018) Sodium alginate adhesives as binders in wood fibers/textile waste fibers biocomposites for building insulation. Carbohyd Polym 184:1–8

Lamaming J, Heng NB, Owodunni AA, Lamaming SZ, Khadir NKA, Hashim R, Sulaiman O, Kassim MHM, Hussin MH, Bustami Y (2020) Characterization of rubberwood particleboard made using carboxymethyl starch mixed with polyvinyl alcohol as adhesive. Compos B Eng 183:107731

Li J, Maplesden F (1998) Commercial production of tannins from radiata pine bark for wood adhesives. Transactions of the Institution of Professional Engineers New Zealand: Electrical/Mechanical/Chemical Engineering Section 25 (1):46

Li J, Wang W, Zhang S, Gao Q, Zhang W, Li J (2016) Preparation and characterization of lignin demethylated at atmospheric pressure and its application in fast curing biobased phenolic resins. RSC Advances 6(71):67435–67443

Li K, Geng X, Simonsen J, Karchesy J (2004) Novel wood adhesives from condensed tannins and polyethylenimine. Int J Adhes Adhes 24(4):327–333

Li RJ, Gutierrez J, Chung Y-L, Frank CW, Billington SL, Sattely ES (2018) A lignin-epoxy resin derived from biomass as an alternative to formaldehyde-based wood adhesives. Green Chem 20(7):1459–1466

Li Y, Wang D, Sun XS (2015) Copolymers from epoxidized soybean oil and lactic acid oligomers for pressure-sensitive adhesives. RSC Advances 5(35):27256–27265

Ling Z, Omura Y, Hori N, Iwata T, Takemura A (2018) In-situ chemical structure analysis of aqueous vinyl polymer solution-isocyanate adhesive in post-cure process by using Fourier transform near infrared spectroscopy. Int J Adhes Adhes 81:56–64

López-Suevos F, Riedl B (2003) Effects of Pinus pinaster bark extracts content on the cure properties of tannin-modified adhesives and on bonding of exterior grade MDF. J Adhes Sci Technol 17(11):1507–1522

Mabrouk AB, Dufresne A, Boufi S (2020) Cellulose nanocrystal as ecofriendly stabilizer for emulsion polymerization and its application for waterborne adhesive. Carbohyd Polym 229:115504

Mandel M (1988) Encyclopedia of polymer science and engineering. In: Mark HF, Bikales N, Overberger CG, Menges G, Kroschwitz JI. Wiley, New York 11

Mansouri H, Navarrete P, Pizzi A, Tapin-Lingua S, Benjelloun-Mlayah B, Pasch H, Rigolet S (2011) Synthetic-resin-free wood panel adhesives from mixed low molecular mass lignin and tannin. Eur J Wood Wood Prod 69(2):221–229

Marra AA (1992) Technology of wood bonding. Van Nostrand Reinhold

Marshall E (1987) EPA indicts formaldehyde, 7 years later. Science 236(4800):381. https://doi.org/10.1126/science.3563516

Medved S, Natasa C, Kunaver M (2011) Sorption and moisture resistance of liquefied wood bonded particle board. In: Proceedings of the international panel and panel product symposium, pp 129–138

Moubarik A, Allal A, Pizzi A, Charrier F, Charrier B (2010) Characterization of a formaldehyde-free cornstarch-tannin wood adhesive for interior plywood. Eur J Wood Wood Prod 68(4):427–433
Müller C, Kües U, Schöpper C, Kharazipour A (2007) 16. Natural Binders. Wood Production, Wood Technology, and Biotechnological Impacts:347
Muller NE (1992) An early example of a plywood support for painting. J Am Inst Conser 31(2):257–260
Nakaya N, Hosoya T, Miyafuji H (2018) Ionic liquids as formaldehyde-free wood adhesives. J Wood Sci 64(6):794–801
Nasir M, Gupta A, Beg M, Chua G, Kumar A (2014) Physical and mechanical properties of medium-density fibreboards using soy—lignin adhesives. J Tropical Forest Sci, 41–49
Nasir M, Hashim R, Sulaiman O, Nordin NA, Lamaming J, Asim M (2015) Laccase, an emerging tool to fabricate green composites: a review. BioResources 10(3):6262–6284
Nasir M, Khali D, Jawaid M, Tahir P, Siakeng R, Asim M, Khan T (2019) Recent development in binderless fiber-board fabrication from agricultural residues: a review. Constr Build Mater 211:502–516
Nath SK, Islam MN, Rahman K-S, Rana MN (2018) Tannin-based adhesive from Ceriops decandra (Griff.) bark for the production of particleboard. J Indian Acad Wood Sci 15(1):21–27
Navarrete P, Kebbi Z, Michenot F, Lemonon J, Rogaume C, Masson E, Rogaume Y, Pizzi A (2013) Formaldehyde and VOCs emissions from bio-particleboards. J Adhes Sci Technol 27(7):748–762
Navarrete P, Mansouri H, Pizzi A, Tapin-Lingua S, Benjelloun-Mlayah B, Pasch H, Rigolet S (2010) Wood panel adhesives from low molecular mass lignin and tannin without synthetic resins. J Adhes Sci Technol 24(8–10):1597–1610
Ni K (2014) Oxidized starch modified urea-formaldehyde resin adhesive. CN,
Nie Y, Tian X, Liu Y, Wu K, Wang J (2013) Research on starch-g-polyvinyl acetate and epoxy resin-modified corn starch adhesive. Polym Compos 34(1):77–87
Nimz H (1983) Wood adhesives chemistry and technology. Marcel Dekker, New York
Nonaka S, Umemura K, Kawai S (2013) Characterization of bagasse binderless particleboard manufactured in high-temperature range. J Wood Sci 59(1):50–56
Norström E, Demircan D, Fogelström L, Khabbaz F, Malmström E (2018) Green binders for wood adhesives. Appl Adhesive Bond Sci Technol, 49–71
Olivares M, Aceituno H, Neiman G, Rivera E, Sellers TJ (1995) Lignin-modified phenolic adhesives for bonding radiata pine plywood. Forest Produc J 45(1):63
Ormondroyd G (2015) Adhesives for wood composites. In: Wood Composites. Elsevier, pp 47–66
Osemeahon S, Maitera O, Hotton A, Dimas B (2013) Influence of starch addition on properties of urea formaldehyde/starch copolymer blends for application as a binder in the coating industry. J Environ Chem Ecotoxicol 5(7):181–189
Özacar M, Soykan C, Şengil İA, (2006) Studies on synthesis, characterization, and metal adsorption of mimosa and valonia tannin resins. J Appl Polym Sci 102(1):786–797
Panel FE (2009) Formaldehyde expert panel report. Part B–Recommendation for listing status for formaldehyde and scientific justification for the recommendation. National Toxicology Program
Pang B, Cao X-F, Sun S-N, Wang X-L, Wen J-L, Lam SS, Yuan T-Q, Sun R-C (2020) The direct transformation of bioethanol fermentation residues for production of high-quality resins. Green Chemistry
Pedieu R, Riedl B, Pichette A (2009) Properties of mixed particleboards based on white birch (Betula papyrifera) inner bark particles and reinforced with wood fibres. Eur J Wood Wood Prod 67(1):95–101
Péguy EAS, Bikoro BAA, Rodrigue ST, Santiago-Medina FJ, Cabaret T, Pizzi A, Bertrand C (2020) The condensed tannins of Okoume (Aucoumea klaineana Pierre): A molecular structure and thermal stability study. Scientific Reports (Nature Publisher Group) 10 (1)
Pichelin F, Nakatani M, Pizzi A, Wieland S (2006) Structural beams from thick wood panels bonded industrially with formaldehyde-free tannin adhesives. Forest Prod J 56(5):31
Pizzi A (1980) Tannin-based adhesives. J Macromol Sci—Rev Macromol Chem 18(2):247–315
Pizzi A (1983) Tannin-based wood adhesives. Wood Adhesives: Chem Technol 1:177–246

Pizzi A (1991) An assessment of the future industrial prospects for panel adhesives from renewable natural materials. Holzforschung Holzverwertung 43:83–87
Pizzi A (1994a) Advanced wood adhesives technology. CRC Press
Pizzi A (1994) Hardening mechanism of tannin adhesives with hexamine. Eur J Wood Wood Prod 52(4):229–229
Pizzi A (2003) Natural phenolic adhesives I: Tannin. Handbook Adhesive Technol 2:573–587
Pizzi A (2006) Recent developments in eco-efficient bio-based adhesives for wood bonding: opportunities and issues. J Adhes Sci Technol 20(8):829–846
Pizzi A (2016) Wood products and green chemistry. Ann for Sci 73(1):185–203
Pizzi A, Salvadó J (2007) Lignin-based wood panel adhesives without formaldehyde. Holz Als Roh-Und Werkstoff 65(1):65
Podschun J, Stücker A, Buchholz RI, Heitmann M, Schreiber A, Saake B, Lehnen R (2016) Phenolated lignins as reactive precursors in wood veneer and particleboard adhesion. Ind Eng Chem Res 55(18):5231–5237
Qiao W, Li S, Guo G, Han S, Ren S, Ma Y (2015) Synthesis and characterization of phenol-formaldehyde resin using enzymatic hydrolysis lignin. J Ind Eng Chem 21:1417–1422
Qiao Z, Gu J, Lv S, Cao J, Tan H, Zhang Y (2016) Preparation and properties of normal temperature cured starch-based wood adhesive. BioResources 11(2):4839–4849
Roffael E, Dix B, Okum J (2000) Use of spruce tannin as a binder in particleboards and medium density fiberboards (MDF). Holz Als Roh-Und Werkstoff 58(5):0301–0305
Rowell RM (2012) Handbook of wood chemistry and wood composites. CRC press
Rüdiger AA (2017) The enzyme mediated autodeposition of casein: from deposition of defined films to nanostructuring with single particles. Universitätsbibliothek
Ruediger AA, Bremser W, Strube OI (2016) The enzyme-mediated autodeposition of casein: effect of enzyme immobilization on deposition of protein structures. J Coat Technol Res 13(4):597–611
Santos J, Antorrena G, Freire MS, Pizzi A, González-Álvarez J (2017) Environmentally friendly wood adhesives based on chestnut (Castanea sativa) shell tannins. Eur J Wood Wood Prod 75(1):89–100
Savov V, Mihailova J (2017a) Influence of the content of lignosulfonate on mechanical properties of medium density fiberboard. Pro Ligno 13(4):252–256
Savov V, Mihailova J (2017b) Influence of the content of lignosulfonate on physical properties of medium density fiberboard. In: International Conference "Wood Science and Engineering in the Third Millennium"–ICWSE, pp 348–352
Savov V, Valchev I, Antov P (2019) Processing factors affecting the exploitation properties of environmentally friendly medium density fibreboards based on lignosulfonate Adhesives. In: Proceedings of the 2nd International Congress of Biorefinery of Lignocellulosic Materials (IWBLCM2019), pp 165–169
Sellers T (1985) Plywood and adhesive technology. CRC Press
Shimatani K, Sano Y, Sasaya T (1994) Preparation of moderate-temperature setting adhesives from softwood kraft lignin. Part 2. effect of some factors on strength properties and chracteristics of lignin-based adhesives. Holzforschung-Int J Biol Chem Phy Technol Wood 48(4):337–342
Sulaiman NS, Hashim R, Amini MHM, Sulaiman O, Hiziroglu S (2013) Evaluation of the properties of particleboard made using oil palm starch modified with epichlorohydrin. BioResources 8(1):283–301
Sun J, Li L, Cheng H, Huang W (2018) Preparation, characterization and properties of an organic siloxane-modified cassava starch-based wood adhesive. J Adhesion 94(4):278–293
Tan H, Zhang Y, Weng X (2011) Preparation of the plywood using starch-based adhesives modified with blocked isocyanates. Proced Eng 15:1171–1175
Tester RF, Karkalas J, Qi X (2004) Starch—composition, fine structure and architecture. J Cereal Sci 39(2):151–165
Tomkinson J (2003) 2.3 adhesives based on natural resources. State-of-the-art report, 46

Tondi G, Wieland S, Wimmer T, Schnabel T, Petutschnigg A (2012) Starch-sugar synergy in wood adhesion science: basic studies and particleboard production. Eur J Wood Wood Prod 70(1–3):271–278

Trosa A, Pizzi A (2001) A no-aldehyde emission hardener for tannin-based wood adhesives for exterior panels. Holz Als Roh-Und Werkstoff 59(4):266–271

Ucar MB, Ucar G, Pizzi A, Gonultas O (2013) Characterization of Pinus brutia bark tannin by MALDI-TOF MS and 13C NMR. Ind Crops Prod 49:697–704

Umemura K, Sugihara O, Kawai S (2014) Investigation of a new natural adhesive composed of citric acid and sucrose for particleboard II: effects of board density and pressing temperature. J Wood Sci 61(1):40–44. https://doi.org/10.1007/s10086-014-1437-8

Valarmathi T, Palanikumar K, Sekar S, Latha B (2020) Investigation of the effect of process parameters on surface roughness in drilling of particleboard composite panels using adaptive neuro fuzzy inference system. Mater Manufact Process, 1–9

Valenzuela J, Von Leyser E, Pizzi A, Westermeyer C, Gorrini B (2012) Industrial production of pine tannin-bonded particleboard and MDF. Eur J Wood Wood Prod 70(5):735–740

Valyova M, Ivanova Y, Koynov D (2017) Investigation of free formaldehyde quantity in production of plywood with modified urea-formaldehyde resin.

Wang P, Cheng L, Gu Z, Li Z, Hong Y (2015) Assessment of starch-based wood adhesive quality by confocal Raman microscopic detection of reaction homogeneity. Carbohydr Polym 131:75–79. https://doi.org/10.1016/j.carbpol.2015.05.044

Wang S, Ma S, Xu C, Liu Y, Dai J, Wang Z, Liu X, Chen J, Shen X, Wei J (2017) Vanillin-derived high-performance flame retardant epoxy resins: facile synthesis and properties. Macromolecules 50(5):1892–1901

Wang Z, Gu Z, Hong Y, Cheng L, Li Z (2011) Bonding strength and water resistance of starch-based wood adhesive improved by silica nanoparticles. Carbohyd Polym 86(1):72–76

Wang Z, Li Z, Gu Z, Hong Y, Cheng L (2012) Preparation, characterization and properties of starch-based wood adhesive. Carbohyd Polym 88(2):699–706

Warmbier K, Wilczyński A, Danecki L (2013) Properties of one-layer experimental particleboards from willow (Salix viminalis) and industrial wood particles. Eur J Wood Wood Prod 71(1):25–28

Wechsler A, Zaharia M, Crosky A, Jones H, Ramírez M, Ballerini A, Nuñez M, Sahajwalla V (2013) Macadamia (Macadamia integrifolia) shell and castor (Rícinos communis) oil based sustainable particleboard: A comparison of its properties with conventional wood based particleboard. Mater Des 50:117–123

Westin M, Simonson R, Ostman B (2001) Kraft lignin wood fiberboards-The effect of kraft lignin addition to wood chips or board pulp prior to fiberboard production. Holz Als Roh-Und Werkstoff 58(6):393–400

Wool R, Kusefoglu S, Khot S, Zhao R, Palmese G, Boyd A, Williams G, LaScala J (1998) Affordable bio-derived plastics and binders for the composite industry. In: Proceedings 2nd European Panel Products Symposium, Bangor, Wales

Xi X, Pizzi A, Gerardin C, Chen X, Amirou S (2019) Soy protein isolate-based polyamides as wood adhesives. Wood Sci Technol, 1–14

Xian-qing X, Ying-ying Y, Yi-ting N, Liang-ting Z (2020) Development of a cornstarch adhesive for laminated veneer lumber bonding for use in engineered wood flooring. Int J Adhes Adhes 98:102534

Xu H-N, Ma S, Lv W, Wang Z (2011) Soy protein adhesives improved by SiO2 nanoparticles for plywoods. Pigm Resin Technol 40(3):191–195

Xu Y, Han Y, Shi SQ, Gao Q, Li J (2020) Preparation of a moderate viscosity, high performance and adequately-stabilized soy protein-based adhesive via recombination of protein molecules. J Clean Product 255:120303

Yang S, Zhang Y, Yuan TQ, Sun RC (2015) Lignin–phenol–formaldehyde resin adhesives prepared with biorefinery technical lignins. J Appl Poly Sci 132(36)

Yotov N, Valchev I, Petrin S, Savov V (2017) Lignosulphonate and waste technical hydrolysis lignin as adhesives for eco-friendly fiberboard. Bul Chem Commun 49:92–97

Yousuf A, Pirozzi D, Sannino F (2020) Fundamentals of lignocellulosic biomass. In: Lignocellulosic Biomass to Liquid Biofuels. Elsevier, pp 1–15

Yuan Y, Guo MH, Liu FY (2014) Preparation and evaluation of green composites using modified ammonium lignosulfonate and polyethylenimine as a binder. BioResources 9(1):836–848

Zhang S, Liu T, Hao C, Wang L, Han J, Liu H, Zhang J (2018) Preparation of a lignin-based vitrimer material and its potential use for recoverable adhesives. Green Chem 20(13):2995–3000

Zhang W, Ma Y, Wang C, Li S, Zhang M, Chu F (2013) Preparation and properties of lignin–phenol–formaldehyde resins based on different biorefinery residues of agricultural biomass. Ind Crops Prod 43:326–333

Zhao XF, Peng LQ, Wang HL, Wang YB, Zhang H (2018) Environment-friendly urea-oxidized starch adhesive with zero formaldehyde-emission. Carbohydr Polym 181:1112–1118. https://doi.org/10.1016/j.carbpol.2017.11.035

Zhou X, Tang L, Zhang W, Lv C, Zheng F, Zhang R, Du G, Tang B, Liu X (2011) Enzymatic hydrolysis lignin derive from corn stover as an intrinstic binder for bio-composites manufacture: effect of fiber moisture content and pressing temperature on board's properties. BioResources 6(1):253–264

Zhu L (2014) Preparation method for high-strength water-resistant starch adhesive for corrugated paperboards. CN

Synthesis Process of Various Types of Bio-adhesives

Muhamad Fareez Ismail, Muhammad Zaki Ramli, Ainil Hawa Jasni, Wei Yun Chew, Mohamad Helmi Abd Mubin, and Li San Chan

Abstract Bioadhesives are gaining more and more interest in various biomedical and drug delivery markets primarily due to the advancement of the synthesis technique and fabrication technology. The performance of bioadhesive systems, derived from either synthetic and biological source, largely depends on the bioadhesion bonding state of the biopolymers which may be varied by their cross-linking properties, chain length, and presence of various functional groups. Due to its biocompatibility, it can get in close contact with the biologic substrate or adhere to the biological surface (mucoadhesion). Thus, bioadhesives, also often called mucoadhesive, can be practically exploited in the various biomedical application (e.g. wound healing, blood restrainer and surgical suture) as well as in drug delivery system (e.g. hydrogel, nano- or microcapsules and membrane bioadhesive). Therefore, in this chapter, bioadhesion mechanism was briefly discussed. We also presented various synthesis methods and recent advances in bioadhesive technologies for abovementioned application.

Keywords Bioadhesive · Mucoadhesion · Synthesis techniques · Microencapsulation · Hydrogel

M. F. Ismail (✉)
Department of Oral Biology and Biomedical Sciences, Faculty of Dentistry, MAHSA University, Saujana Putra Campus, 42610 Jenjarom, Selangor, Malaysia
e-mail: fareezismail88@yahoo.com; muhamadfareez@mahsa.edu.my

M. Z. Ramli
Royal College of Medicine Perak, Universiti Kuala Lumpur (UniKL), Jalan Greentown, 30450 Ipoh, Perak Darul Ridzuan, Malaysia

A. H. Jasni
Faculty of Science and Defence Technology, National Defence University of Malaysia, Kem Sungai Besi, 57000 Kuala Lumpur, Malaysia

W. Y. Chew · M. H. Abd Mubin · L. S. Chan
Centre of Pre-U Studies, MAHSA University, Saujana Putra Campus, 42610 Jenjarom, Selangor, Malaysia

M. Jawaid et al. (eds.), *Eco-Friendly Adhesives for Wood and Natural Fiber Composites*, Composites Science and Technology,
https://doi.org/10.1007/978-981-33-4749-6_2

1 Introduction

Bioadhesive is defined as a glue that is synthetically and biologically produced and designed to facilitate adherence of substances on the skin, mucin or epithelial tissue via interfacial forces, for extended periods (Khanlari and Dubé 2013). Owing to its biocompatibility, bioadhesive has now been recognized as a very promising unidirectional transmucosal drug delivery system (Gong et al. 2017; Zhang et al. 2016) and sutureless tissue sealants (Sani et al. 2019; Tavafoghi et al. 2020). The efficiency of bioadhesive systems lies on the bioadhesion bonding state of the biopolymers which may be varied by their cross-linking properties, chain length, and presence of various functional groups (Kumar et al. 2017). Therefore, current research is replacing the first-generation charged hydrophilic polymer networks to more specific second-generation polymers due to presence of thiol, lectin and other adhesive functional groups which is proven to form stronger covalent bonds with the mucus and the underlying cell layers for improved therapeutic interactions (Andrews et al. 2009). Bioadhesives are currently being used in various biomedical application [i.e. wound healing, body fluid restrainer (e.g., blood) and suture development] and drug delivery system (i.e. hydrogel, micro- or nanocapsules and membrane bioadhesive) among others (Gong et al. 2017; Khurana et al. 2016; Lim et al. 2017; Mendonsa et al. 2017). However, the term has also been used to describe the natural-based polymeric materials that act as adhesives as alternatives to petroleum-based synthetic adhesives in various application including wood and environmental application (Monroy et al. 2019; Zhang et al. 2020b). It is a completely different meaning of bioadhesives highlighted in this chapter. Herein, we presented various synthesis methods and recent advances in bioadhesive technologies for both biomedical and pharmaceutical application. Bioadhesion mechanism was also briefly discussed.

2 Bioadhesion Mechanism

Fundamentally, bioadhesion is the adherence of molecules to biological surfaces. One distinctive characteristic of bioadhesion is that it normally happens while water is present. This approach avoids the unique categorisation of bioadhesion and allows for appropriate mechanisms to be used to describe and interpret the molecular processes that are involved in forming each type of interfacial bond. According to Park et al. (1986), there are three different types of bioadhesion:

Type I: adhesion between two or more biological objects.
Type II: Biological substance adhesion to an artificial substrate.
Type III: Artificial component adhesion to a biological substrate.

For drug administration purpose, some pharmaceutical formulation with active drugs are impregnated into bioadhesive molecules (Kumria et al. 2018). The formulation will be the anchor on/at the biological surface and the drug will be released

Fig. 1 The process of contact and consolidation

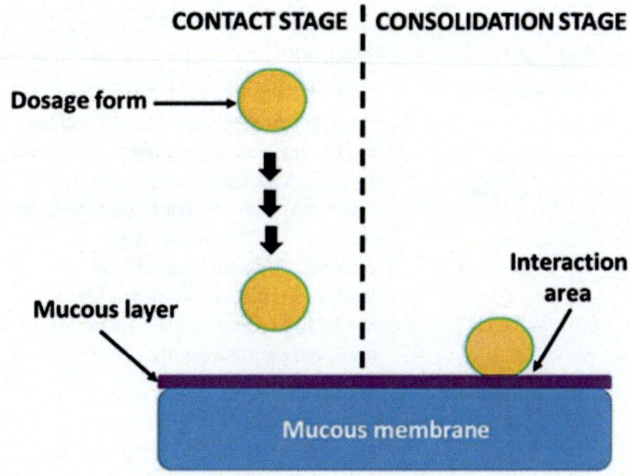

close to the absorptive membrane, with a consequent enhancement of bioavailability (Morales et al. 2017). A more specific term than bioadhesion is mucoadhesion which is when the biological surface is the mucosal tissue. Nevertheless, the interaction between polymers and the biological surface is still not well understood.

The mechanism of mucoadhesion is generally divided into two steps which are the contact stage and the consolidation stage (Fig. 1). The contact stage explains the contact between the mucoadhesive polymer and the mucus membrane, with spreading and swelling of the formulation (Sudheer 2018). In the consolidation stage, here mucoadhesive materials are activated by the presence of moisture, plasticizes the system, allows the mucoadhesive molecules to break free and further bonded by weak Vander Waals and hydrogen bonds (Tangri et al. 2017).

Mucoadhesion is a complex process and numerous theories have been proposed to explain the mechanisms involved. These theories include mechanical interlocking, electrostatic, diffusion interpenetration, adsorption and fracture processes. Table 1 shows the theories involved in mechanisms of mucoadhesion.

3 Synthesis Process of Various Type of Bio-adhesives

3.1 Hydrogel Synthesis

Hydrogels can be synthesized via various techniques. Since a comprehensive discussion on all available techniques is beyond the scope of this chapter, a brief overview is included. Traditionally, physical crosslinking, chemical cross-linking, polymerization grafting, and radiation cross-linking have been used for the preparation of hydrogels (Mavila et al. 2016). In recent years, both physical and chemical methods is actively employed.

Table 1 The theories that involved in mechanisms of mucoadhesion

Theory	Description	References
Electronic theory	The adhesion occurs by means of electron transfer between the mucus and the mucoadhesive happen through their different electronic structures. The electrical double layer of charges at the mucus and mucoadhesive surface will be created. Electrostatic forces are essential for bond adhesion rather than high joint strength	Asati et al. (2019), Kaur and Singh (2019)
Adsorption theory	The adhesion occurs by intermolecular forces (hydrogen bonding) and Vander Waal' forces	Brahmbhatt, (2017), Markov et al. (2019)
Diffusion theory	The adhesion occurs by the interpenetration of both polymer and mucin chains to a sufficient depth to create a semi-permanent adhesive bond. The adhesion force increases with the degree of penetration of the polymer chains	Bruschi et al. (2017)
Wetting theory	The adhesion occurs by the surface scattering property of liquid systems measured by the contact angle. As a general rule, the lower the contact angle, the greater is the affinity. The contact angle should be equal or close to zero to provide adequate spreadability	Akbari et al. (2016), Pandey et al. (2020)
Mechanical theory	The adhesion occurs by the diffusion of the liquid adhesives into the micro-cracks and irregularities present on the substrate surface thereby forming an interlocked structure which gives rise to adhesion	Akbari et al. (2016)
Cohesive theory	The adhesion occurs by the intermolecular interactions amongst like molecules	Tekade et al. (2019)

i. *Physical cross-linking*

In physical cross-linking hydrogels, dissolution is prevented by physical interactions existing between polymeric chains (Xu et al. 2016). It is noteworthy to say that systems synthesized through mild conditions of physical crosslinking methods tend to be frailer in comparison with their counterparts fabricated via the covalently crosslinking method (Mahinroosta et al. 2018). These systems become stable by relatively weak interactions between polymer chains such as hydrogen bonds, hydrophobic or ionic interactions (Mahinroosta et al. 2018).

3.2 *Chemical Cross-Linking*

In chemically crosslinked hydrogels, covalent bonds are present between the polymeric chains. The well-known chemical methods include chemical cross-linking, grafting of monomers, or applying a cross-linker to interconnect two polymer chains are the well-known chemical method (Zhang and Khademhosseini 2017). Chemical crosslinks require a mediator agent for the reaction to occur, but the formed gels offer higher mechanical stability since covalent bonds are stronger (Shen et al. 2016). The most commonly used methods for the chemically cross-linked hydrogels in fabrication are to employ heterogeneous polymerization reactions in the presence of bifunctional or multifunctional cross-linkers (Shen et al. 2016). Cross-linking of natural and synthetic polymers can be attained by the reaction of some functional groups such as hydroxyl, carboxylic, and amine groups with cross-linkers such as aldehydes (Hamedi et al. 2018). Grafting includes polymerization of a monomer. Polymer chains are activated by chemical reagents or operation of high-energy radiation (Pino-Ramos et al. 2017). Accordingly, the grafting is classified into either chemical or radiation types.

3.3 *Micro- and Nanosphere System*

3.3.1 Solvent Evaporation

Solvent evaporation is frequently used to obtain controlled release formulations for microencapsulation of various substance in the pharmaceutical industries (Iqbal et al. 2015). The choice of a process that will provide adequate drug encapsulation usually depends on the hydrophilicity or hydrophobicity of the active molecules. There are various ways of using solvent evaporation techniques for microencapsulation which are single emulsion-solvent evaporation, double emulsion-solvent evaporation and spontaneous emulsification (Li et al. 2008). The advantages and disadvantages of these bioadhesives production system are discussed in Table 2 whilst some of the examples are highlighted in Table 3. In the bio-adhesive industry, the utilization of traditional natural adhesive materials of polysaccharides and proteins had been considered compared to the solvent to water-based or high solid adhesives (Brynjelsen et al. 2014).

Table 2 Pros and cons of various solvent evaporation techniques

Technique	Pros	Cons
Single emulsion-solvent evaporation	It provides high lipophilic agent entanglement The particle size can be modified by adjusting homogenization, stability, organic and aqueous phases viscosity (Hernández-Giottonini et al. 2020)	Hydrophilic drugs were badly entrapped It's challenging to expand the production scale unless integrating with nanoemulsion but it will consume more energy (Hernández-Giottonini et al. 2020) Require purification step
Double emulsion-solvent evaporation	It offers an advantage for the encapsulation of both hydrophilic and hydrophobic active ingredients (Hernández-Giottonini et al. 2020)	Chunky particles (polydisperse) and non-uniform particles Two-step process Leakage in the external aqueous process of the hydrophilic active layer Scale-up is difficult Thermodynamically unstable Require high steering and high-pressure homogenization (Nguyen et al. 2013)
Spontaneous emulsification/ solvent diffusion	It allows thermosensitive substances to be integrated Decreases particle size mean and distribution in a narrow range Strong reproducibility batch-batch No homogenizers Higher lipophilic drug entanglement Non-toxic solvents are used Ease of up-scale, rapid and easy operation Simple and cheap High efficiency due to low energy consumption (Hernández-Giottonini et al. 2020)	Final formulation concentration is important In the final solution, potential organic solvent residues Ineffective hydrophilic substance encapsulation Longer time required for emulsion agitation Require a greater amount of water for the formation of nanoparticles (Chen et al. 2014)

3.3.2 Single Emulsion-Solvent Evaporation

Single emulsion technique involves 2 basic steps of emulsification into a volatile solvent and secondly solvent evaporation. Alternatively, lyophilization which is a patented process is used to completely remove the solvent and to produce submicron particles of less than 500 nm (Kızılbey 2019). For instance, a study by Luo et al., showed that bio-adhesive soybean meal-based's (SM) mechanical properties were improved by combination with the Acrylic Emulation (AE) via Interpenetrating

Table 3 Preparation of bioadhesive using solvent evaporation technique

Bioadhesive	Stabilizer/crosslinker/surfactant/emulsifier	Polymer	Solvent	Purpose	Findings	References
Soybean	Triglycidylamine	Acrylic	Sodium hydroxide	To improve the water-resistance of soybean meal-based (SM) bioadhesive	Improved toughness by 47.9% with acrylic emulsion network	Nguyen et al. (2013)
Seaweed alginate	Renatured DNA	Seaweed polysaccharides	–	To bind silicon and silicon graphite electrodes	The DNA-alginate binder allows homogeneous electrode distribution	Sutthapitaksakul and Sriamornsak (2019)
Lignin	Polyether diamine	Epoxy	Ethyl acetate ethanol, methanol and acetone	To tune network structure and properties of thermosets based on actual lignin fractions	Higher lignin molecular fraction has high thermomechanical strength	Li et al. (2020)
Dried corn starch	Lauryl sodium sulfate (LSS) and alkylphenol ethoxylates (APEO)	Corn starch polysaccharides	Hydrochloric acid (HCl)	To study the mobility and storage stability effects of compound surfactants on wood adhesives with a high starch content	The 7:1 LSS-APEO mixing ratio had the best effect on improvements in HSWA flowability and storage space	Trotta et al. (2001)

Network (IPN) of the mixed plywood with triglycidylamine (TGA). The addition of 8% AE in the bio-adhesive resulted in a 24.6% improvement in water resistance, 44% increase in shear strength to 1.80 MPa, and an increase in the wet shear strength of 47.9% to 1.05 MPa. The enhancements were due to the development of IPN which increased the solid content and increased the adhesive's toughness (Nguyen et al. 2013).

3.3.3 Double Emulsion-Solvent Evaporation

The double emulsion is a dynamic mechanism that involves even the smallest of dispersed phases and is also known as the emulsion of emulsion. There are 2 categories of double emulsion which are water-oil-water (w/o/w) and oil-water-oil (o/w/o) (Hernández-Giottonini et al. 2020). Unlike conventional emulsions with an adequate emulsifier, double emulsions need at least two emulsifiers: the lipophilic emulsifier is needed to stabilize internal water droplets, while the oil droplets have hydrophilic emulsion to guard against coalescence (Iqbal et al. 2015). Double emulations once formed are subjected to several phenomena of coalescence and diffusion which ultimately affect product properties such as the performance of texture or encapsulation. The analysis also indicates that texture changes or the encapsulated active material is lost easily but it cannot indicate which mechanism is responsible (coalescence and/or diffusion). As different emulsifiers adsorb each other on an interface, they react. Changes in dual emulsion stability should be expected. The emulsifier's molecular structure and oil phase affect the distribution between the phases of their emulsifiers. Therefore, to balance the capillary pressure between the inner and outer water phases, the introduction of osmotic active substances to the inner water phase to prevent ripening of Ostwald is normal. Over time, the propagation of osmotic substances through diffusion or coalescence may also occur (Sievens-Figueroa et al. 2012). The adhesive in lithium-ion batteries is one example of this. To ensure a consistent distribution, the adhesive must be compatible with all substances (Sutthapitaksakul and Sriamornsak 2019). Kim et al. (2005) engineered a protein-inspired amphiphilic adhesive that includes both the hydrophobic backbone protein and hydrophilic oligosaccharides, inspired by mucin, an amphiphilic macromolecular lubricant which had expedited the distribution of the electrodes.

3.3.4 Hot Melt Microencapsulation

In the field of drug supply systems, microencapsulation is among the most fascinating fields (Devi et al. 2017). It is an interdisciplinary field that needs to know the pure polymer science, to be familiar with emulsion technology and to have a profound understanding of the stabilization of drugs and proteins (Sayes et al. 2017). This field was considered more art than science in the early 1970s, as most of the research was conducted in drug companies and very few information was discussed in scientific meetings (Rach et al. 2020). In major pharmaceutical companies and universities, as well as research institutes, the topic of microencapsulation is now widely studied. While scientists in the early 1970s primarily dealt with the encapsulation of carbonless paper dyes, scientists today have mastered technology to such a degree as to encapsulate cells as well as delicate proteins and genes (Morales-Narváez et al. 2017).

In one or more polymer coatings, microencapsulation is techniques for capturing solids, liquids or gasses (Suganya and Anuradha 2017). There have been two evolved, chemical and physical classes of encapsulation techniques. Polymerization during preparation of the microcapsules is the first class of encapsulation (Lim 2019). For instance, interface polymerization or in situ polymerization is usually known as this class (Zoppe et al. 2017). The second form includes controlled precipitation of a polymer solution where there are normal physical changes. The number of encapsulation processes is increasing and enormous. Therefore, several new inventions grow only from new ways of generating microspheres. Most researchers are therefore trying to establish a systematic nomenclature for Classification of encoding. The various system forms were listed by literature in Table 4. Encapsulation methods are sometimes difficult to classify because certain methods can be hybrids of two or can

Table 4 Classification of microencapsulation methods

Process	Coating material	Suspended medium
Interfacial polymerization	Water-soluble and insoluble monomers	Aqueous/organic solvent
Complex coacervation	Water-soluble polyelectrolyte	Water
Coacervation	Hydrophobic polymers	Organic Solvent
Thermal denaturation	Proteins	Organic
Salting-out	Water-soluble polymer	Water
Solvent evaporation	Hydrophilic or hydrophobic polymers	Organic or water
Hot Melt	Hydrophilic or hydrophobic polymers	Aqueous/organic solvent
Solvent removal	Hydrophilic or hydrophobic polymers	Organic solvents
Spray-drying	Hydrophilic or hydrophobic polymers	Air, nitrogen
Phase separation	Hydrophilic or hydrophobic polymers	Aqueous/organic

use different mechanisms concurrently, or more processes (Villemin et al. 2019). Some terms have also changed over the years for an example, solvent evaporation is called water drying and double emulsion, which may create confusion (Muschiolik and Dickinson 2017).

To avoid using solvents during the entire process, hot melt encapsulation (Patil et al. 2016) was created. As a modification of the solvent evaporation technique, solvent removal (Subedi et al. 2016) was introduced by using organic solvents as an extraction medium. In spray-drying, a special temperature-controlled cyclone is used to vaporize the solvent. Finally, phase separation is a new process of developing a double-walled microsphere by one-step precipitation of two or more polymers (Zhang et al. 2020a). The process of hot melting is carried out by the extrusion of hot melting or hot melting fusion. Hot melt extruder involves the pre-mixing and a particular excipient, for example, wax or polymer, passed through an extruder at a specified temperature, over or above a wax melting point, to allow the dispersing of drug particles inside a soft or sticky excipient that usually results in a solid dispersion (Jadav and Paradkar 2020). Hot melt or melt granulation is achieved by melting the wax and dispersing the drug into the melt before cooling the mixture (Nart et al. 2017). Table 5 shows the drugs masked in the last three years with hot melting techniques.

A solid dispersal is an inert hydrophilic carrier matrix device in which an active component or more is molecularly dispersed. The formulation of an API in a solid dispersion transforms the waterproof crystalline shape into the amorphous form, thus increasing its solubility. Numerous methods, including melt fusion and solvent evaporation methods, are widely used in preparation for solid dispersions. Hot melt extrusion is a technique which is extremely suitable to shape solid dispersions, and its main advantages are that the method needs no solvent application. Consequently, the related stability risks associated with solvents, since no residual solvent is present, are eliminated during the shelf life of the product. The TSE emerges as the most viable

Table 5 List of drugs masked by hot melting techniques

Drug	Method	Taste masking material
Griseofulvin and caffeine anhydrous	Hot-melt extrusion	Maltodextrin
Mefenamic acid	Hot-melt extrusion	Eudragit E
Ondansetron HCl	Hot-melt fusion	Glycerol monostearate
Cetirizine HCl and verapamil HCl	Hot-melt extrusion	Eudragit L
Caffeine citrate	Hot-melt extrusion	Ethylcellulose
Arbidol hydrochloride	Hot-melt fusion	Octadecanol
Paracetamol	Hot-melt extrusion	Eudragit EPO®, and Kollidon VA64®

option in the pharmaceutical industry for solid dispersion production, based on the equipment's configuration and processing temperature (Aleksovski et al. 2016).

The development of drug systems includes technological designs that maximize the therapeutic effects of drugs by controlling their biodistribution profiles. To this end, an ideal selectivity and specificities for the target tissues or cells are needed for the delivery system (Vizovišek et al. 2018). Different strategies for targeted drug delivery have been developed in recent years and researchers currently concentrate on hot-melt extrusion as a prospective and viable solution. A photosensitiser was given to the colon through a formula developed using hot melt extrusion for targeted drug delivery (Thakkar et al. 2020). Eudragit S 100 was used as a polymer by the researchers, and TEC was extruded to formulation composed of medicine, polymer, photosensitiser and plasticiser (Watts and Williams III 2016). The new treatment choice for multi-medicine resistant species like *Enterococcus faecalis* and *Bacteroides fragilis*, inhabiting a colon, photodynamic antimicrobial chemotherapy (PACT). There was no release of drugs in the acidic medium in the results of this study. For those formulations, the authors demonstrated that the release of photosensitisers would theoretically target the colon and that PACT could destroy the pathogen residing there (Dąbrowski et al. 2016).

The unpleasant bitter taste is present in several active ingredients used for medicinal products. If drugs dissolve in saliva, they interact in tongue with taste receptors and create bitter, sweet, or other feelings of taste by signal transduction from the organs of the taste receiver which are also known as taste buds. These taste buds have highly sensitive nerve endings that produce and transmit electric pulses in the areas of the brain devoted to tasting, via the seventh, the ninth and tenth cranial nerves. Tasting masking is therefore necessary in order to improve the palatability of bitter drugs. There are two ways to conquer foul tastes. Firstly, the solubility of the product in the saliva would be reduced by transforming it from the amorphous into the crystalline. Furthermore, the link between the drug and the gout receptors is altered (Alqurshi et al. 2017). By formulating solid dissipations with a taste masking polymer, hot-melt extrusion was used as a taste masking technique for bitter active ingredients. These solid dispersions avoid bitter drug release in the saliva, thereby avoiding the contact between medicine molecules and taste bud (Liu et al. 2017). The hydrogen bonding between the active ingredient and the polymer matrix will mask the taste through the processing of counter loaded compounds (Saifullah et al. 2019).

3.3.5 Spontaneous Emulsification/Solvent Diffusion

Spontaneous Emulsification refers to immiscible fluid emulsions that develop without external/mechanical support, such as low interfacial tension (Chen et al. 2014). This technique is beneficial in creating micro-and nanodroplets with less energy. Polymeric nanoparticles and pharmaceutical nanosuspensions are often prepared by spontaneous nanoemulsion treatment with the method of solvent diffusion (Iqbal et al.

2015). When immiscible liquids in non-equilibrium conditions are in contact, spontaneous emulsification is triggered by gradients of chemical potential between the two phases, which under certain conditions lead to negative values of free energy of emulsification. Spontaneous emulsification methods have been proposed in various industries, including heavy oil, high temperatures, high-salinity and low-permeability oil tanks, as well as waste-to-water treatments from standard oil reservoirs to hard oil tanks (Aldousary and Kovscek 2019). Lignin is usually used to produce less toxic bio-adhesives (Date et al. 2010). While there is typically an empirical application of lignin in the material synthetic application of chemical complexity and inherent heterogeneity, Gioia et al. (2018) investigated the possibility for regulating lignin at molecular levels for the first time to tailor the thermomechanical characteristics of end materials especially with the integration of epoxy. Kraft lignin was standardized to four distinctive fractions, characterized by low polydispersity and increased molecular weight. The mechanical properties of the materials have shown a good provision with higher fractions of molecular weight (Li et al. 2020).

In another case, the combination of alkylphenol ethoxylates (APEO) with lauryl sulfate (LSS) was used to improve the strength of a high-starch wood adhesive (HSWA) of dried corn starch. Due to the electric charge forces and space, steric impediments induced on the surface of latex particles by the LSS/APEO adsorbed mixture, the addition of LSS/APEO (Mixing Ratio 7:1) mix had prevented latex particles and starch molecules from aggregating significantly. This study can resolve the storage problem of starch-based adhesives efficiently, particularly for high-starch samples which competitively reduce additional production costs and the environmental performance of the wood adhesive industry (Trotta et al. 2001).

3.3.6 Emulsion Reverse Salting-Out

Emulsification Reverse Salting-out is a polymeric nanoparticles preparation method first proposed by Allémann et al. (1992). It is a set of procedures based on preformed polymers (Paliwal et al. 2014). Examples of polymeric nanoparticles formed by using this method are listed in Table 6. Firstly, the emulsion formed by using conventional o/w emulsification (Crucho and Barros 2017). Polymer and drug(s) are added into the water-miscible organic solvent. A viscous solution is prepared separately by

Table 6 Chemicals used in emulsification reverse salting-out method

Polymer	Organic solvent	Salting out agent	Surfactant	References
PLA	Acetone	$MgCl_2$	PVA	Cirstoiu-Hapca et al. (2009)
PLGA	Acetone, DCM, and DMSO	$MgCl_2$	PVA	Van de Ven et al. (2012)
PLA	Acetone	$CaCl_2$	PVA	Vettor et al. (2010)
PLA, PLGA	Acetone	$MgCl_2$	PVA	Zweers et al. (2003)

dissolving salting agent in water with added surfactant (Vettor et al. 2010). Some of the common polymers used are poly(ethylene oxide) (PEO), poly(d, l-lactic acid) (PLA), poly(trimethylene carbonate) (PTMC), and poly(lactic-co-glycolic acid) (PLGA) (Lee et al. 2016; Mendoza-Muñoz et al. 2012; Rao and Geckeler 2011). Acetone, tetrahydrofuran (THF) and acetonitrile are some of the common organic solvent (Akbari et al. 2016; Konan et al. 2002; Mendoza-Muñoz et al. 2012; Rao and Geckeler 2011). Combination of organic solvents is possible as shown by Van de Ven et al. (2012) where they employed combined solvents of Acetone, dichloromethane (DCM) and dimethyl sulfoxide (DMSO) as an organic solvent.

The polymer-solvent emulsified by adding an aqueous solution comprising of salting-out agent and surfactant to form an o/w emulsion (Paliwal et al. 2014). The presence of salting-out agent(s) in high concentration causing an alteration of the miscibility properties between the water and the polymer-solvent. Based on this principle, salt or sucrose with strong salting-out effect in the aqueous phase are recommended as a salting-out agent (Rao and Geckeler 2011; Vauthier and Bouchemal 2009). Some of the well-known examples of salting-out agents are magnesium chloride, magnesium acetate, and calcium chloride (Mendoza-Muñoz et al. 2012; Rao and Geckeler 2011). For the surfactant/colloidal stabiliser, the common examples are Polyvinyl Alcohol (PVA) and Polyvinylpyrrolidone (Lee et al. 2016).

The o/w emulsion undergo the "reverse salting-out effect" by the added large quantity of water to decrease the concentration of the salting-out agents (Zhuang et al. 2017). The organic solvent diffused into the aqueous phase through this process. The polymer encapsulated the drugs will then precipitate; nanoparticles are produced. Next, cross-flow filtration is employed to eliminate the organic solvent and salting-out agent from the nanoparticles (Ibrahim et al. 1992).

The advantage of this preparation method is the o/w emulsion can be achieved without using high shear forces due to an Ouzo-effect. Besides, the preparation does not need to include toxic chlorinated solvents (Crucho and Barros 2017). This preparation is also suitable for heat-sensitive substances because the preparation does not involve heat. However, the intensive purification of the nanoparticles from the solvent and salting-out agent is the drawback of this preparation (Lee et al. 2016).

As for the size of the nanoparticles produced, they can vary between 186 and 1130 nm. The key factors that can affect the size of the nanoparticles formed by this employing this method are the selection of the combination of salting-out agents and organic solvent, the concentration of polymer in the organic phase, stirring speed during the preparation, the concentration of PVA and ratio of aqueous to the organic phase.

3.3.7 Nanoprecipitation

Nanoprecipitation is a procedure based on the preformed polymers (Paliwal et al. 2014). It is a one-step process first developed by Fessi et al. (1989). Commonly, the nanoparticles are formed by the addition of the organic phase into the anti-solvent phase while mixing (Table 7). However, it also can be formed by adding an aqueous

Table 7 Chemicals used in the nanoprecipitation method

Polymer	Solvent	Non-solvent	Surfactant	Particle size (nm)	Remark	References
Poly(hydroxy)urethane (PHU)	Dimethylsulfoxide (DMSO)	Water	Sodium dodecyl sulfate (SDS)	55–179	–	Quérette et al. (2020)
Poly(lactic-co-glycolide) (PLGA)	Acetone	Water	Polyvinyl alcohol (PVA)	80–120	Drug Delivery: Curcumin; niclosamide	Prabhuraj et al. (2020)
Carbon nitride	Acetonitrile	Concentrated sulfuric acid; Methanesulfonic acid (MSA)	–	~40	For photocatalytic application	Gan et al. (2020)
Poly(lactic-co-glycolide) (PLGA)	Acetone; DMSO	Water	Polyvinyl alcohol (PVA); Pluronic F86; Pluronic F127; Tween 20; Tween 80	100–200	Drug Delivery: PKC inhibitor	Shkodra-Pula et al. (2019)
Poly(lactic-co-glycolide) (PLGA)	Acetonitrile; Acetone; Dimethyl sulfoxide (DMSO)	Water	PVA	≤200	Drug Delivery: N-acetylcyteine	Chiesa et al. (2018)

phase into the organic phase (Crucho and Barros 2017). The nanoparticles formed by this procedure requires polymer, polymer-solvent and non-solvent as main components. Addition of the surfactants in the anti-solvent is optional. Surfactant functioned to stabilise the nanoparticles suspension for long storage (Rao and Geckeler 2011).

Formation of the nanoparticles utilised the 'Ouzo effect' which is the rapid mixture of organic and aqueous phases forming a milky emulsion (Crucho and Barros 2017; Lee et al. 2016). The effect is due to the solvent (that is immiscible with water) rapidly diffused into the aqueous phase inducing interfacial deposition of polymer (Rao and Geckeler 2011). A combination of the correct ratio between solvent, aqueous phase, polymer and drugs are required to achieve the effect (Lee et al. 2016).

Rate of injection of organic phase into the aqueous phase (or vice-versa), the mixing rate of the organic phase and aqueous phase, the concentration of polymer, the surfactant, and aqueous phase volume ratio are the important factors that can influence the formation of nanoparticles (Lee et al. 2016). The technical complexity of the preparation is considerably lower than other preparation methods. Hence, it is practical to use this method for large scale production due to its low technical complexity, fast completion, high consistency, and low energy input requirements (Quérette et al. 2020; Rao and Geckeler 2011; Zhuang et al. 2017).

A study by Prabhuraj et al. (2020) successfully demonstrated that two drugs can be loaded into the nanoparticles using this method. The study loaded curcumin and niclosamide into the PLGA nanoparticles resulting in the uniform particle size of 80–120 nm. The study managed to the achieved high encapsulation efficiency of the curcumin (58.09%) and niclosamide (85.36%). Further testing on the cancer cell revealed that the drug release in higher amount in the acidic environment of the cancer cells (pH 6.0) when compared to the normal cells environment (pH 7.4) (Prabhuraj et al. 2020).

Quérette et al. (2020) employing factorial design in the study to identify 3 key parameters influencing the formation of nanoparticles in terms of particles size and polydispersity index (PDI). The key factors identified were the concentration of the polymer, non-solvent volume, and concentration of the surfactant. Poly(hydroxy)urethane (PHU) were selected as polymer and sodium dodecyl sulphate (SDS) as the surfactants in the study.

Surfactant also can prevent aggregation of the nanoparticles allowing nanoparticles to have higher stability. Additionally, the presence of the surfactant can drastically increase the effects of the water volume and concentration of the PHU on the PDI. The lowest PDI achieved in the study was 0.10 with PHU concentration of 5 g/L, the water volume of 50 ml, and SDS concentration of 25 mM. Meanwhile, the study also highlighted that at low concentration of the PHU (1 g/L) in the presence of the SDS (25 mM), the water volume is the main factor of the particle size which can result in a change from 55 to 179 nm (Quérette et al. 2020).

3.4 Fabrication of Nanocomposite

Nanotechnology is applied in bioadhesive of drug delivery system. Nanoparticles are incorporated into the polymer to form nanocomposite polymer that will be used to build a drug delivery system (Soppimath et al. 2001; Vasir et al. 2003). The nanoparticles are added to improve the mechanical properties, bioadhesive properties and hydrophilicity of the polymers. A new function such as wound healing is introduced to the polymers due to the presence of nanoparticles (Akhlaghi et al. 2013; Chamundeeswari et al. 2010; Khanlari and Dubé 2013; Lee and Tsao 2006).

In terms of bioadhesive properties, the nanoparticles in the nanocomposite polymers enhance the release of drugs as the polymers surface increases with the presence of nanoparticles. Thus, the absorption of the drug into the organ improves with a drug delivery system that is built by nanocomposite polymers. As for mechanical properties, it is proven that the nanocomposite polymers delivery system has a suitable strength that is needed during the delivery of the drug to the organ. Nanoclays is a suitable nanofiller in a drug delivery system as it will improve the mechanical properties and the release of a drug (Luo et al. 2020).

In the further development of nanocomposite, Zinc Oxide Eudragit nanohybrids is synthesized for intestinal drug delivery system. In this study, eudragit solution is added to the ZnO nanoparticles to allow the particle to be coated with eudragit solution. Bridges are formed among every eudragit coated nano-ZnO particle and thus, self-polymerization occurs. After the self-polymerization, many "spheres". Each "sphere" is formed with nano ZnO particles as the core while eudragit as the outer layer. The nano-ZnO enhances the absorption of the drug in the intestine. The nano-Zn particles have a characteristic of resisting gastric acid that makes Zinc Oxide Eudragit nanohybrids to be a good drug delivery system for intestine (Luo et al. 2020). There is a wide room for the development of nanotechnology in bioadhesive as the world is heading towards the nanoscale. Various nanocomposite polymers are developed, and this will lead to new bioadhesives.

4 Conclusion and Future Perspective

New generations of bioadhesives are on the move due to a high level of concern to minimize environmental and health impacts and our reliance on the petroleum-based product. The future of bioadhesives development depends heavily depend on the concerted focus of research effort to boost certain specific characteristics (i.e. wet strength) and on how synthetic adhesives can be integrated into the system. A new generation of bio-based adhesives called bio-inspired adhesives has emerged as a comprehensive materials-based platform for the development of sustainable bioadhesives. Bio-inspired adhesives are based on bioadhesives that can be found in nature and it has great potential to replace the existing substrate. Among these bio-inspired adhesives, the main effort was made to develop mussel-mimetic adhesives.

These bio-inspired adhesives can be combined with synthetic compounds for feature enhancements. It will provide a new and promising way to develop a new generation of eco-friendly adhesives, particularly for biomedical and drug delivery applications that partake the highest growing market for bioadhesives.

References

Akbari J, Saeedi M, Morteza-Semnani K, Zarrabi B, Rostamkalaei SS, Kelidari HR (2016) The effect of Plantago major seed mucilage combined with carbopol on the release profile and bioadhesive properties of propranolol HCl buccoadhesive tablets. Pharm Biomed Res 2(2):84–100

Akhlaghi SP, Berry RC, Tam KC (2013) Surface modification of cellulose nanocrystal with chitosan oligosaccharide for drug delivery applications. Cellulose 20(4):1747–1764

Aldousary S, Kovscek AR (2019) The diffusion of water through oil contributes to spontaneous emulsification during low salinity waterflooding. J Pet Sci Eng 179:606–614

Aleksovski A, Vervaet C, Dreu R (2016) Hot-melt extrusion and prilling as contemporary and promising techniques in the solvent free production of solid oral dosage forms, based on solid dispersions. Maced Pharm Bull 62:3–24

Allémann E, Gurny R, Doelker E (1992) Preparation of aqueous polymeric nanodispersions by a reversible salting-out process: influence of process parameters on particle size. Int J Pharm 87(1–3):247–253

Alqurshi A, Chan KA, Royall PG (2017) In-situ freeze-drying-forming amorphous solids directly within capsules: an investigation of dissolution enhancement for a poorly soluble drug. Sci Rep 7(1):1–16

Andrews GP, Laverty TP, Jones DS (2009) Mucoadhesive polymeric platforms for controlled drug delivery. Eur J Pharm Biopharm 71(3):505–518

Asati S, Jain S, Choubey A (2019) Bioadhesive or mucoadhesive drug delivery system: a potential alternative to conventional therapy. J Drug Deliv Ther 9(4-A):858–867

Brahmbhatt D (2017) Bioadhesive drug delivery systems: overview and recent advances. Int J Chem Life Sci 6(3):2016–2024

Bruschi ML, Borghi-Pangoni FB, Junqueira MV, de Souza Ferreira SB (2017) Nanostructured therapeutic systems with bioadhesive and thermoresponsive properties. In: Nanostructures for novel therapy. Elsevier, pp 313–342

Brynjelsen S, Doty M, Kipp JE, Jayswal N, Narayanan K (2014) Preparation of submicron sized nanoparticles via dispersion lyophilization: Google Patents

Chamundeeswari M, Sobhana SL, Jacob JP, Kumar MG, Devi MP, Sastry TP, Mandal AB (2010) Preparation, characterization and evaluation of a biopolymeric gold nanocomposite with antimicrobial activity. Biotechnol Appl Biochem 55(1):29–35

Chen C, Yang W, Wang D-T, Chen C-L, Zhuang Q-Y, Kong X-D (2014) A modified spontaneous emulsification solvent diffusion method for the preparation of curcumin-loaded PLGA nanoparticles with enhanced in vitro anti-tumor activity. Front Mater Sci 8(4):332–342

Chiesa E, Dorati R, Modena T, Conti B, Genta I (2018) Multivariate analysis for the optimization of microfluidics-assisted nanoprecipitation method intended for the loading of small hydrophilic drugs into PLGA nanoparticles. Int J Pharm 536(1):165–177

Cirstoiu-Hapca A, Buchegger F, Bossy L, Kosinski M, Gurny R, Delie F (2009) Nanomedicines for active targeting: physico-chemical characterization of paclitaxel-loaded anti-HER2 immunonanoparticles and in vitro functional studies on target cells. Eur J Pharm Sci 38(3):230–237

Crucho CI, Barros MT (2017) Polymeric nanoparticles: a study on the preparation variables and characterization methods. Mater Sci Eng: C 80:771–784

Dąbrowski JM, Pucelik B, Regiel-Futyra A, Brindell M, Mazuryk O, Kyzioł A, Stochel G, Macyk W, Arnaut LG (2016) Engineering of relevant photodynamic processes through structural modifications of metallotetrapyrrolic photosensitizers. Coord Chem Rev 325:67–101

Date AA, Desai N, Dixit R, Nagarsenker M (2010) Self-nanoemulsifying drug delivery systems: formulation insights, applications and advances. Nanomedicine 5(10):1595–1616

Devi N, Sarmah M, Khatun B, Maji TK (2017) Encapsulation of active ingredients in polysaccharide–protein complex coacervates. Adv Colloid Interface Sci 239:136–145

Fessi H, Puisieux F, Devissaguet JP, Ammoury N, Benita S (1989) Nanocapsule formation by interfacial polymer deposition following solvent displacement. Int J Pharm 55(1):R1–R4

Gan Z, Huang C, Shen Y, Zhou Q, Han D, Ma J, Liu S, Zhang Y (2020) Preparation of carbon nitride nanoparticles by nanoprecipitation method with high yield and enhanced photocatalytic activity. Chin Chem Lett 31(2):513–516

Gioia C, Lo Re G, Lawoko M, Berglund L (2018) Tunable thermosetting epoxies based on fractionated and well-characterized lignins. J Am Chem Soc 140(11):4054–4061

Gong C, Lu C, Li B, Shan M, Wu G (2017) Injectable dopamine-modified poly (α, β-aspartic acid) nanocomposite hydrogel as bioadhesive drug delivery system. J Biomed Mater Res Part A 105(4):1000–1008

Hamedi H, Moradi S, Hudson SM, Tonelli AE (2018) Chitosan based hydrogels and their applications for drug delivery in wound dressings: a review. Carbohydr Polym 199:445–460

Hernández-Giottonini KY, Rodríguez-Córdova RJ, Gutiérrez-Valenzuela CA, Peñuñuri-Miranda O, Zavala-Rivera P, Guerrero-Germán P, Lucero-Acuña A (2020) PLGA nanoparticle preparations by emulsification and nanoprecipitation techniques: effects of formulation parameters. RSC Adv 10(8):4218–4231

Ibrahim H, Bindschaedler C, Doelker E, Buri P, Gurny R (1992) Aqueous nanodispersions prepared by a salting-out process. Int J Pharm 87(1–3):239–246

Iqbal M, Zafar N, Fessi H, Elaissari A (2015) Double emulsion solvent evaporation techniques used for drug encapsulation. Int J Pharm 496(2):173–190

Jadav N, Paradkar A (2020) Solid dispersions: technologies used and future outlook. In: Nanopharmaceuticals. Elsevier, pp 91–120

Kaur L, Singh I (2019) Chitosan-Catechol conjugates—a novel class of bioadhesive polymers: a critical review. Rev Adhes Adhes 7(1):51–67

Khanlari S, Dubé MA (2013) Bioadhesives: a review. Macromol React Eng 7(11):573–587

Khurana JV, Mali AM, Mali RS, Chaudhari AU (2016) Comparative evaluation of healing after periodontal flap surgery using isoamyl 2-cyanoacrylate (bioadhesive material) and silk sutures: a split-mouth clinical study. J Indian Soc Periodontol 20(4):417

Kim B-Y, Jeong JH, Park K, Kim J-D (2005) Bioadhesive interaction and hypoglycemic effect of insulin-loaded lectin–microparticle conjugates in oral insulin delivery system. J Control Release 102(3):525–538

Kızılbey K (2019) Optimization of rutin-loaded PLGA nanoparticles synthesized by single-emulsion solvent evaporation method. ACS Omega 4(1):555–562

Konan YN, Gurny R, Allémann E (2002) Preparation and characterization of sterile and freeze-dried sub-200 nm nanoparticles. Int J Pharm 233(1–2):239–252

Kumar L, Verma S, Vaidya B, Gupta V (2017) Bioadhesive polymers for targeted drug delivery. In: Nanotechnology-based approaches for targeting and delivery of drugs and genes. Elsevier, pp 322–362

Kumria R, Al-Dhubiab BE, Shah J, Nair AB (2018) Formulation and evaluation of chitosan-based buccal bioadhesive films of zolmitriptan. J Pharm Innov 13(2):133–143

Lee BK, Yun Y, Park K (2016) PLA micro-and nano-particles. Adv Drug Deliv Rev 107:176–191

Lee WF, Tsao KT (2006) Preparation and properties of nanocomposite hydrogels containing silver nanoparticles by ex situ polymerization. J Appl Polym Sci 100(5):3653–3661

Li M, Rouaud O, Poncelet D (2008) Microencapsulation by solvent evaporation: state of the art for process engineering approaches. Int J Pharm 363(1–2):26–39

Li Z, Xu D, Yuan Y, Wu H, Hou J, Kang W, Bai B (2020) Advances of spontaneous emulsification and its important applications in enhanced oil recovery process. Adv Colloid Interface Sci 277:102119
Lim F (2019) Biomedical applications of microencapsulation. CRC press.
Lim S, Nguyen MP, Choi Y, Kim J, Kim D (2017) Bioadhesive nanoaggregates based on polyaspartamide-g-C18/DOPA for wound healing. Biomacromol 18(8):2402–2409
Liu M, Yin D, Fu H, Deng F, Peng G, Shu G, Yuan Z, Shi F, Lin J, Zhao L, Yin L, Fan G (2017) Double-coated enrofloxacin microparticles with chitosan and alginate: preparation, characterization and taste-masking effect study. Carbohydr Polym 170:247–253
Luo F, Wang M, Huang L, Wu Z, Wang W, Zafar A, Tian Y, Hasan M, Shu, X (2020) Synthesis of zinc oxide eudragit FS30D nanohybrids: structure, characterization, and their application as an intestinal drug delivery system. ACS Omega
Mahinroosta M, Farsangi ZJ, Allahverdi A, Shakoori Z (2018) Hydrogels as intelligent materials: a brief review of synthesis, properties and applications. Mater Today Chem 8:42–55
Markov PA, Khramova DS, Shumikhin KV, Nikitina IR, Beloserov VS, Martinson EA, Litvinets SG, Popov SV (2019) Mechanical properties of the pectin hydrogels and inflammation response to their subcutaneous implantation. J Biomed Mater Res Part A 107(9):2088–2098
Mavila S, Eivgi O, Berkovich I, Lemcoff NG (2016) Intramolecular cross-linking methodologies for the synthesis of polymer nanoparticles. Chem Rev 116(3):878–961
Mendonsa NS, Thipsay P, Kim DW, Martin ST, Repka MA (2017) Bioadhesive drug delivery system for enhancing the permeability of a BCS class III drug via hot-melt extrusion technology. AAPS PharmSciTech 18(7):2639–2647
Mendoza-Muñoz N, Quintanar-Guerrero D, Allémann E (2012) The impact of the salting-out technique on the preparation of colloidal particulate systems for pharmaceutical applications. Recent Pat Drug Deliv Formul 6(3):236–249
Monroy Y, Rivero S, García MA (2019) Sustainable panels design based on modified cassava starch bioadhesives and wood processing byproducts. Ind Crops Prod 137:171–179
Morales-Narváez E, Sgobbi LF, Machado SAS, Merkoçi A (2017) Graphene-encapsulated materials: synthesis, applications and trends. Prog Mater Sci 86:1–24
Morales JO, Fathe KR, Brunaugh A, Ferrati S, Li S, Montenegro-Nicolini M, Mousavikhamene Z, McConville JT, Prausnitz MR, Smyth HD (2017) Challenges and future prospects for the delivery of biologics: oral mucosal, pulmonary, and transdermal routes. AAPS J 19(3):652–668
Muschiolik G, Dickinson E (2017) Double emulsions relevant to food systems: preparation, stability, and applications. Compr Rev Food Sci Food Saf 16(3):532–555
Nart V, Beringhs AOR, França MT, de Espíndola B, Pezzini BR, Stulzer HK (2017) Carnauba wax as a promising excipient in melt granulation targeting the preparation of mini-tablets for sustained release of highly soluble drugs. Mater Sci Eng: C 70:250–257
Nguyen TA, Liu B, Zhao J, Thomas DS, Hook JM (2013) An investigation into the supramolecular structure, solubility, stability and antioxidant activity of rutin/cyclodextrin inclusion complex. Food Chem 136(1):186–192
Paliwal R, Babu RJ, Palakurthi S (2014) Nanomedicine scale-up technologies: feasibilities and challenges. AAPS PharmSciTech 15(6):1527–1534
Pandey N, Soto-Garcia L, Liao J, Zimmern P, Nguyen KT, Hong Y (2020) Mussel-inspired bioadhesives in healthcare: design parameters, current trends, and future perspectives. Biomater Sci
Park K, Cooper SL, Robinson JR (1986) Bioadhesive hydrogels. CRC Press, Danvers, MA
Patil H, Tiwari RV, Repka MA (2016) Encapsulation via hot-melt extrusion
Pino-Ramos VH, Ramos-Ballesteros A, López-Saucedo F, López-Barriguete JE, Varca GH, Bucio E (2017) Radiation grafting for the functionalization and development of smart polymeric materials. In:Applications of radiation chemistry in the fields of industry, biotechnology and environment. Springer, pp 67–94
Prabhuraj R, Bomb K, Srivastava R, Bandyopadhyaya R (2020) Dual drug delivery of curcumin and niclosamide using PLGA nanoparticles for improved therapeutic effect on breast cancer cells. J Polym Res 27(5)

Quérette T, Bordes C, Sintes-Zydowicz N (2020) Non-isocyanate polyurethane nanoprecipitation: toward an optimized preparation of poly (hydroxy) urethane nanoparticles. Colloids Surf A: Physicochem Eng Asp 589:124371

Rach C, Lukas J, Müller R, Sendler M, Simon P, Salloch S (2020) Involving patient groups in drug research: a systematic review of reasons. Patient Prefer Adherence 14:587

Rao JP, Geckeler KE (2011) Polymer nanoparticles: preparation techniques and size-control parameters. Prog Polym Sci 36(7):887–913

Saifullah M, Shishir MRI, Ferdowsi R, Rahman MRT, Van Vuong Q (2019) Micro and nano encapsulation, retention and controlled release of flavor and aroma compounds: a critical review. Trends Food Sci Technol 86:230–251

Sani ES, Kheirkhah A, Rana D, Sun Z, Foulsham W, Sheikhi A, Khademhosseini A, Dana R, Annabi, N (2019) Sutureless repair of corneal injuries using naturally derived bioadhesive hydrogels. Sci Adv 5(3):eaav1281

Sayes CM, Aquino GV, Hickey AJ (2017) Nanomaterial drug products: manufacturing and analytical perspectives. AAPS J 19(1):18–25

Shen X, Shamshina JL, Berton P, Gurau G, Rogers RD (2016) Hydrogels based on cellulose and chitin: fabrication, properties, and applications. Green Chem 18(1):53–75

Shkodra-Pula B, Grune C, Traeger A, Vollrath A, Schubert S, Fischer D, Schubert U (2019) Effect of surfactant on the size and stability of PLGA nanoparticles encapsulating a protein kinase C inhibitor. Int J Pharm 566:756–764

Sievens-Figueroa L, Bhakay A, Jerez-Rozo JI, Pandya N, Romañach RJ, Michniak-Kohn B, Iqbal Z, Bilgili E, Davé RN (2012) Preparation and characterization of hydroxypropyl methyl cellulose films containing stable BCS Class II drug nanoparticles for pharmaceutical applications. Int J Pharm 423(2):496–508

Soppimath KS, Aminabhavi TM, Kulkarni AR, Rudzinski WE (2001) Biodegradable polymeric nanoparticles as drug delivery devices. J Control Release 70(1–2):1–20

Subedi G, Shrestha AK, Shakya S (2016) Study of effect of different factors in formulation of micro and nanospheres with solvent evaporation technique. Open Pharm Sci J 3(1)

Sudheer P (2018) Mucoadhesive polymers: a review. J Pharm Res 17(1):47–55

Suganya V, Anuradha V (2017) Microencapsulation and nanoencapsulation: a review. Int J Pharm Clin Res 9(3):233–239

Sutthapitaksakul L, Sriamornsak P (2019) Influence of process parameters on the characteristics of hydrophilic drug-loaded microparticles through double emulsion solvent evaporation technique. Paper presented at the key engineering materials

Tangri P, Jawla S, Jakhmola V, Mishra R (2017) Highlights of mucoadhesive drug delivery systems: a review. Indo Glob J Pharm 7(2):112–125

Tavafoghi M, Sheikhi A, Tutar R, Jahangiry J, Baidya A, Haghniaz R, Khademhosseini A (2020) Engineering tough, injectable, naturally derived, bioadhesive composite hydrogels. Adv Healthc Mater 9(10):1901722

Tekade M, Maheshwari N, Youngren-Ortiz SR, Pandey V, Chourasiya Y, Soni V, Deb PK, Sharma MC (2019) Thiolated-Chitosan: a novel mucoadhesive polymer for better-targeted drug delivery In: Biomaterials and bionanotechnology. Elsevier, pp 459–493

Thakkar R, Thakkar R, Pillai A, Ashour EA, Repka MA (2020) Systematic screening of pharmaceutical polymers for hot melt extrusion processing: a comprehensive review. Int J Pharm 576:118989

Trotta M, Gallarate M, Pattarino F, Morel S (2001) Emulsions containing partially water-miscible solvents for the preparation of drug nanosuspensions. J Control Release 76(1–2):119–128

Van de Ven H, Vermeersch M, Vandenbroucke R, Matheeussen A, Apers S, Weyenberg W, De Smedt SC, Cos P, Maes L, Ludwig A (2012) Intracellular drug delivery in Leishmania-infected macrophages: evaluation of saponin-loaded PLGA nanoparticles. J Drug Target 20(2):142–154

Vasir JK, Tambwekar K, Garg S (2003) Bioadhesive microspheres as a controlled drug delivery system. Int J Pharm 255(1–2):13–32

Vauthier C, Bouchemal K (2009) Methods for the preparation and manufacture of polymeric nanoparticles. Pharm Res 26(5):1025–1058

Vettor M, Bourgeois S, Fessi H, Pelletier J, Perugini P, Pavanetto F, Bolzinger M (2010) Skin absorption studies of octyl-methoxycinnamate loaded poly (D, L-lactide) nanoparticles: estimation of the UV filter distribution and release behaviour in skin layers. J Microencapsul 27(3):253–262

Villemin E, Ong YC, Thomas CM, Gasser G (2019) Polymer encapsulation of ruthenium complexes for biological and medicinal applications. Nat Rev Chem 3(4):261–282

Vizovišek M, Vidmar R, Drag M, Fonović M, Salvesen GS, Turk B (2018) Protease specificity: towards in vivo imaging applications and biomarker discovery. Trends Biochem Sci 43(10):829–844

Watts A, Williams R III (2016) Formulation and production strategies for enhancing bioavailability of poorly absorbed drugs. Preclin Drug Dev 187:161

Xu D, Huang J, Zhao D, Ding B, Zhang L, Cai J (2016) High-flexibility, high-toughness double-cross-linked chitin hydrogels by sequential chemical and physical cross-linkings. Adv Mater 28(28):5844–5849

Zhang C, Tang J, Liu D, Li X, Cheng L, Tang X (2016) Design and evaluation of an innovative floating and bioadhesive multiparticulate drug delivery system based on hollow structure. Int J Pharm 503(1–2):41–55

Zhang C, Yang L, Wan F, Bera H, Cun D, Rantanen J, Yang M (2020a) Quality by design thinking in the development of long-acting injectable PLGA/PLA-based microspheres for peptide and protein drug delivery. Int J Pharm 119441

Zhang Y, Chen M, Zhang J, Li J, Shi SQ, Gao Q (2020b) Polysaccharide-based adhesives: a high-performance bio-adhesive using hyperbranched aminated soybean polysaccharide and bio-based epoxide. Adv Mater Interfaces 7(9):2070048

Zhang YS, Khademhosseini A (2017) Advances in engineering hydrogels. Science 356(6337):eaaf3627

Zhuang J, Fang RH, Zhang L (2017) Preparation of particulate polymeric therapeutics for medical applications. Small Methods 1(9):1700147

Zoppe JO, Ataman NC, Mocny P, Wang J, Moraes J, Klok H-A (2017) Surface-initiated controlled radical polymerization: state-of-the-art, opportunities, and challenges in surface and interface engineering with polymer brushes. Chem Rev 117(3):1105–1318

Zweers ML, Grijpma DW, Engbers GH, Feijen J (2003) The preparation of monodisperse biodegradable polyester nanoparticles with a controlled size. J Biomed Mater Res Part B: Appl Biomater: Off J Soc Biomater Jpn Soc Biomater Aust Soc Biomater Korean Soc Biomater 66(2):559–566

Fiberboard Manufacturing from Laccase Activated Lignin Based Bioadhesive

Mohammed Nasir, Mohammad Asim, and Kaushal Singh

Abstract The manufacturing of fiberboard from wood fiber with the help of adhesive is highly developed and widely commercialized technique. However, with an increasing concern over health issues and biomass conservation, the fiber boards from agriculture residues with or without using adhesives are under intensive investigation. In last three decades, several researches have been done various methods and technique dealing the development of bioadhesive based fibre boards. This review article also covered type of adhesives, natural firbes, comparative study of synthetic and bioadhesive based fibreboards. Laccase based adhesive is one of the very promising adhesive for particle boards as far as environment is concern. Chemical reaction of laccase has been studied in detail and also the behavior with wood.

1 Introduction

Wood composite had emerged as an alternative of wood in the early twentieth century, where small logs, non-commercial timber or waste cellulosic materials can be transformed to wood like structure (Isroi et al. 2011). With the ever increasing demand of wood composite, it is sure that future of wood industries will certainly depend on engineered wood product. It exhibits many advantages over solid wood like, smoothness, uniform structure, knots free surface, dimensionally stable, availability in different sizes and thickness, resistance to corrosion and fire, with a good tensile strength and

M. Nasir (✉)
Forest Products Utilization, College of Forestry, Banda University of Agriculture and Technology, Banda 210001, India
e-mail: mnknasir@gmail.com

M. Asim
Institute of Tropical Forestry and Forest Products (INTROP), Universiti Putra Malaysia (UPM), Serdang, 43400 Selangor, Malaysia

K. Singh
Silviculture and Agroforestry, College of Forestry, Banda University of Agriculture and Technology, Banda 210001, India

M. Jawaid et al. (eds.), *Eco-Friendly Adhesives for Wood and Natural Fiber Composites*, Composites Science and Technology,
https://doi.org/10.1007/978-981-33-4749-6_3

easier to work (Nasir et al. 2013). These composite wood products are made from wood-based materials bonded together with a synthetic adhesive using heat and pressure (Nasir et al. 2015). The wood materials may include veneer, strands, particles, chips and fibers whereas adhesives are the most commonly urea formaldehyde or phenol formaldehyde. The construction and furniture market is the most important end-user for plywood, followed by packaging. The furniture industry is the most important user of particleboard and medium density fiberboard (MDF). Laminate flooring is a booming market for MDF and now accounts for nearly more than 40% of all applications (European Panels Federation 2004).

Wood adhesives are essential components in wood composites. As the demand of wood composites are increasing, especially MDF market, there is greater demand of adhesives to convert low value wood to high quality and useful products. At present, formaldehyde-based synthetic adhesives such as phenol–formaldehyde (PF) and urea- formaldehyde (UF) resins are predominantly used (Nasir et al. 2014). These adhesives are synthetically produced from non-renewable resources such as petroleum and natural gas (Moubarik et al. 2010). The major drawbacks of formaldehyde-based resins are they emit formaldehyde which is harmful to human health (Li et al. 2009). Agency for Research on Cancer has classified it as a carcinogenic to human (IARC 2004). Furthermore, the decreasing petroleum resources and increasing price of fossil fuel has been a concern to the future cost and continues supply of synthetic adhesives. Number of research has been done to reduce or replace formaldehyde contents in adhesive preparation but none of them is commercially applicable till now (Mozaffar et al. 2004; Khan and Ashraf 2006).

The increasing health and environment concern, and to maintain continues supply of adhesive, bio-based adhesives are a growing interest among researchers. Number of natural products like tannin, lignin, and more recently proteins based adhesives are under intense study to produce a bio-based natural adhesive (Pizzi 2006; Mansouri et al. 2010). Since tannin and lignin both are polymer of phenol compound, they are primarily viewed as substituting option for synthetic phenolic resins. Tannin based adhesives have received more improvement and it is being used commercially for the last 20 years in southern hemisphere of the world (Pizzi 2003). However, a huge availability of lignin as a byproduct of pulp and paper mills has made it an attractive raw material for adhesives (Pizzi 2003). In the past few decades number of patents has been done dealing with the application of pulp lignin as a wood adhesive, where lignin is cross-linked by condensation reactions. The commercial use of lignin is still growing but the progress is very slow (Pizzi 2006).

2 Medium Density Fiberboard (MDF) from Natural Adhesive

MDF is manufactured by combining the wood fiber with a synthetic thermosetting resin such as urea–formaldehyde (UF) or phenol- formaldehyde (PF) and hot pressed into various shape and size (Huttermann et al. 2001). In this entire process, basically, only one foreign component is added which is not found in the natural wood, i.e. the synthetic adhesive. The major drawback of these adhesives is that they contain volatile organic compounds that are harmful to human health and the final products are difficult to recycle (Nasir et al. 2014). Due to this reasons, many developed countries from Europe and North America have already imposed regulations, limiting the emission of formaldehyde from building materials (Kavvouras et al. 1998).

Therefore to fabricate a completely natural MDF, it should be either prepared with no synthetic adhesive or supplemented with a natural adhesive such as lignin based, soy based adhesives etc. Fiber modification is also one of the approaches to prepare MDF without formaldehyde based adhesive. It includes the enzyme treatment, steam treatment, microwave treatment or direct pressing at high temperature to modifying the thermoplastic behavior of cellulosic fiber and initiates the self-bonding (Nasir et al. 2017). Being a semi-crystalline molecule, lignin is the principal component which starts to plasticize at lower temperature (above 200 °C) than the cellulose (Lora and Glasser 2002). To achieve strong bonding between the crystalline layers, a semi-crystalline thermoplastic lignin matrix must have sufficient time to melt and diffuse in between existing layers. Crystallization kinetics is a complex process and plays an important role in determining the bonding strength of product made up of consolidation of semi crystalline material (Tierney and Gillespie 2004). Felby et al. (1997, 2002, 2004) and Kharazipour et al. (1997) have revealed that wood fibers can be enzymatically activated in vitro by laccase enzyme, and this treated fiber can be used to produce wood composites with enhanced self-bonding between fibers. Table 1, enlists the important developments in MDF preparation by self-bonding of fiber. A fiber board prepared from laccase treatment shows an improved tensile strengths but the biggest problem is lower water resistance (Huttermann and Haars 1981). However, in order to meet the international standard, the final composite product is supposed to be a water-proof.

3 Natural Fiber

A woody plant can either be gymnosperms (softwoods) or angiosperms (hardwoods), classified on the basis of their anatomical differences. But woody plants have many anatomical differences in species and even within species based on growing conditions. Gymnosperms are primarily composed of longitudinal tracheid which is simple as compared to the structure of angiosperm and length ranges from 3 to 4 mm. Angiosperms are composed of a complex structure made up of 4 major cells, longitudinal parenchyma, ray parenchyma, vessel elements and fiber tracheids. The fiber length of angiosperms is smaller than softwoods, at less than 1 mm (Haygreen and Bowyer 1996).

Table 1 Important development in MDF preparation by self-bonding of fiber

Treatment	Method	Advantage	Limitation	References
Ferric sulfate	Treatment before pressing	no adhesive	Not resistant to water	Linzell (1950)
No treatment	High pressure (100 g cm^{-3}) and high temperature (200 °C) for longer time	Water resistant	Energy consuming, harsh condition	Klauditz and Stegmann (1955)
Hydrocloric acid + ferrous sulfate	Chemical were sprayed on the wood fiber	High mechanical strength	Corrosive nature	Stofko and Zavarin (1977)
Lignin plasticizing	Heating above glass transition temperature (Tg)	Completely eco-friendly	Required a high temperature	Wagenführ (1988)
Laccase treatment	Enzyme activates the lignin content of fiber	Completely eco-friendly	Not water resistant	Felby et al. (1997), Huttermann et al. (2001)
Laccase/HBT	Laccase mediated compound is used to improve self-bonding of fiber (HBT = 1-hydroxyl benzotriazole)	Improved wet strength in paper	Not water resistant	Wong et al. (2000)
Improved milling	Normal MDF preparation technique	Eco-friendly	Not comparable to commercial	Velasquez et al. (2003)
Steam treatment	Steam exploded method plasticizing native lignin	ecofriendly	energy requirement, not water resistance	Salvado et al. (2003)

3.1 Physical Properties of Natural Fibers

The properties of MDF depend on natural fiber that may varies by species, shapes, sizes, orientations, and thickness of the cell walls of individual fibers (Rafiqah, Abdan et al. 2020). These fibers may differ, in coarseness, strength, stiffness and roughness or smoothness of the fiber surface of cell walls (Siakeng, Jawaid et al. 2020). Thus it is better to know the length, width type and source of the fibers which play an important role to determine the properties of MDF. Being a cellulosic material, wood fiber is a hygroscopic, dimensionally unstable and prone to biological attack. When MDF prepared from such fiber comes into contact of water, it swells up and sometime swelling is not recoverable. A thickness swelling markedly weakens panel products and reduces the appearance. Thus a MDF quality improves by improving the dimensional stability of the panel product.

3.2 Chemical Composition of Fiber

Cell walls of both, gymnosperm and angiosperm are chemically composed of cellulose, hemicellulose, lignin, and some amounts of extractives (5–10% by weight) (Asim, Paridah et al. 2020). A simplified structure of a typical hardwood fiber and its orientation in different layer is depicted in Fig. 1. A woody cell wall can be divided into three distinct layers, i.e. middle lamella, primary cell wall and secondary cell wall. The middle lamella contains a high proportion of amorphous material which holds neighboring fibers together. The primary cell wall is approximately 0.03–1.0 μm thick, and also contains a high percentage of lignin. The secondary cell wall can be further divided into three layers, namely S1 S2 and S3 from outward to the inward. The S2 layer is the most dominating layer and generally constitute 1–5 μm whereas, S1 and S3 layers are thin and made up of 0.1–0.3 μm. S2 layer is primarily responsible for the all the mechanical properties of individual fibers. The micro fibrils of secondary cell wall are arranged in a set pattern and wound helically around the fiber axis whereas it is oriented randomly in the primary walls.

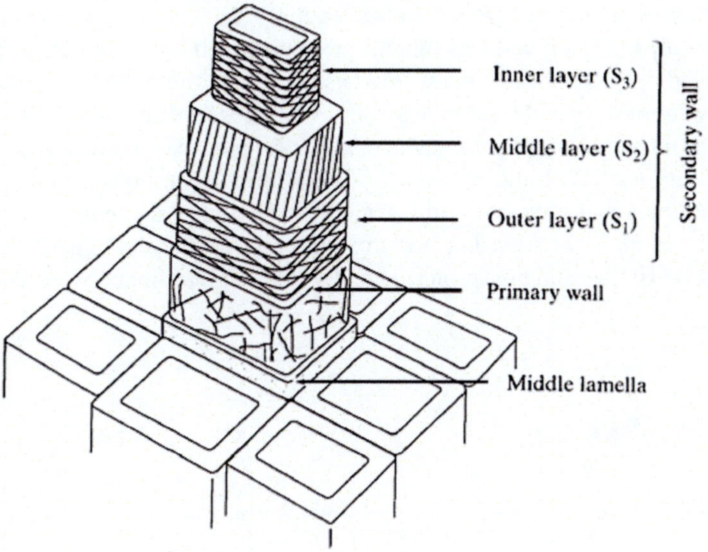

Fig. 1 Typical structure of cell wall with its micro-fibril arrangement in different layer (Adopted from Dinwoodie (1989))

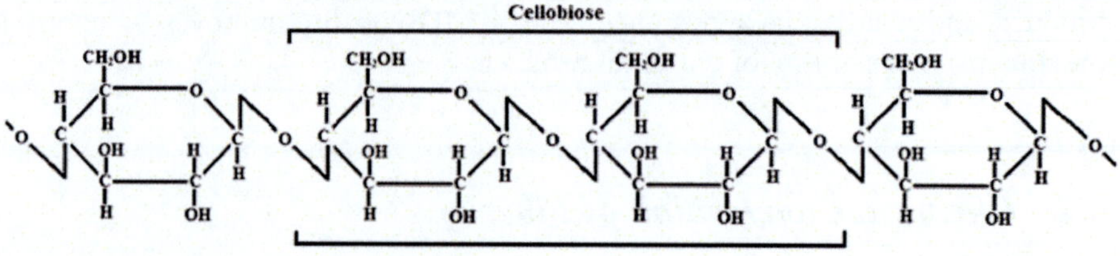

Fig. 2 Long cellulose chain depicting, a structural unit Glucose and a functional unit cellobiose (Adopted from Sjostrom (1993)))

3.3 Cellulose

Cellulose is one of the most abundant natural polymer on this planet (Asim et al. 2015). It contributes 40–45% of the dry weight of wood and located primarily in the secondary cell wall. Celluloses are long linear chains of D-glucose connected with ß-1,4-glycosidic bonds (Fig. 2) (Lamaming et al. 2015; Osong et al 2016). The degree of polymerization ranges from 1000 to 10, 000 or more in native wood. The hydroxyl groups present in its basic structural unit link laterally by strong hydrogen bonding networks giving rise to a crystalline structure (Gumuskaya et al. 2003). Cellulose exists in four different crystalline forms (polymorphs) of cellulose: I, II, III, and IV. Cellulose I is the native cellulose exist in natural state, and exhibits two allomorphs, I_α(triclinic) and I_β (mono clinic), rest all the polymorphs are the result of some chemical modification (Lavoine et al. 2012).

A crystalline region cellulose chains are interrupted by some occasional folds or links called defects and amorphous regions at every 60 nm which makes cellulose, a semi-crystalline molecule (Ioelovich 2008). The degree of crystallinity of cellulose is one of the most important parameters to study the physical and mechanical behavior of cellulose fiber. With the increase of crystallinity to amorphous ratio, rigidity of cellulose fibers increases but at the same time flexibility decreases (Ishikawa et al. 1994; Mo et al. 1994). Furthermore, crystallinity of cellulose plays an important role in accessibility and digestion (longevity) of cellulose fiber (Awadel-Karim et al. 1999).

3.4 Hemicellulose

Hemicelluloses are the second most abundant biopolymer, which lend support to the cell wall and constitutes 20–30% of the dry weight of wood. It comprises the backbone of different monosaccharides such as glucose, xylose, arabinose, galactose, fructose, mannose, galacturonic acid, and glucuronic acid, linked by β (1 → 4) bond (Ebringerova and Heinze 2000). Figure 3 showed the monomers of Hemicellulose. It is branched, hetero-polysaccharides with lower molecular weights and lower degree of polymerization (80–200) as compared to cellulose. It is an amorphous polymer

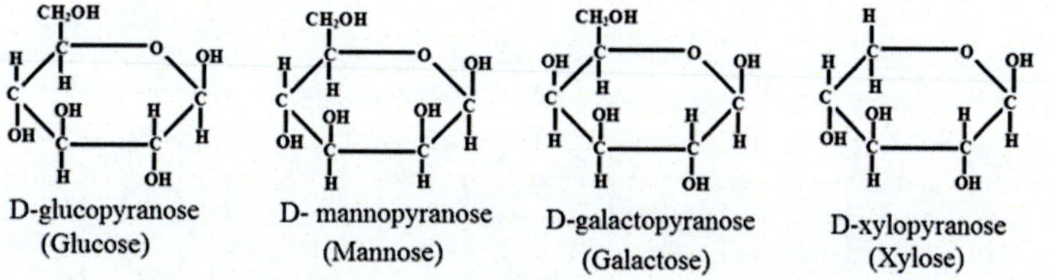

Fig. 3 Structure of hemicellulose monosaccharides (Adopted from Sjostro (1993))

having monosaccharides as side chains. Due to these factors hemicelluloses are very susceptible to chemical attack. It is usually found attached with various other cell-wall components such as cellulose, lignin, and cell-wall proteins by covalent bonding, hydrogen bonding, or by ionic interaction (Sun et al. 2000).

3.5 Lignin

Lignin is the third major polymeric components found in the cell walls of wood, contributes 20–30% of total dry weight of the wood (Nasir et al. 2015). It acts as an adhering material in the cell wall and binds up the cellulose and hemicellulose component together, provide strength and rigidity to the plant stem necessary for vertical growth (Feldman 2002). Higher lignin content in cell walls increases the rigidity, improves compressive strength and lowers the porosity. Apart of providing strength, it performs as a resistance against biological attack and makes the vessel tissue, water proof and helps in water transport. The highest concentration of lignin is found in the middle lamella and primary cell wall (Saka and Goering 1985), whereas the highest amount is found in secondary cell wall as it contribute major portion of the cell wall (Harris 1990).

A lignin molecule can be described as an amorphous, highly irregular, complex polymer of phenyl propane monomers (Nasir et al. 2013). Figure 4 depicts the lignin monomer units, based on methoxy group present in phenyl propane; it is divided into types, namely p-coumaryl alcohol (p-hydroxy phenyl), coniferyl alcohol (Guaiacyl), and synapyl alcohol (syringyl) (Dence and Lin 1992; Sjostrom 1993).

These three phenylpropane are found together in all plants but in different proportions. Softwoods are composed of approximately p-coumaryl alcohol 14% coniferyl alcohol 80%, and only 6% of sinapyl alcohols, whereas a hardwood are composed of p-coumaryl 4%, coniferyl 56% and considerably large amount of sinapyl alcohols i.e. 40% (Jung and Fahey 1983). Phenylpropane monomers linked together by radical coupling of carbon to carbon or ether reaction and make a network like structure (Glasser and Jain 1993). In native state, lignin shows various linkage pattern, among them the most common linkage are β-O-4, α-O-4, β-5, 5-5, 4-O-5, β-1, and

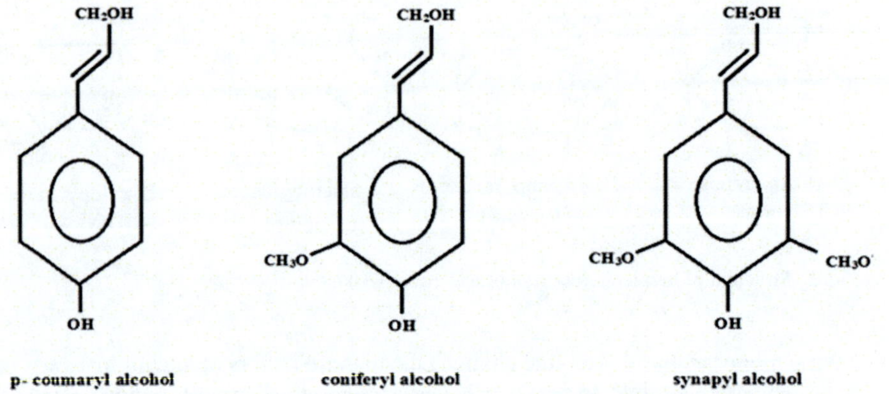

Fig. 4 A simplified structure of phenyl propane units of lignin (Adopted from Dence and Lin (1992))

β-β (Dence and Lin 1992; Sjostrom 1993). Figure 5 show a simplified structure of lignin units, depicting various bonding pattern of monolignols units.

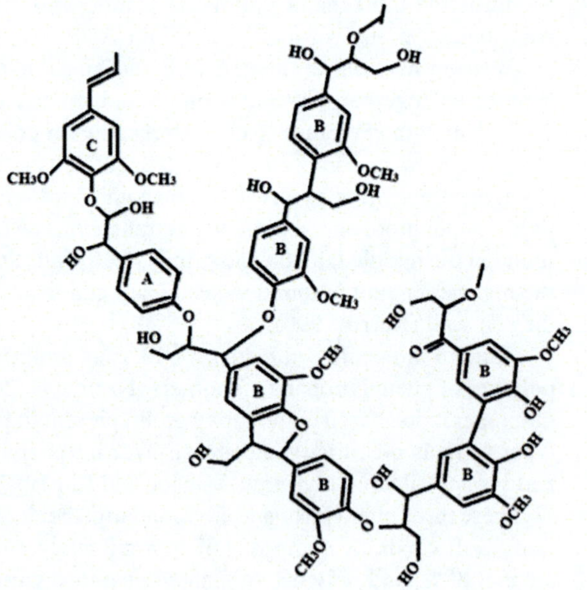

Fig. 5 A simplified structure of lignin polymer. p Hydroxy phenyl (**a**), guaiacyl (**b**) and syringyl (**c**) are linked together by β-O-4, α-O-4, β-5, 5-5 and 4-O-5 (Adopted from Dence and Lin (1992))

3.6 Extractive

These are the low molecular weight compound which often gives colour to wood. It contributes 5–10% of the total dry weight of the wood and varies by species to species and location within individual trees grows. Its role in the plant is very diverse such as it may provide energy, improves resistance against microbial attack and changes the hydrophobicity behavior. Some examples of extractives are terpenes, resin acid, triglycerides and various phenolic compounds.

4 Synthetic Adhesive

Urea–formaldehyde (UF) adhesive is the most common adhesive used because of it several positive aspects such as non-flammable, low cost, rapid curing rate, and a light color. However, it has some strong negative aspects also such as, poor water-resistant and formaldehyde emission form the product (IARC, 2004; WHO, 2004). But still UF adhesive is commercially predominant adhesives used for interior grade plywood, particleboard and MDF.

Phenol–Formaldehyde (PF) resins performs better than UF resin is a product of addition and condensation reactions of phenol and formaldehyde either it is an acidic or basic condition. Methylol from formaldehyde, has a very high tendency of addition on Phenol and generally occurs at ortho and para position, and form 2, 4, 6 methylol-substituted phenols. This methylol-substituted phenols, further goes under condensation reaction and forms a mixture of linear or branched polymers of different size. Polymerization is the last reaction starts during the setting process when it is heated to definite temperature (Koch 1987).

Resorcinol–Formaldehyde (RF) resins are prepared from same reaction mechanism as PF resin i.e. addition and condensation reaction of resorcinol and formaldehyde in basic condition. Resorcinol contains one more hydroxyl group than phenol (Asim, Saba et al. 2018), at meta-position which makes it more reactive towards formaldehyde. Resorcinol is very reactive towards formaldehyde and starts polymerization reaction quickly even at room temperature (Gierenz and Karmann 2001). In wood composites, RF resin exhibits a very strong bonding strength compared to PF resin. However, still it is not preferred over PF resin because of its high cost, short pot-life and dark colour.

Melamine contains three reactive amino groups which have an addition reaction with formaldehyde and forms Melamine–Formaldehyde (MF) resins. Furthermore, these MF monomers undergo a condensation reaction and forms a cross linked network structure. MF resin is resistant to moisture and generally preferred for exterior grade of composite. However, MF resins are less popular than UF and PF resin because of high cost of melamine.

Epoxy resins are chemical compound, contains epoxy functional groups that starts polymerization in presence of catalyst and hardener (Gheith, Aziz et al. 2019).

Bisphenol A diglycidyl ether (BADGE) is the most common and widely used epoxy resin worldwide, it is a derivative of bisphenol A (a condensation product of phenol and acetone). Epoxy resin exhibits an excellent physical and chemical property like superior mechanical strength, good weather resistance, low cure shrinkage, excellent chemical and corrosion resistance (Asim, Jawaid et al. 2019) (Ratna 2003). Epoxy resin shows a wide range of application and forms a good bonding to various materials including wood, plastics, metals, concrete and ceramics. It is less popular in wood working because of its high cost and durability limitation; in some cases its cost become higher than the cost of wood itself (Karayannidou et al. 2006).

Polyvinyl Acetate (PVA) is prepared by emulsion polymerization of vinyl acetate. It is generally of white colour and sold in aqueous emulsion form. It is the most commonly used glue in wooden joints. PVA glue has excellent advantage to set up in short time, at ambient temperature and gives a high strength. The only disadvantage of PVA is, not resistant to water and heat. However, various cross linking monomers such as N-(hydroxymethyl) have been introduced that can polymerase with PVA and improve its water and heat resistance (Liu et al. 2005).

5 Natural Adhesive

They are obtained from renewable raw materials and could be origin from proteins, natural rubber, starch, or cellulose etc. Based on their source of raw material it is further divided as follows.

5.1 Animal Glues

Animal glues are prepared from the skin and bones of animal such as sheep cattle etc. These glues are basically protein based adhesives formed by hydrolysis of collagen protein. Since proteins are amphoteric compound it contains many free reactive sites of carboxylic acid group (–COOH), amide group (–CONH–) and amino groups (–NH–, –NH_2) which interacts with each other within protein chains as well as with wood also (Blomquiest et al. 1981). Animal glues were being used for wooden works from very ancient time but now it has been replaced by synthetic polyvinyl acetate because of their low moisture resistance, prone to biological attack and its high price. Furthermore, proteins from different source have different structure and number of functional group, thus the properties also differs significantly, and preparing uniform glue from different source is difficult (Eckelman 1997).

5.2 *Casein-Based Adhesives*

Casein is a milk protein, separated when milk is treated with acid. A casein protein contains very high amounts of carboxylic acid groups thus have a high solubility in an alkaline medium such as sodium hydroxide solution or lime solution. This carboxylic acid of casein protein can react with polyvalent metal ions and forms a crosslink network. It shows a better water resistance then animal glue and suitable for interior grade panels (Sellers 2001). However, its high viscosity makes it difficult to apply in wood furnishing industries (Yang et al. 2006). Another problem of casein glue is the high amount of tannic acid gives a dark stains to wood.

6 Blood-Based Adhesives

Bloods a byproduct of slaughterhouse can be used as an adhesive. It contains high amount of albumen of protein (Lina and Gunasekaran 2010). Bloods are given alkali treatment to unfold the protein and formaldehyde or PF resins are used to improve the water resistance property. Blood-based adhesives shows a better water resistance as compare to casein adhesive but having a lower strength compared to casein based adhesive. It can be used efficiently, in combination with PF resin or soy based resin in panel industries. However, commercially it is not being used any more (Lina and Gunasekaran 2010).

7 Soy-Based Adhesives

Soybean is an abundant, renewable, and inexpensive agricultural product readily available to all over the world. The traditional soybean market as food and animal feed has been saturated and nowadays, soybean meal is mainly used for livestock and poultry. A non-food industrial market for oversupplied soybean is still need to be developed. In the beginning of twentieth century, soy-based adhesive was being used in the commercial production of composite, but replaced by urea formaldehyde because of the superiority over soy-based adhesive (Liu 1997). However, soybean can be an ideal raw material for wood adhesives because of its advantage such as low cost, easy handling, low pressing temperature. Although some inferior properties such as low strength and low water-resistance of the resulting wood composite limits the application of soy-based adhesive.

In recent years, there is an increasing interest of researchers can be seen to develop a soy-based adhesive with improved strength and water resistance. A novel method to modify soy-protein with urea, sodium dodecyl sulfate (SDS) and guanidine hydrochloride was observed to increase the strength and water- resistance in plywood panels (Huang and Sun 2000). It is proven that the alkali treatment or protease enzymes pretreatment to soybean protein significantly improved the strength and water- resistance of plywood samples bonded with modified soy proteins adhesive (Nasir et al. 2014). Liu and Li (2004) demonstrated the chemical modifications of soybean like a mussel adhesive protein that can be used as a strong and water- resistant wood adhesive. A resin prepared from soy protein treated with polyamidoamine-epi-chlorohydrin (PAE) has been proven an excellent curing agent (Li et al. 2004).

8 Lignin Based Adhesive

Lignin is considered as one of the potential and attractive substitutes of formaldehyde based adhesive due to its phenolic based composition and abundance of availability. Approximately 5×10^6 million metric ton of lignin is produced as a byproduct annually from different processes such as soda kraft, hydrolysis, or enzymatic pulping (Vishtal and Kraslawski 2011). Since lignin molecule is tightly attached to the holocellulose, lignin solution contains impurities such as glucan, xylan, mannan, galactan, and traces of ash when separated from cell wall (Mansouri et al. 2010). Such lignin solution is called organosolv lignin (OSL) if obtained through organic solvent process or enzymatic hydrolysis lignin (EHL) if obtained from enzyme hydrolysis. Most of the lignin obtained from pulp mills is either burned or disposed wastefully and only 15–20% of total lignin is used for various purposes (Pizzi 2003).

An extensive work has been done on lignin for utilizing it as a wood adhesives but none of them is being applied yet at commercial scale (Nasir et al. 2013). The most success full attempt to utilize lignin in industrial application was done to replace phenol partially from urea formaldehyde or phenol formaldehyde (Sulaiman et al. 2018) (Khan and Ashraf 2007). The lower reactivity of lignin with formaldehyde was compensated by the addition of UF or PF adhesives up to some extent. Such adhesives (lignin-UF or lignin-PF) can be utilized in plywood but not in MDF or particle boards where rate of pressing is a critical parameter determines the cost and profit of both process and product (Stephanou and Pizzi 1993).

Based on the concept of lignin biosynthesis in plant which is a free radical mechanism where destruction and chain oxidation takes place (Moldes et al. 2010). It is hypothesized that enzyme activated lignin polymerisation may overcome the low reactivity of problem of lignin (Suurnakki et al. 2010). The laccase enzyme reacts with the lignin part of the cell component and generate free radicals of lignin, lignans, and different lipophilic extractives (Moya et al. 2011) and allows a free radical reaction, between carbon–carbon as well as carbon–oxygen very quickly at low activation energy (Pizzi 2003).

9 LACCASE

Laccase (EC 1.10.3.2) enzyme is glycoproteins found in nature in white-rot fungi, higher plants, insects and bacteria (Nasir et al. 2019). It is a well-studied oxido-reductase enzyme act on lignin and its allied molecule (such as phenols, polyphenols, anilines, aryldiamines, methoxy- substituted phenols, hydroxyindols, benzenethiols etc.) (Kunamneni et al. 2007; Van de Pas et al. 2011). Further, it not only removes the amorphous lignin content, but also changes the hemicellulose content and ultimately the cellulose crystallinity (Kim and Holtzapple 2005, 2006). Although several researchers have studied the enzymatic hydrolysis of lignocellulosic fiber, the changes in physical and mechanical properties of fiber are not fully understood (Setliff and Eudy 1980).

Laccase acts as an oxido-reductase agent; on one side it removes the lignin molecule from cellulosic material and on other hand it helps in vivo polymerization of lignin through free radical reactions (Dashtban et al. 2010). Laccase catalyze one-electron substrate oxidation along with four-electron reduction of O_2 (Zhou et al. 2009). It is believed that four moles of phenoxy hydroxyl groups are oxidized to four moles of phenoxy radical and reduce oxygen to two molecules of water (Camarero et al. 2007). This efficiency of the laccase enzyme to oxidize lignin is being exploited commercially in pulp and paper industries for the eco-friendly removal of lignin from pulp (Suurnakki et al. 2010). A diagrammatic representation of laccase activity is presented in Fig. 6.

The treatments usually involve the application of laccase enzymes to activate lignin on fibers (one-component system) or the addition of another component with laccase to act as a potential cross-linking agent (two component system) (Gochev and Krastanov 2007). Since the laccase enzymes are too large to penetrate fibers (50–100 kDa) treatments result only in a surface modification (Kunamneni et al. 2007). Therefore, the treatments helps in increasing the board strength and free phenolic groups on the fiber surface create potential reactive sites as phenoxy radicals. Based on this theory, there is a growing interest that laccase can be employed to generate

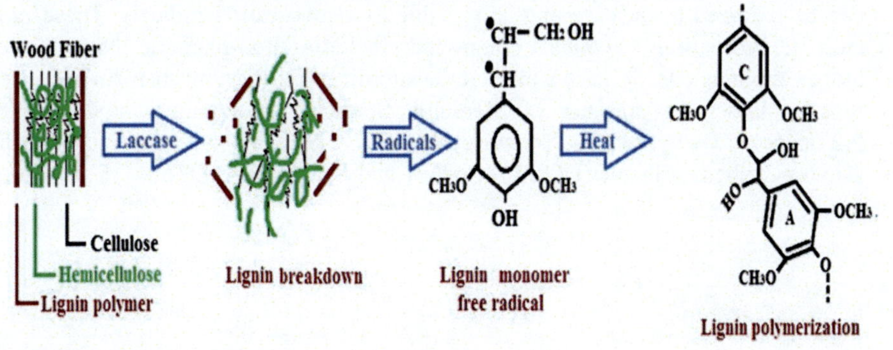

Fig. 6 A hypothetical diagrammatic of laccase reactivity

reactive quinonoid structures in lignin-rich fibers that could be reacted with amino acids to generate, enhanced fiber charge. A chronological development in laccase application in fiber treatment is listed in Table 2.

10 Laccase: Reaction Mechanism

Laccase is specific to the phenolic compounds leading to the Cα oxidation, Cα-Cβ cleavage and aryl–alkyl cleavage. A simplified laccase enzyme reaction mechanism is drawn in Fig. 7 without mediator. The laccase catalysis mechanism involves three major steps of reaction. Initially the Cu I, is reduced by a reducing substrate and oxidized itself. The electron generated at Cu I is transferred to Cu II internally and finally to Cu III which is arranged in a triangular structure of each other (Fig. 7) (Kunamneni et al. 2008).

Laccase utilize oxygen as the electron acceptor and remove protons from the phenolic hydroxyl groups. Thus the free radicals at phenolic compounds are formed that can spontaneously rearrange, and lead to fission at C–C or C–O bonds of the alkyl side chains, or causes cleavage to aromatic rings. In this whole process oxygen molecule is reduced to water at the tri-nuclear cluster. It is observed that the O_2 molecule binds to the tri-nuclear cluster of Cu for asymmetric activation and it is assumed that this O_2–bond pocket restricts the entrée any other oxidizing agents except O_2 (Petri and Laine 2002).Thus a laccase can reduce one molecule of oxygen to two molecules of water through one electron oxidation of various aromatic compounds,such as phenols, polyphenols, anilines, aryl diamines, methoxy- substituted phenols, hydroxyindols, benzenethiols etc. (Bourbonnais et al. 1997, 1995). This oxidation reaction generates an oxygen-centered free radical, which can then be transformed in a second enzyme-catalyzed reaction to quinine (Moilanen et al. 2011).

A laccase can extend its substrate range from phenolic to non-phenolic organic substrate if a mediator supplemented with laccase (Gochev and Krastanov 2007). A mediator is a group of organic compounds (generally low molecular-weight) that can be oxidized by laccase and form a highly active cation radicals. These radicalsreacts to various chemicals compounds including non-phenolic that a laccase alone cannot oxidase. Figure 8 present the simplified reaction mechanism of laccase reaction under the presence of Mediator. Some of the common mediators are 2,2′-azinobis 3-ethylthiazoline-6-sulfonat (ABTS), N-hydro-xyphthalimide (NHPI), 1-hydro- xybenzotriazole (HOBT) (Gochev and Krastanov, 2007).

Table 2 Chronological development of laccase application fiber treatment and lignin modification

References	Contribution/Application/Novelty
Yoshida (1883)	Credited to discover laccase first time in latex from the Japanese lacquer tree (*Rhus vernicifera*) hardened in the presence of air
Huttermann and Haars (1981)	In-vitro treatment of lignin, generates quinonoid intermediates, and subsequently lignin polymerization
Kawai et al. (1988)	Laccase polymerized lignin by radical coupling, alkyl–aryl cleavage or Ca oxidation
Vaheri et al. (1991)	Patented first the pretreatment for fiber pulping, it reduces energy consumption during mechanical pulping
Bao et al. (1993)	Reported laccase in plants, and its role in lignin biosynthesis
Heinzkill and Messner (1997)	Isolated laccase from many sources such as plants, bacteria, and insects
Bourbonnais et al. (1997)	Studied laccase alone is not so effective oxidizing agent but needs some mediator to enhance its activity
Murphy et al. (1997)	studied the structural comparison all these laccase
Xu et al. (1998)	Studied the reaction mechanism and various parameters of enzyme reaction in pulp,
Ikeda et al. (1998)	Laccase induced radical polymerization of phenoxy redical of lignin with or without mediator
Gianfreda et al. (1999)	Explained laccase is a glycoprotein with average molecular mass of 60–70 kDa
Bajpai (1999)	Explained it as a nonspecific enzyme, can be utilized in various application such as delignification, bioleaching etc.
De Souza and Peralta (2003)	Explained, that majority of fungal laccase are a extracellular having molecular size 60–70 kDa with isoelectric point (pI) around pH 4.0
Felby et al. (2004)	Treated wood fiber for MDF manufacturing, it was concluded that Laccase treatment causes a significant fiber surface modification by enriching with lignin and extractives precipitation on to the surface
Xu (2005), Nyanhongo et al. (2002)	Studied the nonspecific nature of laccase, its substrate has wide range such as: phenols, polyphenols, anilines, aryl diamines, methoxysubstituted phenols, hydroxyindols, benzenethiols and many inorganic/ organic metal compounds
Alcalde (2007)	Studied laccases molecules are monomeric extracellular enzymes contains four copper atoms bound to 3 redox sites. Also studied the laccase-mediator system (LMS) such as 2,2′-azino-bis-(3-ethylbenzothiazoline-6-sulphonic acid) (ABTS) and 1-hydroxybenzotriazole (HBT)

(continued)

Table 2 (continued)

References	Contribution/Application/Novelty
Suurnakki et al. (2010)	Laccase is highly reactive towards fiber containing more accessible and reactive surface lignin. Also observed that a small amount of enzyme is more relevant in larger scale activation of pulp fiber

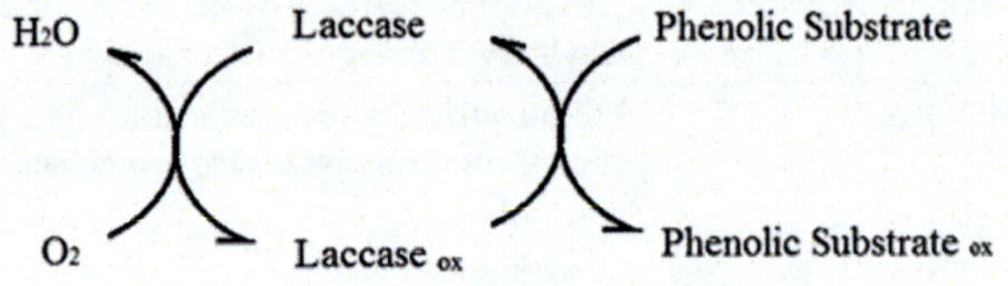

Fig. 7 Laccase reaction specific to substrate (Adopted from Gochev and Krastanov (2007))

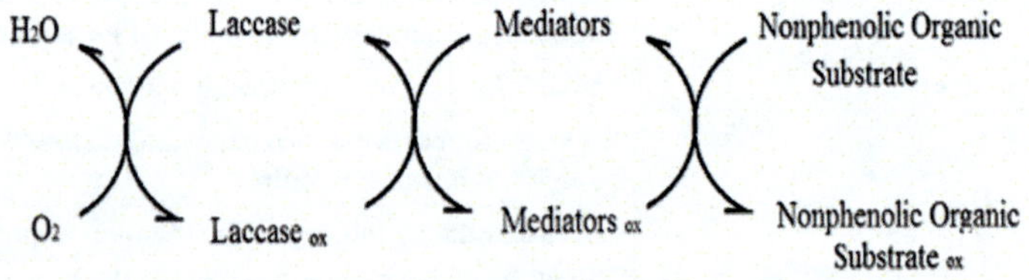

Fig. 8 Laccase reaction under mediator system increase the substrate range (Adopted from Gochev and Krastanov (2007))

11 Factor Affecting Laccase Activity

11.1 Influence of pH on Laccase Activity and Stability

Laccase activity is highly dependable to the pH of substrate or medium of reaction (Kim and Nicell 2006). The mediator such as ABTS also can change the pH optima of the reaction and it is observed from the range 3.0–5.0 (Kunamneni et al. 2007). Laccase exhibits a bell shaped trend of enzyme activity against pH and varies considerably (Slomezynski et al. 1995). Laccase obtained from *Trametes modesta* was observed the best activity at pH 4.5 and decreased to great extent when pH was brought to 3.0 (Edens et al. 1999). It is found that at high pH value a redox potential deference is created that can increase the oxidation of substrate at Cu I clusterbut the OH^- binding at Cu II and Cu IIIcauses an inhibition of the laccase activity because of the interference of the internal electron transfer between the Cu I and Cu II/III centers. These two reasons play an important role in deciding the optimal pH of the bi-phasic laccase enzymes (Slomezynski et al. 1995).

11.2 Temperature on Laccase Activity and Stability

The optimal temperature of laccase varies to a wide range and depends on the source of origin. The stability of the enzyme can be increased by immobilizing the enzyme on glass powder by air drying. Laccase obtained from *Myceliophtera thermophila* can perform up to 50 °C (Felby et al. 2002), whereas the laccase obtained from *Trametes villosa* had a standard temperature 30 ± 5 °C (Felby et al. 2004). In general, a high temperature (>65 °C) or a moderately high temperature (47–50 °C) for a duration of 24 h causes the inactivation of laccase enzyme (Snajdr and Baldrian 2007). In general the optimum temperature of laccase activity is observed between 25 and 35 °C which is best suited for the cultivation and production of enzyme also (Pointing and Vrijmoed 2000).

11.3 Influence of Inhibitors on Enzyme Activity

Many ions such as halides, azide, thiocyanide, cyanide, hydroxide and fluoride bind to the Cu II and Cu III, of laccase causes the interruption of internal electron transfer, therefore check of activity. Some other inhibitors include fatty acids, metal ions (e.g. Hg + 2), hydroxyglycine, sulfhydryl reagents, desferal, cationic quaternary ammonium detergents, and kojic acid, alsocauses inhibition to amino acid residue modifications, or Cu chelation (Kiiskinen et al. 2004). Enzymatic oxidation is highly depend on configuration of the Cu atoms, any change in configuration may leads to the loss of activity (Kunamneni et al. 2007).

11.4 Fiber Modification

Fiber modification can be discussed in two aspects, first as a physical modification and second as chemical modification.

11.4.1 Physical Modification of Fiber

A close observation of MDF revels that an adhesive alone is not only responsible for fiber bonding but a mechanical entanglement also plays major role. A catalyst or enzyme that initiate an oxido-reductas reaction in the fiber can be used for cross-linking or self-bonding of wood fibers to prepared a MDF without using any external adhesive (Nasir et al. 2013). An enzyme modified fiber can improve the inter-bonding strength of fiber in MDF by many ways such as mechanical interlocking, adsorption behaviour, chemical bonding or combination of all (Siakeng, Jawaid et al. 2019).

Mechanical interlocking is a type of physical force in which two component of distinct interface are held together. It is a like a dovetail joints where the surface of one component is embedded into another. In a laccase hydrolysis process with breakdown of lignin, precipitation and adsorption of lignin is also occurred at same time (Maximova et al. 2004). A lignin adsorbed surface can change the sorption characteristics, dimensional stability and intermolecular adhesion of the fiber. Since Laccase are too large to penetrate fibers (55–100 kDa), it modify the amorphous extracellular component only (Kunamneni et al. 2007). Hence, due to a size exclusion reasons, laccase treatment is only a surface modification and it alters extracellular lignin only without affecting the cellulose fiber.

11.4.2 Chemical Modification of Fiber

A chemical modification involves a direct chemical reaction of components at the interface. It can be free radical reaction, ionic reaction, hydrogen bonding or carbonyl bonding (Asim, Jawaid et al. 2018) (Hill and Ceti 2000). Figure 9 showed, some of the possible self-bonding reactions may occur at the time of hot pressing. Laccase is a well-established approach to initiate the phenoxy radicals from lignin by oxido-reduction process (Mai et al. 2004; Huttermann et al. 2001). To modify the wood fiber

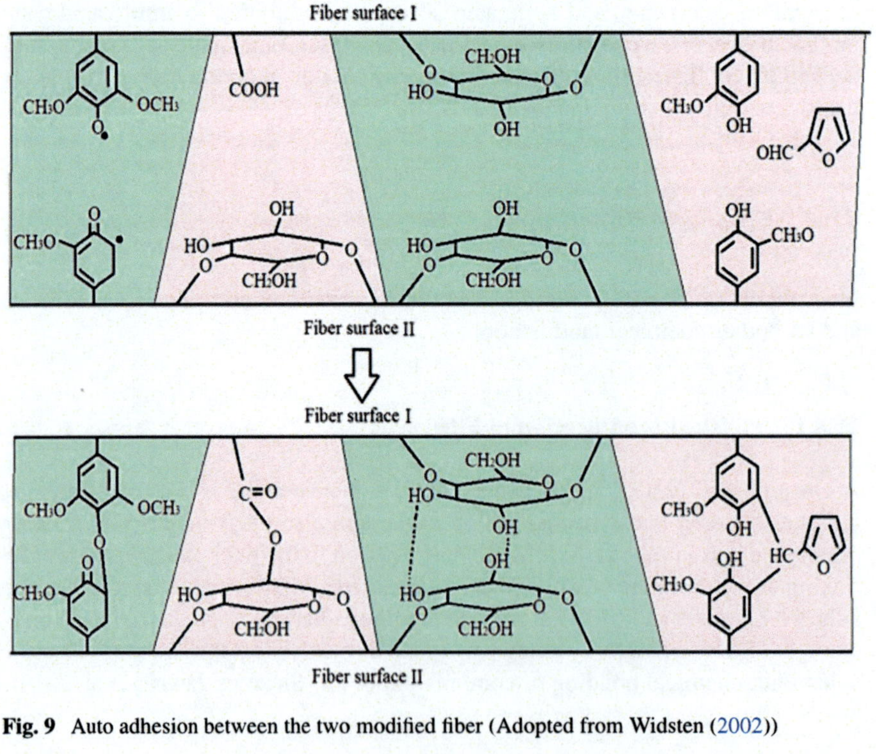

Fig. 9 Auto adhesion between the two modified fiber (Adopted from Widsten (2002))

surface it is crucial to understand all the major components i.e. cellulose, hemicellulose and lignin of the fiber (Asim et al. 2019). In order to improve the self-bonding of fibers many hypothetical ideas are under investigation such as: (1) How to enrich a lignin component at the surface wood fiber, (2) How to plasticized the fiber matrix (lignin and hemicellulose) in-situ and utilized as an adhesive for fiber bonding, And (3) How to activate the native lignin to start the free radical polymerization within and between the fibers.

The first approach to modify wood fiber is based on the hypothesis that laccase enzyme can generate a free radicals by an oxido-reductase reaction into phenoxy radicals of lignin molecule (Gradwell 2004). These free radicals further undergoes to polymerization reaction and form network of polymer by coupling (like thermoset adhesives). In recent study, it is well established that laccase enzyme obtained from fungi is a best suited for activation of native lignin and accelerate oxido-reductase coupling of fiber (Felby et al 2004; Nasir et al. 2014, 2015).

Another approach to improve the self-bonding of the fiber is to bring out the copolymer matrix (lignin and hemicellulose) onto the surface of the fiber that will take part in auto adhesion of fiber when the fiber will be pressed at high temperature. Laccase enzyme is a specific enzyme to a wide range of substrate (Phenolic compound) (Felby et al. 1997). It can oxidize the number of organic compounds present in plant cell wall such as lignin, ortho and para diphenols, aminophenols, polyphenols, aryl diamines polyamines, and some inorganic ions (Yaropolov et al. 1994). These copolymer of the cell wall exhibits an amphiphilic nature and serves as both adsorbing surface and adsorbable amphiphiles (Akiyoshi et al. 1995). It is evident that a ligno-cellulosic composite can be formed successfully if the wood surface is coated with the thermoplastic cell matrix such as lignin and hemicellulose (Akiyoshi et al. 1992). Thus enzymatic treatment to wood fibers has been emerged as a potential tool to solving multi-direction problem.

11.5 Lignin Modification

Being a polyphenol compound, lignin structure has a very close similarity with phenol–formaldehyde (PF) resins. To transform a lignin into insoluble adhesive, it should have to be additionally cross-linked. The lower number of free position in aromatic nuclei and lower rate of reactivity limits the utility as an adhesive. Furthermore, the presences of methoxy or methoxy-equivalent groups on the aromatic ring of lignin are considerably less reactive toward hydroxybenzyl alcohol groups than the hydroxyl group found in phenol (Pizzi 2003). Because of these reasons, lignin cannot be effectively utilized as a potential adhesive as synthetic PF resins.

Therefore, the potential cross-linking agents such as polyisocyanates epoxides, polyols, polyethyleneimine, polyacrylamides, maleic anhydride, aldehydes, proteins, amines, hydrazine or melamine are required to achieve desire result. So far, all these procedures, for different reasons, have not led to any major practical application. The laccase treatment to activate lignin molecule is relatively new technique and a new hope of success (Wu et al. 2011; Huttermann et al. 2001. Laccase activity to initiate cross-linking, is same as the reaction mechanism of hydrogen peroxideas reported by Nimz (1986) but it is little milder.

11.6 Lignin: Chemistry

Lignin is an amorphous polphenolic material formed by the polymerization of three types of phenyl-propanoid monomers, therefore, the chemical structure cannot be drawn specifically (Sakakibara 1980; Hwang et al. 1989). It is chemically bonded to polysaccharides of cell wall, hence its isolation and purification is also a difficult (Koshijima and Watanabe 2003).

The chemistry of lignin is complicated compared to other biopolymers like proteins or carbohydrates that are linear chains or at the most branch polymers. Lignin is composed of a three dimensional network, lacking the regular and ordered repeating units like other biopolymers such as cellulose (Gonzalez et al. 2005). There are very few information is available about the crosslinking between lignin and cell wall carbohydrates. Ester and ether linkages have been reported for ferulic acid and saccharide molecules (Ralph et al. 2008).

The most accepted theory on lignin polymerization, is radicals coupling reactions of three basic monolignols, p-coumaryl, coniferyl and sinapyl alcohols (Fig. 10). The process of lignin polymerization is known as lignifications in which phenolic monomers, produces radicals, and couples together with other monomer radicals and forms oligomers and ultimately a phenyl-propanoid polymer (Brunow 1998). The majority (approximately two-third) of chemical bonds in the native lignin polymeric network are of the C–O–C ether linkage type between the phenylpropane units. Such bonding can be, β-O-4, β-5, 5-5, 5-3 or α-4, among these β-O-4 and 5-5 are the most common and contribute two-third of the total bonding (Fig. 11).

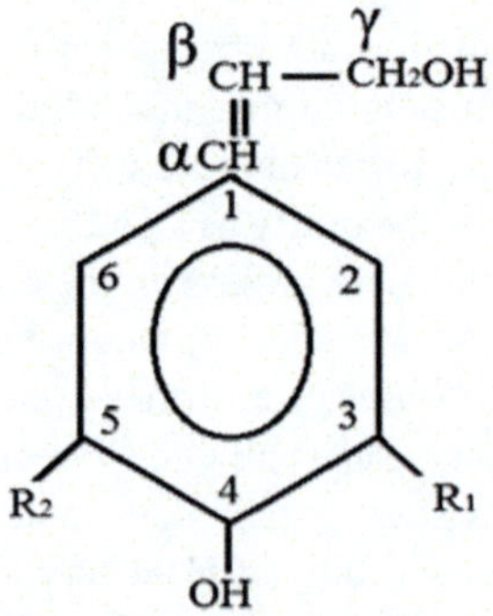

1) R_1 = H R_2 = H

2) R_1 = OCH_3 R_2 = H

3) R_1 = OCH_3 R_2 = OCH_3

Fig. 10 Structural formula of lignin unit. Where (1) represents the p-hydroxyphenyl (H) (2) guaiacyl (G) and (3) syringyl (S) (Adopted from Ralph et al. (1999))

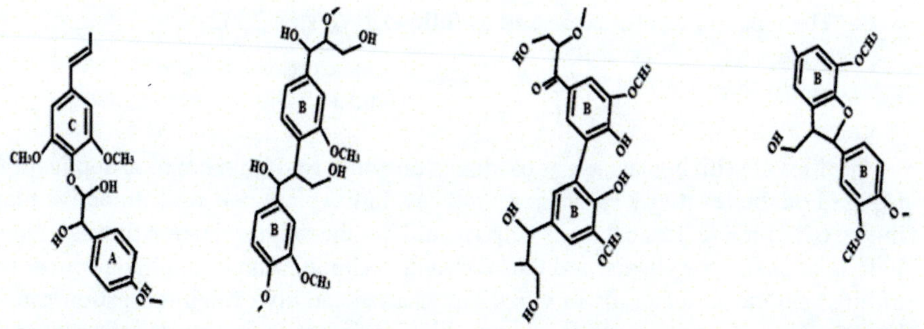

(β-O-4) Bond (β-4) Bond (5-3) Bond (β-5 α-O-4) Bond

Fig. 11 Type of bonding formed during polymerization (Glasser et al. 2000)

Table 3 Type of linkage in percent found in lignin polymer, separately in soft wood and hardwood plant (Adopted from Sjostrom (1993))

Linkage types	Percent total linkage	
	Softwood Lignin	Hardwood Lignin
β-O-4	50	60
β-5	9–12	6
5-5	10–11	5
α-O-4	2–8	7
4-O-5	4	7
β-1	7	7
β-β	2	3

Table 3. exhibits various linkages in lignin chain with percent distribution in soft wood and hard wood species. Furthermore lignin includes branched and cross-linked structures also. The more the lignin is condensed the more difficult it is to degrade and to get it dissolved in the pulping or fractionation processes.

11.7 Lignin Polymerization

Number of theories has been proposed to explain the lignin polymerization. Dean and Eriksson (1994), Higuchi (2000) studied the lignin polymerization by using chemicals and observed that phenolic monomers are polymerized by the catalytic action of phenol oxidases or peroxidases. A similar reaction was observed when lignin was oxidized by laccase enzyme. The enzymatic oxidation of lignin with an oxido-reductases laccase and peroxidase has been shown to increase the bonding strength of fibers in MDF (Felby et al. 1997 and Kharazipour et al. 1997). Laccase catalyzes a one-electron oxidation and transformed four phenolic hydroxyl groups in to four phenoxy radicals while reduces oxygen molecule, in to two molecule of water. The reaction can be expressed as follows (Widsten 2002):

$$4PhOH + O_2 \overset{Laccase}{\rightleftharpoons} 4PhO + 2H_2O$$

The idea of utilizing laccase as oxidizing enzymes for lignin bonding applications is based on the reactivity of phenoxy radicals in the plant cell wall. In native plant in-situ oxido-reductases catalysis activity initiates the polymerization through cross-linking of phenoxy radicals, and thus the same technique can be applied for bonding of lingo-cellulosic materials in vitro. The laccase activity of delignification can be improved by adding a redox mediator which increases the range of substrate from phenolic to non-phenolic compound (Bourbonnais et al. 1997). Furthermore, the laccase mediated oxidation reaction mechanisms; to generate the oxidized radical in the substrate is different from laccase alone. It enables the oxidative transformation at high redox potential of substrates which is not possible with laccase enzyme carry a low redox potential (Bourbonnais et al. 1997). Thus the substrate range and ultimately industrial applicability of laccase can be improved to a great extent by the use of a laccase-mediator system (LMS). But in the fiber treatment for MDF fabrication needs a mild and short term treatment just to improve the quality of fiber and activate the lignin molecule, thus this technique is not applied in this research. Furthermore using a mediator in enzyme reaction may increase the chances of laccase reaction with some non-phenolic compound also which may leads to reduce the fiber quality.

Lignin polymerization is begun with oxidation of the phenyl-propane phenolic hydroxyl groups (Kuzina et al. 2011). The lignin precursors undergo dimerization through enzymatic dehydrogenation, which is initiated by an electron transfer and yields resonance-stabilized phenoxy radicals. Figure 12 exhibits the phenoxy radical formed from lacaase oxidations and different form of resonance. Through, in the resonance process radicals changes many position to stabilize the oxidised phenolic compound, it forms different coupling reaction with another radical in any of the positions of the unpaired electron. Such mono-lignol, having free radical can undergo radical coupling reactions and producing number of dimers, called as di-lignols (van

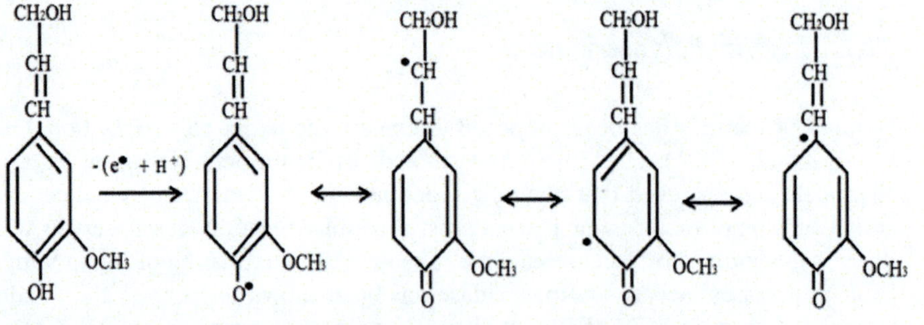

Fig. 12 Generation of resonating phenoxyl radicals by enzymatic dehydrogenation of coniferyl alcohol (Adopted from Freudenberg and Neish (1968))

Parijs et al. 2010). The bonding type β-O-4 and β-5 leads to the linear polymer. However, a branched polymer may form when a nucleophilic compound such as alcohols or phenolic hydroxyl groups, water attacks on the benzyl carbon of the quinone methide intermediate.

In one-electron transfer reaction, lignin molecules are converted into small precursors of lignin called lignols. These lignols react further with additional lignol radicals in a typical chain reaction to form the bilignols. The dilignols then undergo further endwise polymerization, instead of combining with one another. Figure 13 present the polymerization reaction of lignin units. Goring 1989 studied, these additions of more monomers to the already-existing lignin molecule, by a simple radical reaction catalyzed by either laccase or peroxidase, finally leads to the very complex structure of lignin, which forms an infinite random three-dimensional network in the middle lamella of woody plants.

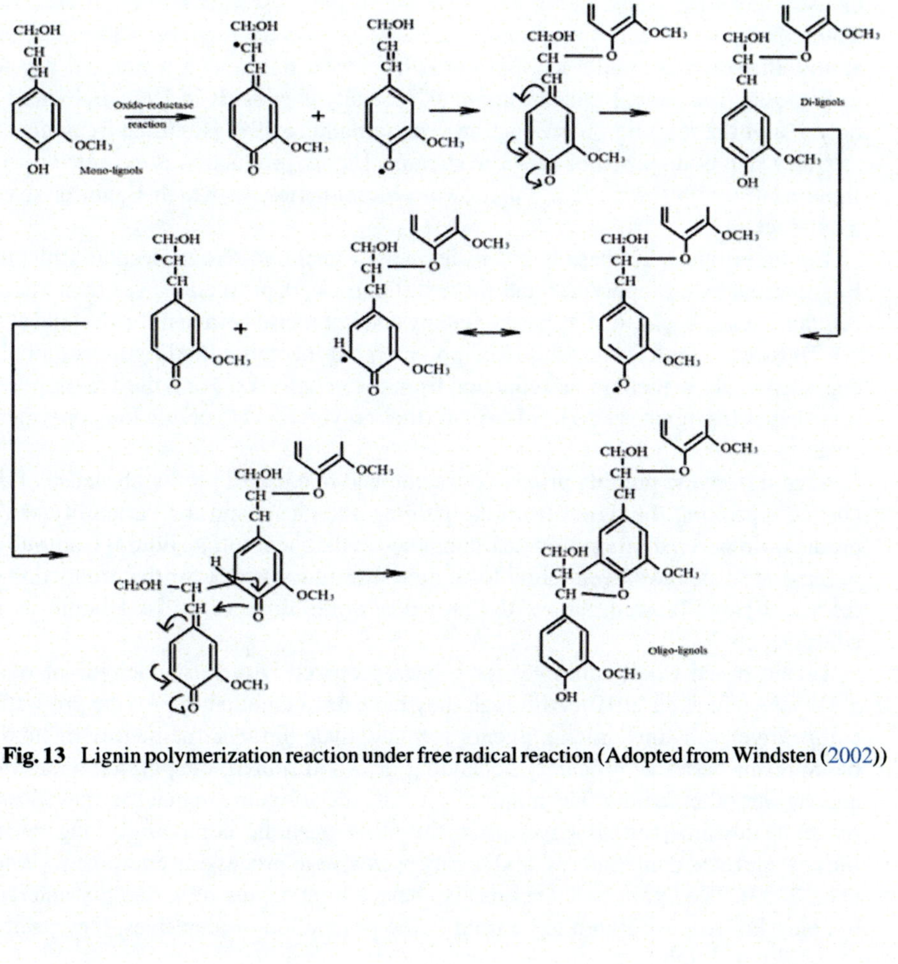

Fig. 13 Lignin polymerization reaction under free radical reaction (Adopted from Windsten (2002))

12 Development in Lignin Based Adhesive

The idea of using natural products in adhesives is not new. Animal blood and casein adhesives are one of the oldest materials used as wood adhesives. They were displaced by higher performance synthetic polymers in the 1950s. However, due to cost, environmental and concerns, the natural products are now regaining a prominent position. The natural adhesive/bio-adhesive could be any material develops from natural resources as a starting point which will provide similar or improved properties to their petroleum-based counterparts. Number of patent has been filed on natural adhesive such as protein based tannin based, soy based, lignin based like Organosolve ligin (OSL) and Enzyme hydrolysis lignin (EHL) etc. The research is going but none of them is claiming to be a better replacement of commercially available adhesive (Moubarik et al. 2010; Kim 2009). One of the naturally occurring polymers that are achieving significant attention is lignin—a waste product of the pulp and paper industry. Over the years, lignin has been one of the most intensely researched raw materials for wood adhesives, mainly because of its very low cost and vast availability. In fact, paper mills and pulp producers treat lignin as a waste and pay for its disposal. The annual sales of lignin as specialty chemicals in 1998 amounted to only 1% of the total lignin production. The remaining 99% is burned in an energy recovery step or disposed of in waste streams. Lignin production is expected to rise significantly with the current shift toward biorefineries, in which lignin is also a byproduct.

Lignin provides biomass based resins with a phenolic-like molecular structure. It is unusual as a biopolymer and more difficult to adapt to adhesives than starch, cellulose or soy because of its heterogeneity and imprecise composition (Tejado et al. 2007). Native lignin is a cross-linked polymer, but the delignified lignin is partially degraded to allow them to be separated from the cellulosics. For adhesive purposes, these degraded lignin resins need to be further polymerized to obtain long polymeric chain.

Lignin from the pulping process does not lead to a useful product because of the cost of separating the lignin from the pulping chemicals and the variability of the product. However, lignin sulfonates, contained in the spent sulfate liquids from sulfite pulping of wood have been found to be more useful feedstock for the production of reactive lignin. Table 4, shows the previous works done to utilized lignin as an adhesive.

Lignin based wood adhesives have been prepared with formaldehyde or other aldehydes (Jin et al. 2010). Although they have been claimed to provide properties competitive with conventional phenol formaldehyde binders, the quality of lignin-based resins varies significantly depending on wood source, the pulping operation, and various other factors. The major disadvantages of lignin are that they have lower reactivity towards formaldehyde than the other phenolic compound, long curing times at elevated temperatures, and highly corrosive to processing equipment (Wang et al. 2009). The lignin based resins also have a high degree of variability and dark in color. For reactive lignin to be used in the production of adhesives, they need to

Table 4 Development in lignin based adhesiveto prepare a composite

Method and application	Limitation	References
Lignin polymerisation by oxidizing with H_2O_2	Lignin more reactive towards H_2O_2	Nguyen et al. (1981)
Lignin was applied as a cementing material when polymerized	Low degree of polymerization	Sjostrom et al. (1993)
50 wt.% phenol of PF was substituted from organosolv lignin obtained from GNSL	Low reactivity of phenol obtained from lignin	Khan and Ashraf (2006)
Kraft lignin and soda lignin were used to prepared a phenol formaldehyde	Less number of hydroxyl group	Mansouri et al. (2010)

be modified in some manner. This has met with little significant commercial success principally due to the disadvantages noted above.

With the advancement of laccase based fiber treatment and subsequent removal of lignin are under intensive investigation. The laccase treated fibers get modified as discussed and can be utilized in MDF making in many ways (Fig. 14a). Laccase treated fiber will have a modified fiber with smooth surface as the lignin is deposited on fiber surface Crystallinity index of fiber is improved (Nasir et al. 2017). The obtained lignin solution contains mainly a physiologically active fraction of lignin that can be directly utilized or modified to some lignin based adhesive (Huttermann et al. 2001). In the previous study it was observed that a laccase treated fiber alone is not enough to prepared MDF comparable to commercial (Nasir et al. 2014). Therefore, a cumulative approach will have to develop where a modified fiber is added with enzyme activated lignin solution or with lignin based natural adhesive. It is assumed that a fiber surface enriched with lignin on treatment with laccase generated phenoxy

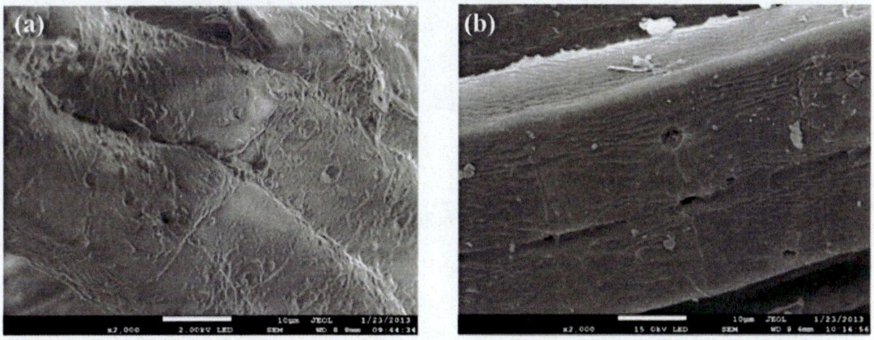

Wrinkled and uneven distribution of lignin Smooth and uniform layer of lignin

Fig. 14 SEM picture of untreated (**a**) and treated fiber (**b**) at 2k magnification (Nasir et al. 2014)

radicals. These phenoxy radicals are highly reactive and forms a network by coupling reaction (like thermoset adhesives).

When the fiber is treated with laccase enzyme, in addition of removing lignin it also breaks down the aliphatic and aromatic ring molecules and makes the fiber surface smooth (Fig. 14b) (Bledzki et al. 2010). Unlike the native fiber lignin, the fragmented lignin molecules are soluble in water solution. As the reaction proceeds, the dissolved fragmented lignin molecules precipitate back on to the fiber surface and form a uniform smooth layer around the fiber surface (Hansson and Hartler 1969). That layer of lignin deposition on the fiber surface is anticipated to be helping in fiber bonding (Maximova et al. 2001; Koljonen et al. 2004). Thus, it is concluded that an untreated fiber has a high lignin-content but with uneven distribution and rough surface. However, a laccase treated fiber gives a smooth surface with a thin layer of lignin deposition.

The MDF board prepared from 10% Con-EHL exhibited an improved mechanical strength when compared to binderless boards meet the ASTM D 1037 standard (Fig. 15). The FESEM of MDF reveals that Con-EHL boards are having many voids which provide more free surfaces to the water molecule to adhere that leads to the water absorption and thickness swelling. It was expected that if the solid content (or viscosity) of the Con-EHL would increase, its voids among the fibers will reduce. This can be achieved by either further concentrating the EHL or by adding of some foreign additive into EHL.

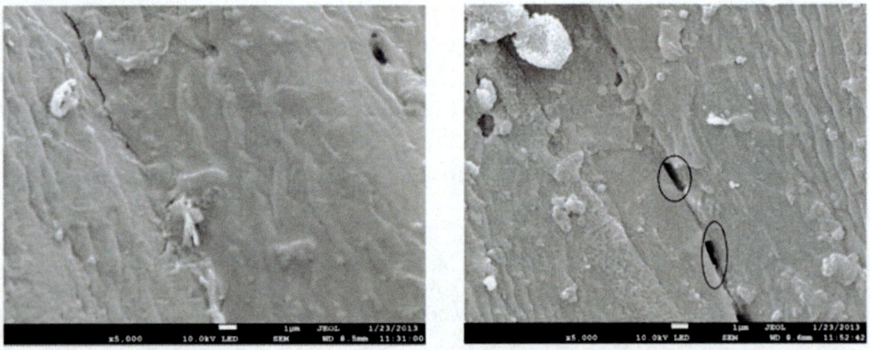

(a) Close interfacial bonding.

(b) Circle shows the presence of voids in interfacial bondings.

Fig. 15 FE-SEM of MDF boards at 5 k magnification. **a** Prepared from UF 10%, **b** Prepared from laccase treated fiber and Con-EHL 10%

13 Conclusion and Recommendation

Laccase oxidizes the lignin into a stable phenoxy radical that act as a potential cross-linking agent. It was assumed that a thin uniform layer of lignin on wood fiber surface was plasticized at high temperature and free radicals of lignin form cross-linked joints between the phenoxy molecules. From the literature it is concluded that the binderless board prepared from laccase treatments are not strong enough to meet the required minimum ASTM standard. Therefore, the technique can be used in combination with some other synthetic/natural adhesive, to reduce the total amount of adhesive required for board making. The obtained enzyme hydrolysis lignin (EHL) can also be modified to concentrated-EHL and used as an adhesive to prepare MDF board. The board prepared from modified EHL showed the improved mechanical strength as compared to binderless boards and pass the required minimum ASTM standard. It was also observed that the laccase treated fibers were having the higher crystallinity index as compared to the untreated fibre that also helps in increasing the mechanical strength of the boards. The enzyme reaction parameters (like temperature, enzyme amount, pH and time etc.) when optimized is given the best results. The maximum crystallinity index was observed maximum up to $\approx$76%, at optimum reaction condition. Fiber treated at optimum condition exhibited superiority in thermal stability, physical properties and mechanical properties over untreated fibers. Further, an optimum manufacturing process of wood composite boards may further improve the fiberboard strength.

A wood composite prepared from-lignin based natural adhesive will have no formaldehyde emission, as it does not contain any formaldehyde based compound. In-depth study on laccase treated enzyme hydrolysis lignin (EHL) as an adhesive will lead the way to utilize the waste product of pulp industries such as sulfite lignin and kraft lignin can also be utilized.

References

Akiyoshi K, Deguchi S, Tajima H, Nishikawa T, Sunamoto J (1995) Self-assembly of hydrophobized polysaccharides. Proc Jpn Acad, Ser B 71:15–19

Akiyoshi K, Nagai K, Nishikawa T, Sunamoto J (1992) Self-aggregates of hydrophobized polysaccharide as a host for macromolecular guests. Chem Lett 9:1727–1730

Alcalde M (2007) Laccase: biological functions, molecular structure and industrial applications. In: Polaina J, MacCabe AP (eds) Industrial enzymes: structure, function and applications. Springer, New York, pp. 461–476

Asim M, Abdan K, Jawaid M, Nasir M, Dashtizadeh Z, Ishak MR, Hoque ME (2015). A review on pineapple leaves fibre and its composites. Int J Polym Sci 950567

Asim M, Jawaid M, Abdan K, Nasir M (2018) Effect of Alkali treatments on physical and mechanical strength of pineapple leaf fibres. IOP Conf Ser: Mater Sci Eng 290:1

Asim M, Nasir M, Jawaid M, Tahir PM (2019) Thermo-gravimetric analysis of various ratio of blended phenolic and epoxy composites. Int J Innovative Technol Explor Eng 9:5435–5439

Asim M, Saba N, Jawaid M, Nasir M, Pervaiz M, Alothman YO (2018) A review on phenolic resin and its composites. Curr Anal Chem 14(3):185–197

Asim M, Tahir PM, Chandrasekar M, Shahroze Rao M, Jawaid M, Nasir M, Riakeng S (2020) Thermal stability of natural fibers and their polymer composites. Iran Polym J 174:175

Awadel-Karim S, Nazhad MM, Pazsner L (1999) Factors affecting crystalline structure of cellulose during solvent purification treatment. Holzforschung 53(1):1–8

Bajpai P (1999) Application of enzymes in the pulp and paper industry. Biotechnol Prog 15:147–157

Bao W, O'Malley DM, Whetten R, Sederoff RR (1993) A laccase associated with lignification in loblolly pine xylem. Science 260:672–674

Bledzki AK, Mamun AA, Jaszkiewicz A, Erdmann K (2010) Polypropylene composites with enzyme modified abaca fiber. Compos Sci Technol 70:854–860

Blomquiest RF, Christiansen AW, Gillespie RH, Myers GE (1981) Adhesive bonding of wood and other structural materials. Pennsylvania State University, Madison WI

Bourbonnais R, Paice MG, Freiermuth B, Bodie E, Borneman S (1997) Reactivities of various mediators and laccases with kraft pulp and lignin model compounds. Appl Environ Microbiol 63:4627–4632

Bourbonnais R, Paice MG, Reid ID, Lanthier P, Yaguchi M (1995) Lignin oxidation by laccase isozymes from *Trametes versicolor*and role of the mediator ABTS in kraft lignin depolymerization. Appl Environ Microbiol 61:1876–1880

Brunow G (1998) Oxidative coupling of phenols and the biosynthesis of lignin. In: Lewis NG, Sarkanen S (eds) Lignin and lignan biosynthesis. American Chemical Society, Washington, DC, pp 131–147

Camarero S, Ibarra D, Martinez AT, Romero J, Gutierrez A, Del Rio JC (2007) Paper pulp delignification using laccase and natural mediators. Enzym Microb Technol 40:1264–1271

Dashtban M, Schraft H, Syed TA, Qin W (2010) Fungal biodegradation and enzymatic modification of lignin. Int J Biochem Mol Biol 1:36–50

De Souza CGM, Peralta RM (2003) Purification and characterization of the main laccase produced by the white-rot fungus Pleurotus pulmonarius on wheat bran solid state medium. J Basic Microbiol 43:278–286

Dean JFD, Eriksson KEL (1994) Laccase and the deposition of lignin in vascular plants. Holzforschung 48(S1):21–33

Dence CW, Lin SY (1992) Methods in Lignin Chemistry. In: Lin SY, Dence CW. Springer, Berlin, pp 3–19

Dinwoodie JM (1989) Wood: nature's cellular, polymeric fibre-composite, vol 420. The Institute of Metals, p 13

Ebringerova A, Heinze T (2000) Xylan and xylan derivatives-biopolymers with valuable properties, 1. Naturally occurring xylans structures, isolation procedures and properties. Macromol Rapid Commun 21:542–556

Eckelman CA (1997) A brief survey of wood adhesives. Forest Nat Resour 154:10

Edens WA, Goins TQ, Dooley D, Henson JM (1999) Purification and characterization of a secreted laccase of *Gaemannomyces graminis* var. tritici. Appl Environ Microbiol 65:3071–3074

European Panels Federation (2004) Annual Report 2003–2004

Felby C, Hassingboe J, Lund M (2002) Pilot-scale production of fiberboards made by laccase oxidized wood fibers: board properties and evidence for cross-linking of lignin. Enzyme Microb Technol 31:736–741

Felby C, Nielsen BR, Olesen PO (1997) Identification and quantification of radical reaction intermediates by electron spin resonance spectrometry of laccase-catalyzed oxidation of wood fibers from beech (Fagus sylvatica). Appl Microbiol Biotechnol 48:459–464

Felby C, Thygesen LG, Sanadi A, Barsberg S (2004) Native lignin for bonding of fiber boards—evaluation of bonding mechanisms in boards made from laccase-treated fibers of beech (Fagus sylvatica). Ind Crops Prod 20:181–189

Feldman D (2002) Lignin and its polyblends-a review. In: Hu TQ (ed) Chemical modification, properties, and usage of lignin. Springer, New York, pp 81–100

Freudenberg K, Neish AC (1968) Constitution and biosynthesis of lignin. In: Norman AG (ed) Molecular biology, biochemistry, and biophysiscs. Springer, New York

Gheith MH, Aziz AM, Ghori W, Saba N, Asim M, Jawaid M, Alothman OY (2019) Flexural, thermal and dynamic mechanical properties of date palm fibres reinforced epoxy composites. J Mater Res Technol 8(1):853–860

Gianfreda L, Xu F, Bollag JM (1999) Laccases: a useful group of oxidoreductive enzymes. Bioremed J 3:1–25

Gierenz G, Karmann W (2001) Adhesives and adhesive tapes. Wiley-VCH Verlag Gmb. H

Glasser WG, Jain RK (1993) Lignin derivatives. I. Alkanoate. Holzforschung 47:225–233

Glasser WG, Northey RA, Schultz TP (2000) Classification of lignin according to chemical and molecular structure. In: Lignin: historical, biological, and materials perspectives. ACS symposium series No. 742. American Chemical Society, Washington, D.C., pp 216–238

Gochev VK, Krastanov AI (2007) Fungal laccases (review). Bul J Agr Sci 13:75–83

Gonzalez K, Arias E, Carnicero A, Perestelo F, Falcon M (2005) Improvements in an industrial lignin (Indulin at) with laccase and laccase-mediator systems evaluated by spectroscopic and chromatographic techniques. In: ITALIC 3 Wood derivatives and agroindustrial waste valorisation 23–24 June

Goring DAI (1989) The lignin paradigm. In: Glasser WG, Sarkanen S (eds) New polymeric materials from lignin. ACS Symposium Series, no. 397. American Chemical Society, Washington, DC, pp 2–10

Gradwell S (2004) Surface modification of cellulose fibers: towards wood composites by biomimetics. CR Biol 327:945–953

Gumuskaya E, Usta M, Kirci H (2003) The effects of various pulping conditions on crystalline structure of cellulose in cotton linters. Polym Degrad Stab 81:559–564

Hansson JA, Hartler N (1969) Sorption of hemicellulose on cellulose fibers. Svensk Papperstid. 17:521–530

Harris JH (1990) Plant cell wall structure and development. In: Akin DE, Ljungdahl LG, Wilson JR (eds) Microbial and plant opportunities to improve lignocellulose utilization by ruminants. Harris PJ, London, pp 71–90

Haygreen JG, Bowyer JL (1996) Forest products and wood science: an introduction, 3rd edn. Lowa State University Press, Ames, p 484

Heinzkill M, Messner K (1997) The ligninolytic system of fungi. In: Anke T (ed) Fungal biotechnology. Chapman & Hall, Weinheim, pp 213–226

Higuchi T (2000) The present state and problems in lignin biosynthesis. Cellul Chem Technol 34(1–2):79–100

Hill C, Cetin N (2000) Surface activation of wood for graft polymerization. Int J Adhes Adhes 20:71–76

Huang W, Sun XM (2000) Adhesive properties of soy proteins modified by sodium dodecyl sulfate and sodium dodecylbenzene sulfonate. J Am Chem Soc 77:705–708

Huttermann A, Haars A (1981) In Der Wald als Rohstoflquelle. In: Hiittermann A (ed) Sauerllnder's, Frankfurt, pp 184–194

Huttermann A, Mai C, Kharazipour A (2001) Modification of lignin for the production of new compounded materials. Appl Microbiol Biotechnol 55:387–384

Hwang, RH. Kennedy, JF. Melo, EHM. and White, CA. 1989. A probable lignin structure by conformation analysis. Carbohyd Polym 10:15–30

IARC (2004) Press release #153. International Agency for Research on Cancer

Ikeda R, Tanaka H, Uyama H, Kobayashi S (1998) Laccase-catalyzed polymerization of acrylamide. Macromol Rapid Commun 19:423–425

Ioelovich M (2008) Cellulose as a nanostructured polymer: a short review. BioResources 3:1403–1418

Ishikawa A, Sugiyama J, Okano T (1994) Fine structure and tensile properties of ramie fibers in the crystalline form of cellulose I, II and III1. Wood Res 81:16–18

Ria Millati I, Syamsiah S, Niklasson C, Nur Cahyanto M, Lundquist K, Taherzadeh MJ (2011) Biological pretreatment of lignocelluloses with white-rot fungi and its applications: a review. Bioresources 6:1–36

Jin Y, Cheng X, Zheng Z (2010) Preparation and characterization of phenol-formaldehyde adhesives modified with enzymatic hydrolysis lignin. Biores Technol 101:2046–2048

Jung HG, Fahey GC (1983) Nutritional implications of phenolic monomers and lignin: a review. J Anim Sci 57:206–219

Karayannidou EG, Achilias DS, Sideridou ID (2006) Cure kinetics of epoxy–amine resins used in the restoration of works of art from glass or ceramic. Eur Polymer J 42:3311–3323

Kavvouras PK, Koniditsiotis D, Petinarakis J (1998) Resistance of cured urea–formaldehyde resins to hydrolysis: a method of evaluation. Horzforschung. 52:105–110

Kawai S, Umezawa T, Higuchi T (1988) Degradation mechanisms of phenolic beta-1 lignin substructure model compounds by laccase of Coriolus versicolor. Arch Biochem Biophys 262:99–110

Khan MA, Ashraf SM (2006) Development and characterization of groundnut shell lignin modified phenol formaldehyde wood adhesive. Indian J Chem Technol 13:347–352

Khan MA, Ashraf SM (2007) Studies on thermal characterization of lignin substituted phenol formaldehyde resin as wood adhesives. J Therm Anal 89:993–1000

Kharazipour A, Hüttermann A, Lüdemann HD (1997) Enzymatic activation of wood fibers as a means for the production of wood composites. J Adhes Sci Technol 11(3):419–427

Kiiskinen LL, Kruus K, Bailey M, Ylosmaki E, Siika-aho M, Saloheimo M (2004) Expression of Melanocarpus albomyces laccase in Trichoderma reesei and characterization of the purified enzyme. Microbiology 150:3065–3074

Kim S (2009) Environment-friendly adhesives for surface bonding of wood-based flooring using natural tannin to reduce formaldehyde and TVOC emission. Bioresour Technol 100:744–748

Kim S, Holtzapple MT (2005) Lime pretreatment and enzymatic hydrolysis of corn stover. Biores Technol 96:1994–2006

Kim Y-J, Nicell JA (2006) Impact of reaction conditions on the laccase-catalyzed conversion of bisphenol A. Bioresour Technol 97(12):1431–1442

Kim S, Holtzapple MT (2006) Effect of structural features on enzyme digestibility of corn stover. Biores Technol 97:583–591

Klauditz W, Stegmann G (1955) Beiträge zur Kenntnis des Ablaufes und der Wirkung thermischer Reaktionen bei der Bildung von Holzwerkstoffen. Holz Roh-Werkstoff 13:434–440

Koch GS (1987) Adhesives for the composite wood panel industry. Park Ridge, NJ, Noyes Data Corp

Koljonen K, Osterberg M, Kleen M, Fuhrmann A, Stenius P (2004) Precipitation of lignin and extractives on kraft pulp: effect on surface chemistry, surface morphology and aper strength. Cellulose 11:209–224

Koshijima T, Watanabe T (2003) Association between lignin and carbohydrates in wood and other plant tissues. Springer series in wood science

Kunamneni A, Ballesteros A, Plou FJ, Alcalde M (2007) Fungal laccase—a versatile enzyme for biotechnological applications. In: Mendez-Vilas A (ed) Communicating current research and educational topics and trends in applied microbiology, pp 233–245

Kunamneni A, Camarero S, García-Burgos C, Plou FJ, Ballesteros A, Alcalde M (2008) Engineering and applications of fungal laccases for organic synthesis. Microb Cell Fact 7:32

Kuzina SI, Shilova IA, Mikhailov AI (2011) Chemical and radiation-chemical radical reactions in lignocellulose materials. Radiat Phys Chem 80:937–946

Lamaming J, Hashim R, Peng C, Sulaiman O, Sugimoto T, Nasir M (2015) Isolation and characterization of cellulose nanocrystals from parenchyma and vascular bundle of oil palm trunk (Elaeis guineensis). Carbohydr Polym 134:534–540

Lavoine N, Desloges I, Dufresne A, Bras J (2012) Microfibrillated cellulose—its barrier properties and applications in cellulosic materials: a review. Carbohyd Polym 90:735–764

Li K, Peshkova S, Geng X (2004) Investigation of soy proteinKymene adhesive systems for wood composites. J Oil Chem Soc 81:487–491

Li X, Li Y, Zhong Z, Wang D, Ratto JA, Sheng K, Sun XS (2009) Mechanical and water soaking properties of medium density fiberboard with wood fiber and soybean protein adhesive. Bioresour Technol 100:3556–3562

Lina H, Gunasekaran S (2010) Cow blood adhesive: characterization of physicochemical and adhesion properties. Int J Adhes Adhes 30:139–144

Linzell HK (1950) The Insulite Wallboard Mill at Kymi Finland. Paper Trade J 131:10

Liu K (1997) Soybeans–chemistry, technology, and utilization. International Thomson Publishing, New York

Liu Y, Li K (2004) Modification of soy protein for wood adhesives using mussel protein as a model: the influence of a mercapto group. Macromol Rapid Commun 25:1835–1838

Liu YM, Zong CZ, Xu Y (2005) Manufacture of the self-crosslinking PVAc emulsion under room temperature. Zhongguo Jiaonianji. 14:9–11

Lora JH, Glasser WG (2002) Recent industrial applications of lignin : a sustainable alternative to nonrenewable materials. J Polym Environ 10:39–48

Mai C, Kues U, Militz H (2004) Biotechnology in the wood industry. Appl Microbiol Biotechnol 63:477–494

Mansouri HR, Navarrete P, Pizzi A, Tapin-Lingua S, Benjelloun-Mlayah B, Pasch H, Rigolet S (2010) Synthetic-resin-free wood panel adhesives from mixed low molecular mass lignin and tannin. Eur J Wood Wood Prod 69:221–229

Maximova N, Osterberg M, Koljonen K, Stenius P (2001) Lignin adsorption on cellulose fiber surfaces: effect on surface chemistry, surface morphology, and paper strength. Cellulose 8:113–125

Maximova N, Osterberg M, Laine K, Stenius P (2004) The wetting properties and morphology of lignin adsorbed on cellulose fibers and mica. Colloids Surf A 239:65–75

Mo Z, Yang B, Zhang H (1994) The degree of crystallinity of multi- component polymers by waxd. Chin J Polym Sci 12:296–301

Moilanen U, Kellock M, Galkin S, Viikari L (2011) The laccase-catalyzed modification of lignin for enzymatic hydrolysis. Enzym Microb Technol 49:492–498

Moldes D, Cadena EM, Vidal T (2010) Biobleaching of eucalypt kraft pulp with a two laccase-mediator stages sequence. Biores Technol 101(18):6924–6949

Morozova O, Shumakovich G, Shleev S, Yaropolov Y (2007) Laccase-mediator

Moubarik A, Allal A, Pizzi A, Charrier F, Charrier B (2010) Characterization of a formaldehyde-free cornstarch-tannin wood adhesive for interior plywood. Eur J Wood Wood Prod 68:427–433

Moya R, Saastamoinen P, Hernandez M, Suurnakki A, Arias E, Mattinen ML (2011) Reactivity of bacterial and fungal laccases with lignin under alkaline conditions. Bioresour Technol 102:10006–10012

Mozaffar AK, Sayed Marghoob A, Ved Prakash M (2004) Development and characterization of a wood adhesive using bagasse lignin. Int J Adhes Adhes 24:485–493

Murphy MEP, Lindley PF, Adman ET (1997) Structural comparison of cupredoxin domains: domain recycling to construct proteins with novel functions. Protein Sci 6:761–770

Nasir M, Gupta A, Beg MDH, Chua GK, Kumar A (2014) Physical and mechanical properties of medium-density fibreboards using soy—lignin adhesives. J Trop For Sci 26, 41–49

Nasir M, Gupta A, Beg MDH, Chua GK, Kumar A (2013) Fabrication of medium density fibreboard from enzyme treated rubber wood (Hevea brasiliensis) fibre and modified organosolv lignin. Int J Adhes Adhes 44

Nasir M, Hashim R, Sulaiman O, Gupta A, Khan TA, Jawaid M, Asim M (2017) Natural fiber improvement by laccase; optimization, characterization and application in medium density fiberboard. J Nat Fibers 14

Nasir M, Hashim R, Sulaiman O, Nordin NA, Lamaming J, Asim M (2015) Laccase, an emerging tool to fabricate green composites: a review. BioResources 10

Nasir M, Khali DP, Jawaid M, Tahir PM, Siakeng R, Asim M, Khan TA (2019) Recent development in binderless fiber-board fabrication from agricultural residues: a review. Constr Build Mater 211:502–516

Nasir M, Gupta A, Beg MDH, Chua GK, Jawaid M, Kumar A, Khan TA (2013) Fabricating eco-friendly binderless fiberboard from laccase-treated rubber wood fiber. BioResources 8:3599–3608

Nasir M, Gupta A, Hossen Beg MD, Chua GK, Asim M (2014) Laccase application in medium density fibreboard to prepare a bio-composite. RSC Adv 4

Nguyen T, Zavarin E, Barrell EMII (1981) Thermal analysis of lignocellulosic materialas. Part I Unmodified materials. J Macromol Sci Chem 20:1–65

Nimz HH (1983) Lignin-based adhesives. In: Pizzi A (ed) Wood adhesives: chemistry and technology, vol 1. Marcel Dekker Inc., New York, USA

Nyanhongo GS, Gomes J, Gubitz G, Zvauya R, Read JS, Steiner W (2002) Production of laccase by a newly isolated strain of Trametes modesta. Biores Technol 84:259

Osong SH, Norgren S, Engstrand P (2016) Processing of wood-based microfibrillated cellulose and nanofibrillated cellulose, and applications relating to papermaking: a review. Cellulose 23:93–123

Pizzi A (2003) Natural phenolic adhesives II: Lignin. In: Pizzi A, Mittal KL (eds) Handbook of adhesive technology, 2d edn. Marcel Dekker Inc., New York, USA

Pizzi A (2006) Recent developments in eco-efficient bio-based adhesives for wood bonding: opportunities and issues. J Adhes Sci Technol 20:829–846

Pointing SB, Vrijmoed LLP (2000) Decolourisation of azo and triphenylmethane dyes by Pycnoporus sanguineus producing laccase as the sole phenoloxidase. World J Microbiol Biotechnol 16:317–318

Rafiqah A, Abdan K, Nasir M, Asim M (2020) Effect of extraction on the mechanical, physical and biological properties of pineapple leaf fibres. Springer, Pineapple Leaf Fibers, pp 41–54

Ralph J (1999) Lignin structure: recent developments. US Dairy Forage Research Center, USDA-Agricultural Research Service, pp 1–16

Ralph J, Kim H, Lu F, Grabber JH, Leple JC, Berrio-Sierra J, Derikvand MM, Jouanin L, Boerjan W, Lapierre C (2008) Identification of the structure and origin of thioacidolysis marker compound for ferulic acid incorporation into angiosperm lignin (and an indicator for cinnamoyl CoA reductase deficiency). Plant J 53:368–379

Ratna D (2003) Modification of epoxy resins for improvement of adhesion: a critical review. J Adhes Sci Technol 17:1655–1668

Saka, S. and Goering, D.A.I. 1985. *Localization of lignins in wood cell walls.* In: Higuchi T, Orlando FL (eds) Biosynthesis and Biodegradation of wood components. Academic Press, pp 51–62

Sakakibara A (1980) Structural model of softwood lignin. Wood Sci Technol 14:89–100

Salvado J, Velasquez JA, Ferrando F (2003) Binderless fiberboard from steam exploded Miscanthus sinensis: optimization pressing and pretreatment conditions. Wood Sci Technol 37:279–286

Sellers T Jr (2001) Wood adhesive innovations and applications in North America. For Prod J 51:12–22

Setliff, E.D. and Eudy, W.W. 1980. *Screening white-rot fungi for their capacity to delignify wood.* Lignin Biodegradation: Microbiology, Chemistry, and Potential Application, Vol. I, T. K Kirk, H.-M. Chang, and T. Higuchi (eds.), CRC Press, Boca Raton, FL, USA. 135–149.

Siakeng R, Jawaid M, Ariffin H, Sapuan SM, Asim M, Saba N (2019) Natural fiber reinforced polylactic acid composites: a review. Polym Compos 40(2):446–463

Siakeng R, Jawaid M, Tahir PM, Siengchin S, Asim M (2020) Improving the properties of pineapple leaf fibres by chemical treatments. Springer, Pineapple Leaf Fibers, pp 55–71

Sjostrom, E. 1993. *Wood chemistry: fundamentals and applications.* 2nd ed. San Diego. CA: Academic Press Inc.

Slomezynski D, Nakas JP, Tanenbaum SW (1995) Production and characterization of laccase from Botrytis cinerea 61–34. App Enviro Microbiol 61:907–912

Snajdr J, Baldrian P (2007) Temperature affects the production, activity and stability of ligninolytic enzymes in Pleurotus ostreatus and Trametes versicolor. Folia Microbiol 52:498–502

Stephanou A, Pizzi A (1993) Rapid Curing Lignin-Based Exterior Wood Adhesives. Part I. Diisocyanates Reaction Mechanisms and Application to Panel Products. Holzforschung 47:439–445

Stofko, J. and Zavarin. 1977. U.S. *Patent 4,007,312.*

Sulaiman NS, Hashim R, Sulaiman O, Nasir M, Amini MHM, Hiziroglu S (2018) Partial replacement of urea-formaldehyde with modified oil palm starch based adhesive to fabricate particleboard. Int J Adhes Adhes 84:1–8

Sun RC, Fang JM, Tomkinson J (2000) Characterization and esterification of hemicelluloses from rye straw. J Agric Food Chem 18:1247–1252

Suurnakki A, Oksanen T, Orlandi M, Zoia L, Canevali C, Viikari L (2010) Factors affecting the activation of pulps with laccase. Enzyme Microb Technol 46:153–158

Tejado A, Pena C, Labidi J, Echeverria JM, Mondragon I (2007) Physico-chemical characterization of lignins from different sources for use in phenol-formaldehyde resin synthesis. Biores Technol 98:1655–1663

Tierney JJ, Gillespie JW Jr (2004) Crystallization kinetics behavior of PEEK based composites exposed to high heating and cooling rates. Compos Appl Sci Manuf 35:547–558

Vaheri M, Salama N, Ruohoniemi KM (1991) Procedure for the production of pulp. European Patent, EP: 429422

Van de Pas D, Hickson A, Donaldson L, Lloyd-Jones G, Tamminen T, Fernyhough A, Mattinen ML (2011) Characterization of fractionated lignins polymerized by fungal laccases. BioResources 6:1105–1121

Van Parijs FRD, Morreel K, Ralph J, Boerjan W, Merks RMH (2010) Modeling lignin polymerization. I. Simulation model of dehydrogenation polymers. Plant Physiol 153:1332–1344

Velasquez JA, Ferrando F, Salvado J (2003) Effects of kraft lignin addition in the production of binderless fiberboard from steam exploded Miscanthus sinensis. Ind Crop Prod 18:17–23

Vishtal A, Kraslawski A (2011) Challenges in industrial applications of technical lignins. BioResources 6:3547–3568

Wagenführ A (1988) Praxisrelevante Untersuchungen zur Nutzung biotechnologischer Wirkprinzipien bei der Holzwerkstoffherstellung. Dissertation, Technsiche University, Dresden

Wang M, Leitch M, Xu C (2009) Synthesis of phenol–formaldehyde resol resins using organosolv pine lignins. Eur Polymer J 45:3380–3388

Widsten P (2002) Radical formation on laccase treatment of wood. Nord Pulp Pap Res J 17:2

Wong KKY, Richardson JD, Mansfield SD (2000) Enzymatic Treatment of Mechanical Pulp Fibers for Improving Papermaking Properties. Biotechnol Prog 16(6):1025–1029

World Health Organisation (WHO), 2004 Annual report.

Wu J, Zhang X, Wan J, Ma F, Tang Y, Zhang X (2011) Production of fiberboard using corn stalk pretreated with white-rot fungus Trametes hirsute by hot pressing without adhesive. Biores Technol 102:11258–11261

Xu F (2005) Applications of oxidoreductases: recent progress. Industrial Biotechnol. 1:38–50

Xu F, Berka RM, Wahleithner JA, Nelson BA, Shuster JR, Brown SH, Palmer AE, Solomon EI (1998) Site-directed mutations in fungal laccase: effect on redox potential, activity and pH profile. Biochem J 334:63–70

Yang I, Kuo M, Myers DJ, Pu A (2006) Comparison of protein-based adhesive resins for wood composites. J Wood Sci 52:503–508

Yaropolov AI, Skorobogatko OV, Vartanov SS, Varfolomeyev SD (1994) Laccase properties, catalytic mechanism and applicability. Appl Biochem Biotechnol 49:57–79

Yoshida H (1883) Chemistry of lacquer (urushi). Part I. J Chem Soc 43:472–486

Zhou G, Li J, Chen Y, Zhao B, Cao Y, Duan X, Cao Y (2009) Determination of reactive oxygen species generated in laccase catalyzed oxidation of wood fibers from Chinese fir (*Cunninghamia lanceolata*) by electron spin resonance spectrometry. Biores Technol 100:505–508

Biobased Thermoplastic Polyurethanes and Their Capability to Biodegradation

Atika Alhanish and Mustafa Abu Ghalia

Abstract Extensively used petrochemical polymers possess harmful influence on the ecosystem, therefore the utilizing of bio-based materials should be essentially important to employ in wide range of applications, due to positive affect to the environment. The advantage of these materials significantly enhances the carbon footprint and has a helpful impact on the life cycle assessment (LCA) of plastic products; consequently, the development of biobased thermoplastic polyurethane (bio-TPU) from renewable resources is gain much interested from various polymers industries. In addition, recent advances in biotechnology dedicate toward produce a new class of TPU to be developed from plant-based biomass. Thus, this review focuses on recent advances of bio-TPU formulations. Factors effecting the TPU formulation are presented in view of the potential applications in various fields. Biodegradation of TPU and some aspects accelerating their biodegradability are also described.

Keywords Thermoplastic polyurethane · Biodegradation · Bio-based · Natural materials · Automotive · Elastomer

1 Introduction

Polymers identified as eco-friendly materials assistance to enhance the environment by at least one property that reduces environmental issues. Ecofriendly polymers are evaluated not only by their origin from biobased sources, nonetheless also by simple way to disposal them, recycling such as reuse of waste, biodegradation or composting, health-safe and reducing impact on human health, light weightiness, energy efficiency, reduction of pollution, and material usage (Nguyen et al. 2010).

A. Alhanish
Department of Chemical Engineering, Faculty of Petrolum and Natural Gas, University of Zawia, Zawia, Libya

M. Abu Ghalia (✉)
Department of Chemical Engineering, Ryerson University, Toronto M5B 2K3, Canada
e-mail: mabughal@ryerson.ca; abughalia76@gmail.com

M. Jawaid et al. (eds.), *Eco-Friendly Adhesives for Wood and Natural Fiber Composites*, Composites Science and Technology,
https://doi.org/10.1007/978-981-33-4749-6_4

Recently, much attention is dedicated to the concept of sustainable development. Therefore, it is essential to integrate social, economic, and ecological aspects and make several changes. Substantial advances have been achieved in the development of alternative resources of raw materials. The use of biobased raw materials considerably enhance the carbon footprint and thus has a positive influence on the LCA of final products (Rydz et al. 2015).

For decades, thermoplastic polyurethanes (TPUs) have been extensively explored. They exhibit outstanding properties. They have good hydrolytic stability, abrasion properties, suitable mechanical properties with superior chemical resistance, dimensional stability, stiffness, high impact resistance, high strength, and good process ability. TPUs are one of the most important synthetic polymers produced in large quantities for high-performance engineering thermoplastics, which expected a growth of ~3.5% (CAGR) by 2024 in the global market (Khalifa et al. 2020). In this context, the progress of TPU from renewable resources becomes vital in medical field and modern industry. Furthermore, the biobased monomers and their polymerization process is becoming progressively significant form the economic and ecological prospective. Consequently, the development of environmentally friendly and sustainable production from biomass-based TPU is highly requested (Rydz et al. 2015). Currently, a major part of bio-based raw materials obtained from plant oils and natural fats is devoted to polyurethane industry. Nevertheless, most of the reported bio-TPU are chemically cross-linked networks (thermosets), except for the very recent renewable-sourced TPU Pearlthane® ECO from Merquinsa (Rusu et al. 2011).

Fully biodegradation rate of TPU still being challenge as the hard segment (isocyanate) is not degradable in various environment conditions. Based on that, incorporate biobased materials into the isocyanate chemistry will help to accelerate biodegradation process. Recent studied by Zhang et al. (2017), developed a novel polyurethane (PU) derived from lignin and poly(caprolactone) (PCL), PCL was incorporated as a biodegradable soft segment to the lignin by the bridge of hexamethylene diisocyanate (HDI) with long flexible aliphatic chains. The degradation rate of PU was conducted in soil from 0 to 48 weeks were the degree of. The weight loss vs degradation time was represented the degradation rate, which was up to 89.99 wt% after 48 weeks. This is explained that the PU have ester bonds which were vulnerable to hydrolysis because of microbial attack, consequently, the degradation was basically occurs from the break-down of ester bonds.

In this chapter, the recent researches and expansion the applications of bio-TPUs are outlined. Environmental impact of TPU is described in prospective applications in various fields. Biodegradable TPU growth much attention in sustainable polymer development. Thus, microbial and enzymatic biodegradation of TPU and some other factors that affect their biodegradability are also discussed.

2 Sustainable Thermoplastic Polyurethane

TPU is a potentially versatile elastomer with exceptional properties that offer both superior performance and processing flexibility. The synthesis of TPU classified into soft and hard segments, the soft part is usually bio-based polyether or polyester while the hard part is a composed of a diisocyanate and chain extenders (Datta and Kasprzyk 2018). Therefore, the manufacture of a TPU is mainly based on petroleum feedstock chemicals like ethylene and propylene. In addition, the automotive industry is presently facing challenges to reduce the environmental impact from the current TPU. Meanwhile, recent advances in biotechnology dedicate toward produce a new class of TPU to be developed from plant-based biomass. This is good news for prospective automotive and footwear regarding sustainability and light weights (Zhao et al. 2020).

The TPU market size was estimated at 53.94 USD billion in 2015 and in general has prospered over the last 5 years. Growth is estimated at 6–7% CGAR by 2025. The market is competitive, oligopolistic and dependent on product development and innovation for applications. 60% of market share is dominated by key top players including Covestro, BASF and The Dow Chemical Company. Market player contribution is expected to grow as the industry grows and as many major internationals form strategic alliances with smaller regional players (Globe Newswire 2020).

Recent industries plans are to synthesis, develop, produce, and commercialize bio-TPU that's made from natural resources and end-product formulations for various verticals and product line categories whereas still maintaining mechanical specifications comparable to chemicals-based products currently used. Likewise, the TPU were synthesized using novel formulations synthesis whereby incorporate bio-based ingredients (Zhao et al. 2020).

3 Thermoplastic Polyurethane: Biobased Synthesis, Properties, and Applications

A TPU is a binary system from soft and hard segment which formulated of three raw ingredients that combined in a specific way: (1) polyol or long-chain diol, (2) chain extender or short-chain diol, and (3) diisocyanate.

Figure represented the fundamental chemistry of TPU (Fig. 1). The soft phase or soft segment (SS), built out of a polyol, is dedicated to the flexibility and elastomeric character of the TPU. The hard phase or hard segment (HS), built up from an isocyanate and chain extender, provided a TPU its stiffness and overall mechanical performance.

One of the unique privileges of TPU that is combines the processability of thermoplastic and the elasticity and softness of rubber (Wang et al. 2019). Figure 2 highlight the cross-linking network of a thermoplastic elastomer against a thermoset rubber. It

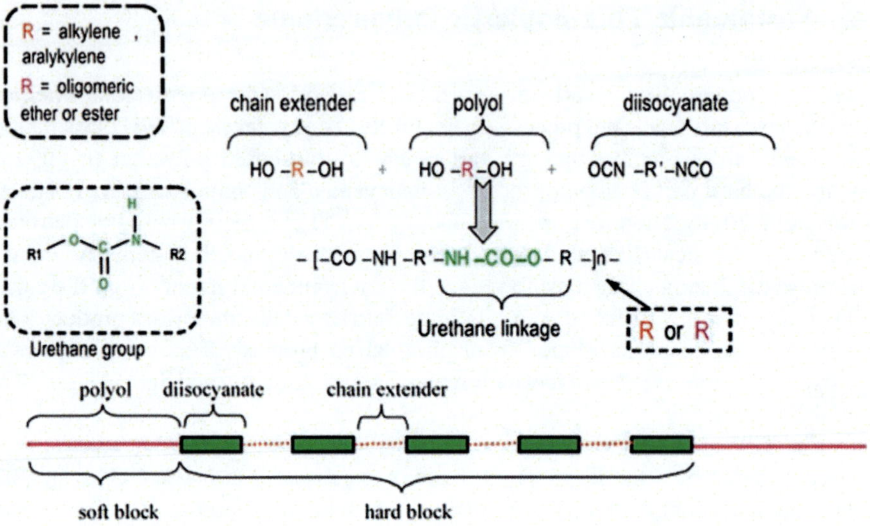

Fig. 1 Basic chemistry of TPU

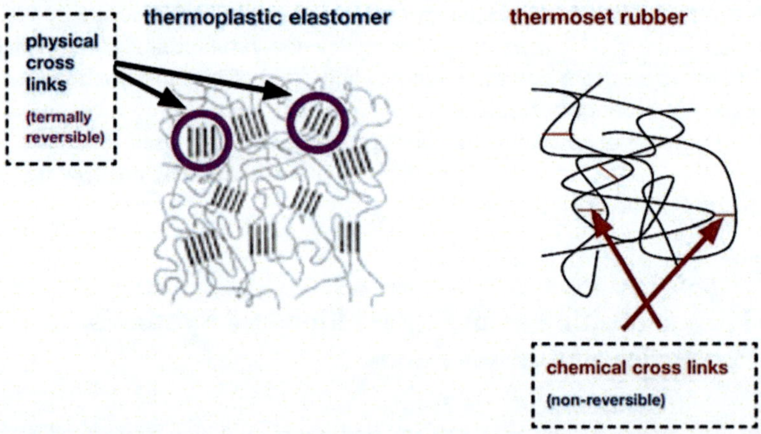

Fig. 2 Cross-linkers structures of thermoplastic elastomer versus thermoset rubber

demonstrates that there are absent of chemical cross-links in TPUs, which is different in thermoset rubbers or casted polyurethane systems.

The crystalline rigid blocks are formed by the aggregation of portions of the long polar HS through hydrogen bonding, which separated from SSs. Other long portion of the HS is mixed with the SSs. Furthermore, short HSs are amorphous dissolved in the SSs. This structure is common among segmented copolymers (Liu et al. 2017).

Due to the crystallization and thermodynamic incompatibility of the HS and SS, a microphase separation morphology is obtained (Xiang et al. 2020). This morphology is a good indicate for understand the processing and maintain mechanical properties (Christenson et al. 2007). It was reported that TPU with low HS (30–40 Vol.%) will display microphase separation, were the HS form isolated domains dispersed within the SS (Datta and Kasprzyk 2018). However, the covalent bond between the two segments prevent the microphase separation as might exist in polyol and isocyanate blends (Yilgör et al. 2015). It worth to mention that a proper choice of the type and symmetry of diisocyanate allow the synthesizing of TPU without the use of chain extender, where a high homogenous accomplished with improved mechanical properties (Marcano et al. 2019).

Although, other factors play vital role in the formation of TPU morphology such as the ratio of HS and SS, synthesis approach, processing method, molecular weight of polyol, chemical structure of basic raw materials i.e. polyol, isocyanate and chain extender (Kasprzyk and Datta 2019; Xiang et al. 2020).

Commonly, TPU can be synthesized via two routes: one step method were all reactants at desired ratios are mixed and the polymerization reaction take place spontaneously; or through two steps method (prepolymer method). In this route, isocyanate terminated prepolymer is obtained by the reaction of polyol with small quantity of isocyanate. Then, the TPU is produced via the reaction of the prepolymer with chain extender (Whelan 2017). The two steps route is more common than one step route due to several merits. In instance, two step polymerization produces less random block distribution of the SS and HS in the polymer compared to the one produced through single step (Yilgör et al. 2015). Also, through two steps approach a processible polymer with high molecular weight and low polydispersity can be obtained (Ojha et al. 2009). It was reported also that TPU produced by two steps route has better mechanical properties compared to the one produced by one step (Table 1).

In addition to the polymerization approach, processing method also has a vital influence on the morphology of TPU, consequently, affects the mechanical properties (Xiang et al. 2020). In industry, TPUs are produced in bulk using extrusion and injection processes through one and two steps approaches. Generally, TPUs are produced through (Yamasaki et al. 2007):

(1) A batch process technique were all the raw materials are mixed vigorously in an open container for several seconds. Then, the mixture is poured into casting trays and polymerized in ovens under static conditions.

Table 1 Properties of TPUs produced via one and two steps methods (Whelan 2017)

Property	One step	Two steps
Elongation at break (%)	550	559
Hardness shore A	80	82
Modulus at 100% (MPa)	5.6	6.6
Tensile strength at break (MPa)	14.0	11.9
T_g (°C)	−35	−34

(2) Continuous procedure or band casting technique were the raw materials fed individually to a mixing head, where the raw materials mixed in stream onto plastic heated conveyor belt. Then, under static condition TPU is polymerized.
(3) Reactive extrusion which is another continues technique, were the raw materials through one or two steps are fed into twin screw extruder and then the reaction take place under dynamic conditions at high temperature. Finally, the TPU is pelletized at die face.

Furthermore, a third polymerization process called quasi-prepolymer method was reported. In this approach, two similar volumetric flow ratios are prepared from polyol + chain extender mixture (curative) and isocyanate terminated prepolymer before being pumped into the extruder or reactor (Yilgör et al. 2015).

The current trend is producing TPUs by using bio-based precursors via the conventional routs, which offers various merits from economic and ecological prospectives, such as low cost, low pollutions, and biodegradability (Wang et al. 2019). According to the literature, bio-based polyol: polyether (Kasprzyk and Datta 2019; Kasprzyk et al. 2020), polyester (Tuan Ismail et al. 2019; Parcheta et al. 2020; Pattamaprom et al. 2020); and poly carbonate (Kim et al. 2017; Oh et al. 2015); bio-based isocyanate (Li et al. 2014); and bio-based chain extenders (Rashmi et al. 2013; Oh et al. 2015; Puszka 2018) derived from different renewable resources had been used to obtain bio-TPU with even better performance in some cases compared to their fossil analogues. It worth to mention that Covestro Company developed a bio-TPU under trade name of Desmopan 37385A® that have mechanical properties at the level of the traditional TPUs with similar hardness and even exceed some of them. They declare that Desmopan 37385A® had tensile strength of 36 MPa and elongation at break almost 660%, APILON®52 Bio grades from TRINSEO are other bio-TPUs which offers similar performance of traditional TPUs and covers a wide range of applications (Donato and Miji 2020).

Moreover, successful synthesizing of non-isocyanate based TPU with enhanced properties had been reported (Unverferth et al. 2013; Poussard et al. 2016). For more details on bio-based substrates employed in synthesis of bio-TPU, the authors suggest to read the notable review (Datta and Kasprzyk 2018). A bio-TPU with good performance was synthesized successfully with no solvent involved (Pattamaprom et al. 2020) and in other work without catalyst (Oh et al. 2015).

TPU behave as thermoplastic in molten state and as elastomer in wide range of use temperatures due to their segmented structure (Rusu et al. 2011). Fundamentally, the unique structure of TPU manifest itself in diverse and special properties which can be easily tuned into rubber behavior under certain conditions, the elongation at break and Young's modulus for example can be varied from MPa to GPa (Kojio et al. 2020). On one hand, the SSs describe to softness, elasticity, elongation, water absorption, and biodegradability. On the other hand, the HSs contribute to tensile strength, hardness, impact resistance, modulus and stiffness (Krasowska et al. 2015). Furthermore, TPUs Young's modulus which is between plastics and rubber provide TPUs with properties such as resistance to oxidation, wear and to solvents, excellent toughness, high elongation and tear strength (Puszka 2018). Recently, bio-TPUs exhibit shape memory characteristic (Alagi et al. 2017), electrical/dielectric conductivity (Xie et al. 2019) and self-healing (Kim et al. 2017) have been well developed.

Another important aspect is introducing different functional groups into TPU structure to enhance different properties. According to the literature (Puszka 2018), introducing sulfur atom into the polymers improved their anti-bacterial properties, refractive index and adhesive strength, while introducing carboxylic and sulfonic enhanced the polymer biocompatibility. In recent study, Rougulska et al. (2020) used a chain extender containing sulfur "4,4′-(ethane-1,2-diyl)bis(benzenethiohexanol)" in synthesizing TPU. It was found that incorporating the sulfur atom had enhanced the adhesive properties to copper compared to TPU based on butane-1,4-diol chain extender.

In addition to the unique structure, the diversity of TPUs properties relays on other different factors such as the synthesis route, processing method, characteristics of the basic raw materials used, among others. Altering the ratio, type and the molecular weight of the SS and HS permit the production of versatile TPU. In instance, the amount of HS i.e. isocyanate and chain extender controlling the physical crosslinking in the polymer backbone were increasing the amount of HS leads to increasing of hardness, tensile strength, and decreasing of elongation at break (Datta and Kasprzyk 2018). In the other hand, elastomeric behavior controlled by chain mobility of the polyol used, which is dependent on their nature and the length of SSs. A more flexible TPU could be obtained by increasing the molar mass of the SS, furthermore enhancing resilience and elongation to break (Ionescu 2005). The degree of phase separation and the packing density of the hard segments have vital role in self-healing efficiency of TPUs (Kim et al. 2017).

The thermal properties of TPUs are determined by the two segments. The SS affect the low temperature properties, while the HS affect the high temperature properties (Xiang et al. 2020). Thus, the key for modulating the properties of TPU for a desired application is adjusting the raw materials employed.

It well known that sustainable PU generally derived from bio-based polyols and chain extenders are more common. Here we will address briefly through the chapter some recent efforts utilizing bio-based polyols and chain extenders in synthesizing TPU and their effect on the properties obtained.

Fully bio-based polyols are not widely used in polyurethane (PU) production, although various studies investigated applying it. In recent work, Parcheta et al. (2020) employed a fully bio-based polyester, namely, poly(propylene succinate) which synthesized at different temperatures. These bio-based polyols were used to synthesize bio-TPU with two different chain extenders from a natural origin. The study established that by using this fully bio-polyol, TPU can be produced without the usage of catalyst. Also, the study showed that the type of chain extender had a major influence on the properties of TPU more than the synthesizing conditions of polyols.

Utilizing carbon dioxide (CO_2) in the production of many chemicals and polymers has attracted considerable attention. CO_2 based polyol has also been reported for producing bio-TPUs. LCA studies showed that CO_2 based polyols reduced the global warming impact to 11–19% lesser than the traditional polyether-based polyol and can be achieved in lower production cost than polyether-based polyols. CO_2 based polyols are now commercialized by various companies such as Sumitomo, China Blue, and Covestro (Alagi et al. 2017). Recently, a non-isocyanate route to synthesis a biodegradable TPU using CO_2 based polyol was developed by Ye and his team (2020). The product showed a remarkable performance including mechanical properties and biodegradability compared to commercial poly(butylene adipate-co-terephthalate) (PBAT). In other study, Alagi and his co-authors (2017) employed CO_2 based polyol to synthesized TPU with superior shape memory characteristic (100% shape recovery) and anticorrosion characteristic regardless of the environment. According to the literature, bio-TPUs demonstrate an excellent performance as a component in blends (Rashmi et al. 2017; Tan et al. 2015) and in composites (Taha 2017; Khalifa et al. 2020).

These tunable properties of TPU made it useful for wide spectrum of fields such as food packing, automotive, medical applications, furniture, electronic devices, and so on. In recent years, TPU in general attracted attentions in emerging and high performance applications such as FDM 3D printing (Donato and Mija 2020).

Commercially, bio-TPUs had been employed in wide range of applications as replacement of their traditional analogues (Fig. 3). In instance, bio-TPUs® products by Lubrizol (bio-content ranging from 30 to 70%) had been used in textile coatings, 3D printing, consumer electronics, and automotive.

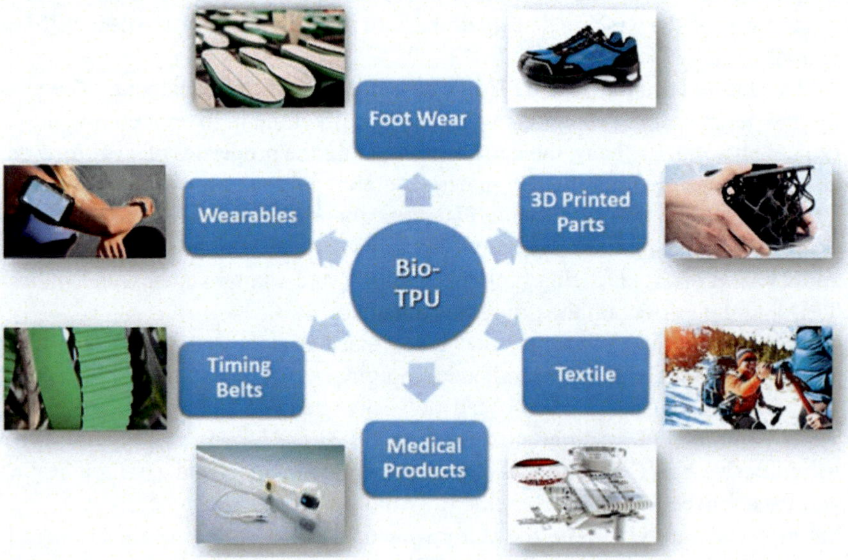

Fig. 3 Some of bio-TPU's commercial applications

4 Thermoplastic Polyurethane for Automotive Industries

For many years, TPUs with their excellent performance and processability were used in automotive industry. The exceptional combination of aging and scratch properties makes it a good choice for automotive industries.

The superior physical properties of TPU such as flexibility, toughness, resistance to abrasion, ozone, weather, oil, solvents and gasoline, allowed the use of TPU in elastomeric wheels and tires, cushions, hard plastic parts, electric potting compounds and flexible foam seating (Ghosh et al. 2020). The mechanical properties of TPUs also satisfied the design specification of car parts such as automotive anti-roll bar (Mastura et al. 2018).

The recent ecological and economical challenges such as sustained depletion of fossil resources and their prices, together with the recent regulations for controlling gas emissions and management of vehicles end life, has encouraged the automotive manufactures to adapt and develop eco-friendly plastics in general for their modern cars. Nowadays, big chemical companies such as BASF, Covestro, and Lubrizol produces different products of bio-TPUs that employed in synthesizing different automotive components (Fig. 4).

Currently, automobile manufacturer implemented eco-friendly routes in order to mitigate major issues, which are: huge and heavy material usage, higher fuel consumption, durability and environmental solution (Ghosh et al. 2020). As bio-TPU have comparable properties to the traditional TPU, they were employed in automotive industry to address some or even all the above-mentioned challenges. Yet, there are many commercial bio-TPU available for automotive applications, such as Pearlbond® Eco and Pearlthane® Eco (from Merquinsa) with bio-content ranging

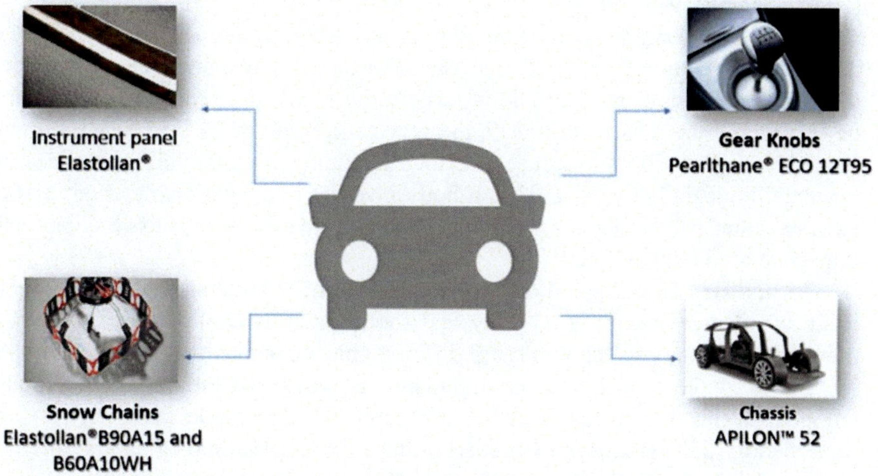

Fig. 4 Commercial bio-TPU used in automotive applications

from 40 to 95% based on polyols derived from fatty acids and vegetable oils. In addition, their unique biocomposites and alloys with styrene-ethylene/butylene-styrene triblock copolymers (SEBS) were promising in several automotive application such as interior trim, covers, skins, and soft-touch applications (Rusu et al. 2011).

Also, the type of raw materials used in synthesizing bio-TPU determine the prospective use in automotive application. A study made by Ismail and his co-authors (Tuan Ismail et al. 2019) suggest that bio-TPU based on crystalline azelate polyol is good candidate for dynamic applications such as wheels and roller due to the low heat up build up, while bio-TPU based on amorphous azelate polyol is suitable for adhesives and coatings applications.

Regardless of the promising that bio-TPU show in this industry, other criteria should be considered in automobile design. As an example, controlling degradation either photochemical or thermal is vital for using TPUs as elastomers or coating for automotive applications. During processing the colour of TPU should not change (Bueno-Ferrer et al. 2012). In the other hand, evaluating the biodegradation of bio-TPU used in this industry, helps to control the massive wastes that generates from it.

5 Biodegradation

Biodegradation of polymers can be induced by thermal, photo and biological influences, which result in changing in the properties such as discoloration, surface erosion, phase separation, cracking, and crystallinity changes (Goodby 2014). The disruption of polymer chain accelerate the initial aging, and then external aspects such as chemicals compositions and temperature enhance the rate of degradation (Pathak and Navneet 2017).

As any other organic polymers, TPU is not stable under all conditions. When using TPU in manufacturing devices; degradation rates in specific environment for a certain application must be known, where the major outcome of degradation to be considered the most is to reduce the service lifetime of TPU material. Furthermore, the degradation may cause a serious problems in certain circumstances, for example biodegradation of TPU in a body involves chemical degradation, hydrolysis, enzymatic degradations, calcification and oxidation where toxic compounds may generated (Xie et al. 2019).

The first step in biological degradation is the initial break down of the polymer chain which may cause by: (1) physical forces such as cooling, heating, drying, thawing and wetting which will cause different mechanical damages, and (2) biological forces by the growth of microorganism that will break the polymer to smaller compounds and even release gases such as methane (Krasowska et al. 2015).

Generally, TPU is known to be susceptible to biodegradation by natural microorganisms (Howard 2011). Based on the published literature, there are three types of TPU biodegradation, namely bacterial biodegradation, degradation by TPU enzymes, and fungal biodegradation (Xie et al. 2019). Since 1980s, biodegradation of TPUs by fungi and bacteria has been studies, and different enzymatic activities have been acknowledged, although the underlying mechanisms of biodegradation are still unknow (Álvarez-Barragán et al. 2016).

It was demonstrated that degradation of TPU with fungi is high effective than with bacteria because of the remarkable enzymatic arsenals (Magnin et al. 2019). However, the activities of fungi and bacteria against PUs has been established under laboratory conditions with certain identified microbial strains, and not in natural conditions (Xie et al. 2019).

The synergistic action of different factors in the natural environmental increase the polymers degradation. Each environment contains various microorganisms, microorganisms, and enzymes, also different chemical and physical parameters which have impact on rates of microbial activity, subsequently affect the rate of degradation. Thus, it is crucial to estimate the degradability of TPU under natural environmental conditions. Nevertheless, the natural environment is sophisticated for investigating the degradation such as sea and compost because variety of microorganisms, wide population of living organisms, salt, animals, sunlight, rain, fluctuation of water, etc., all play a part in the process of degradation (Krasowska et al. 2015).

Generally, there are two types of TPU; degradable and un-degradable TPUs. Both of them can be synthesized through a proper selection of the raw materials (Krasowska et al. 2015). In spite of their mechanical, biocompatible and physicochemical properties, TPU is reported to degrade under various conditions such as oxidation, hydrolysis, environmental stress cracking, enzymatic degradation, etc. in vivo (Tatai et al. 2007). However, most of TPUs are not biodegradable. Thus, enhancing the biodegradability of TPU and minimize the reliability on fossil based materials have become urgent in the field of polymer engineering (Wang et al. 2019).

Improving the biodegradation of TPU have been well developed by introducing a biodegradable raw material. Table 2 shows the common raw materials used to synthesis partially degradable and biodegradable TPU.

The abovementioned materials show different degradation behavior in different conditions and based on the target application the raw material is selected. For wide range of applications, aliphatic polyesters have been used to prepare biodegradable TPUs due to their good biodegradability and can be synthesized by fully biobased polyester based on raw materials such as bio-succinic acid, sebacic acid and 1,3 propanediol, 1,4 butanediol. Despite, the hydrolytic biodegradation it is likely to occur but with low rate than the desired rates for biomedical applications. By using polar, hydrolytically labile aliphatic polyester in TPUs the hydrolytic degradation of the designed material is accelerated. In the meantime, additives have hydrophobic character such as poly carbodiimide, silica and amide group can significantly control hydrolytic stability (Tang 2011).

The TPUs based on polyester were found to be susceptible to microbial attack under aerobic environments, while those based on polyether are resistant to microbial attack, with negligible biodegradations in aerobic and anaerobic conditions (Xie et al. 2019). Also, TPUs based on polyester are highly susceptible to fungal degradation such as Chaetomium, Pullularia, Penicillium, and Trichoderma species, whereas TPUs based on polyether reveals higher resistance (Schmidt et al. 2017). Nonetheless,

Table 2 Common ingredients used in bio-TPU and their degradability

Polyol	Isocyanate	Chain extender	Degradation character	Application	Reference
Polycaprolactone	Methylene-bis(cyclohexyl isocyanate HMDI	Putrschine (diamines)	Biodegradable	Cardiac tissue material	Rodriguez et al. (2013)
Modified cater oil with lignin	(MDI)	Propylene glycol (PPG)	Bio-based polyurethane	Automotive	Tavares et al. (2016)
Sodium liqnosulfonate-Glyercol	Methylenediphenyldiisocyante (PMDI)	Propylene glycol (PPG)	Biobased Polyurethane foam	Insole materials	Katarzyna et al. (2016)
Polycaprolactone	Hexamethylene diisocyanate (HDI)	Piperazine (diamines)	Biodegradable	Shape memory polymers	Rodriguez et al. (2013)
Macrodiol (PEG 1000)	HDI	Bio-based 1,2 propan doil	Partially Bio-TPU	Internal automotive	Rashmi et al. (2013)
Diglycidyl ether of cardanol	HMDI	Isophorone diamine (IPDA)	10% degradable	Coating applications	Mukesh et al. (2014)
Alginic acid, PEG	HDI	Glycerin	N/A	Cushioning, insulating, electrical	Seong et al. (2007)
Polycaprolactone	Diphenylmethane diisocynate (MDI)	1,4 BDO	Long term degradable	Drug delivery	Rodriguez et al. (2013)

(continued)

Table 2 (continued)

Polyol	Isocyanate	Chain extender	Degradation character	Application	Reference
Soy-based Polyol (Soyol)	MDI	Wood fiber as chain extender	Biodegradable	Spray foam insolution	Mustafa et al. (2011)
Castor oil	Polymethylene polyisocyanates (PM-200)	N/A	Biodegradable	Sound insulating	Shuai et al. (2018)
Prepolymer long-chain diamine	Biobased nonisocyanate polyurethanes	1,4-butanediamine	Bibased material	Automotive or adhesives	Camille et al. (2016)
Soybean straw	MDI	Glycerol	Biobased	PU foam	Shengjun et al. (2012)
Diethylene Glycol	TDI	Glacial acetic acid	Biobased	Flexible PU foam	Piotr et al. (2012)

further fungal strains isolated from environmental samples reveals the ability to degrade PU produced by both polyesters and polyether (Álvarez-Barragán et al. 2016). Subsequently, the nature of the basic raw material considered the key to determining the ability to degrade in a certain environment. Petrović et al. (2010) studied the biodegradation of two different bio-based polyesters in compost. It was reported that polyricinoleic acid based polyester TPU and castor oil-based polyester TPU biodegrade up to 30% and 42%, respectively in 60 days (composting).

It was also reported that the presence of ester linkage in TPU makes them degradable in water (hydrolysis) as ester bonds are sensitive to hydrolysis, while polyether based TPU possess excellent hydrolytic stability (Christenson et al. 2007). In the other hand, the presence of ether make TPU susceptible to oxidative degradation (Rodriguez et al. 2013). These demerits can be overcome when using hydrolytically and oxidatively stable polycarbonate instead of polyester and polyether as soft segments (Rogulska et al. 2020).

Researchers investigated using block copolymers as the SSs to regulate the biodegradability and the elasticity of TPU (Mi et al. 2017). It has been reported that copolymerizing polyethylene glycol (PEG) with PCL and/or PLA accelerated the degradation of TPU due to the high hydrophilicity and PEG act as plasticizer (Rodriguez et al. 2013).

Degradation rate can be amended basically by altering the compositions and component of the raw materials. The most used SS in synthesizing bio-TPU is PCL, although it has a low degradation rate (Mi et al. 2017). In instance, PU based on PCL(polyester) as SS showed a slow degradation rate (15% weight loss in 11 weeks) (Rodriguez et al. 2013).

Hence, the raw materials are not the only factors controlling the degradation rate. In essence, many factors had an influence on the biodegradation of TPU, which will be discussed in the next section.

5.1 Factors Affecting the TPU Biodegradation

Degradation of bio-TPUs is determined by various factors; and the main factor is the nature of the raw materials used and the type of the environment is also important. Incorporation of biodegradable polyols, chain extenders, and isocyanate is well known as the main factor to determine the ability to degrade, specifically polyols. A wide range of different bio-polyols were investigated as SS for synthesizing bio-TPU which showed an enhancement in degradation. For example, TPU based on lignin polyols showed a high fungal degradation, also improve tear strength and hardness (Krasowska et al. 2015). Also, a novel bio-TPU was synthesized through the addition of chitin i.e. an amino polysaccharide to the SS (hydroxy terminated polybutadiene) in the prepolymer first step, beside the use of combination of curcumin and 1,4-butanediol as chain extender. The results revealed the synthesized of TPU with enhanced antibacterial, antioxidant, and biocompatibility with good thermal properties (Amjed et al. 2020). In other study (Wang et al. 2019), a combination of five bio-based aliphatic monomers (fumaric acid, succinic acid, sebacic acid, 1,4-butanediol, and 1,3-propanediol), were used as soft segment in order to synthesize bio-TPU. Various contents of the biodegradable polyester diol (SS) were used. The results revealed that the degradation rate enhanced as the SS content increased, and the sample with 80% SS exhibit the highest weight loss of 36.7% after 14 days in lipase solution, compared with some conventional TPU which do not show a significant degradation in lipase solution. The authors attributed this result to the structure of the polyol, which consist of carbon–carbon double bond and six kinds of polyester repeat units, so the main chain was simply attacked and decomposed by lipase. Also, as the SS content increased, the HS decreased, resulting the flexibility and subsequently decreasing the physical crosslinking allowing the lipase to penetrate easily into molecule interior.

The modification of the HS by altering the diisocyanate and chain extender also has been explored to influence the degradation of TPU. Tatai et al. (2007) studied the effect of introducing degradable chain extenders containing hydrolysable ester linkage (DL-lactic acid and ethylene glycol) into the HS of TPU in the in-vitro degradation (phosphate buffered saline) during one year. The study found that the presence of degradable chain extender showed a sustainable weight loss (samples with 100% hard segment showed 100% mass loss in 180 days), also the samples containing chain extender showed high weight loss compared with other samples were no chain extender included. The LDI was used as degradable diisocyanate where the type of the diisocyanate also had an impact on biodegradations rate. It was demonstrated that TPU based on polyester and aliphatic diisocyanate show higher biodegradation rate than those composed of aromatic diisocyanate, and the rate growing in this order: 4,4′-MDI < 4,4′-H12MDI < HDI (Xie et al. 2019).

Another important factor is the content of the HS and SS. As biodegradation is mainly proceeds though the SSs and it is well known that HSs degrade slower than the SSs. Thus, a higher SS displays a higher biodegradation. Furthermore, TPU with dispersed morphology (the HSs domains dispersed in the soft matrix) shows a faster biodegradation than those of co-continues morphology (Petrović et al. 2010). In contrast, the morphology has opposite influence on degradation rate in case of TPU blends. In instance, a study found that co-continues morphology in PLA/TPU blend revealed higher degradation than blends with dispersed (droplet) morphology. In addition, it was demonstrated that the addition of TPU to the blend decreases the degradation rate while improving stability (Jašo et al. 2015).

In the other hand, hydrophilicity has been shown to influence the biodegradability of TPU, where using hydrophilic polyols such as PEG has been proved to enhance the hydrophilicity and the degradation rate of TPU (Goodby 2014). Incorporating of bio-fillers is also reported to alter the biodegradation behavior. The addition of antimicrobial bio-fillers such as chitosan to TPU had enhanced the antimicrobial properties which are preferred in the certain applications (Javaid et al. 2018). Despite the privileges of using renewable materials, there is no assurance that bio-TPU will be fully biodegrade, this attributed to the alteration of crosslink density, functional groups, and incompatibility with non-renewable co-monomers (Khalifa et al. 2020).

6 Future of Biobased-Thermoplastic Polyurethane

Bo-TPU is consider as a bio-based polymer that's produced with renewable material (content ranging from 30 to 70%). This material had found a way to wide range of applications with high demand in the field of medical, automobile, footwear, and aerospace applications, due to their unique structure and the superior properties. However, bio-TPUs transcend the traditional TPUs by being sustainable, relatively cheaper and had tendency to biodegradation. Nowadays, bio-TPUs are produced from partially bio-based raw materials for commercially with various trademarks and grades. Yet, efforts are sustained to investigate new precursors and improving their biodegradability is the most concerns. However, in many cases a good degradation rate companied a poor property which dependent on the raw materials used. Introducing polyether as SS conveys high resistance to hydrolysis but imparts low bio-based content in contrast with PU using polyester. While crosslinked agent linked to polyester is similar thermally stable to polyether, although showing minimal physical or chemical degradation with adequate mechanical and heat resistance (Asensio et al. 2019). Thus, more investigations of the structure–property relationship for different bio-based raw materials it will be a valuable addition to this field.

7 Conclusion

TPU is one of the distinguished PUs classes which is versatile with good processability. As the most industries in the world are forwarding to use more sustainable materials, bio-TPUs were extensively investigated. Nowadays, TPUs synthesized from bio-based with low cost, low carbon footprint, eco-friendly and with comparable properties to their fossil analogies. However, 100% bio-TPU has not achieved yet at least at industrial scale. Commercial bio-TPUs are available in the global market, many companies produce various grads of bio-TPUs derived from different renewable materials with various contents. Despite the huge production of PUs, their disposal still a challenge as PU degrades slowly and the recycling processes are limited. Enhancing the biodegradation of TPU to satisfy the LCA through employing bio-based materials which possess the characteristic of biodegradation was a proper solution to reduce the negative affect not only of the slow degradation but also the depending on fossil resources. However, detailed studies on the use of new bio-materials as raw materials or fillers are needed to exploited their potential to biodegrade in different environment.

References

Aimin X, Huafeng TA.Varada R, Shuai et al (2018) Water-Blown castor oil-based polyurethane foams with soy protein as a reactive reinforcing filler. J Poly Env 26:15–22.https://doi.org/10.1007/s10924-016-0914

Alagi P, Ghorpade R, Choi YJ et al (2017) Carbon dioxide-based polyols as sustainable feedstock of thermoplastic polyurethane for corrosion-resistant metal coating. ACS Sustain Chem Eng 5:3871–3881. https://doi.org/10.1021/acssuschemeng.6b03046

Anagha S, Dawid D'M, Mukesh K (2014) Isocyanate free polyurethanes from new CNSL based bis-cyclic carbonate and its application in coatings. Euro Poly J 57:99–108. https://doi.org/10.1016/j.eurpolymj.2014.05.008

Álvarez-Barragán J, Domínguez-Malfavón L, Vargas-Suárez M et al (2016) Biodegradative activities of selected environmental fungi on a polyester polyurethane varnish and polyether polyurethane foams. Appl Environ Microbiol 82:5225 LP–5235. https://doi.org/10.1128/AEM.01344-16

Amjed N, Bhatti IA, Zia KM et al (2020) Synthesis and characterization of stable and biological active chitin-based polyurethane elastomers. Int J Biol Macromol 154:1149–1157. https://doi.org/10.1016/j.ijbiomac.2019.11.097

Asensio M, Costa V, Nohales A et al (2019) Tunable structure and properties of segmented thermoplastic polyurethanes as a function offlexible segment. Polymers (Basel) 11:1–22. https://doi.org/10.3390/polym11121910

Bueno-Ferrer C, Hablot E, Garrigós MDC et al (2012) Relationship between morphology, properties and degradation parameters of novative biobased thermoplastic polyurethanes obtained from dimer fatty acids. Polym Degrad Stab 97:1964–1969. https://doi.org/10.1016/j.polymdegradstab.2012.03.002

Caixia W, Yebo L, Shengjun et al (2012) Production and characterization of biopolyols and polyurethane foams from crude glycerol based liquefaction of soybean straw. Bioresour Tech 103:227–233. https://doi.org/10.1016/j.biortech.2011.09.125

Christenson EM, Anderson JM, Hiltner A (2007) Biodegradation mechanisms of polyurethane elastomers? Corros Eng Sci Technol 42:312–323. https://doi.org/10.1179/174327807X238909

Cristiane R, Stilhano VB, Gabriel RS, Tavares et al (2016) Bio-based polyurethane prepared from Kraft lignin and modified castor oil. Exp Poly Let 10: 927–940. https://doi.org/10.3144/expresspolymlett.2016.86

Datta J, Kasprzyk P (2018) Thermoplastic polyurethanes derived from petrochemical or renewable resources: a comprehensive review. Polym Eng Sci 58:E14–E35. https://doi.org/10.1002/pen.24633

Donato RK, Mija A (2020) Keratin associations with synthetic, biosynthetic and natural polymers: an extensive review. Polymers (Basel) 12:1–64. https://doi.org/10.3390/polym12010032

Doo S L, Nguyen et al (2010) Injectable biodegradable hydrogels. Macromol Biosci. 10:563–579. https://doi.org/10.1002/mabi.200900402

Ghosh M, Ghosh A, Roy A (2020) Renewable and sustainable materials in automotive industry. Elsevier Ltd.

GLOBE NEWSWIRE (2020) The polyurethane (PU) foam market: global industry trends, share, size, growth, opportunity and forecast 2020–2025.

Goodby A (2014) Biodegradable thermoplastic polyurethanes. Aston University, Birmingham, England

Howard GT (2011) Microbial biodegradation of polyurethane. In: Fainleib A, Grigoryeva O (eds) Recent developments in polymer recycling. Transworld Research Network, Kerala, pp 215–238

Hugo Z, Stéphane D, Camille et al. (2016) Synthesis and characterization of advanced biobased thermoplastic nonisocyanate polyurethanes with controlled aromatic-aliphatic architectures. Euro Poly J 84: 759–769. https://doi.org/10.1016/j.eurpolymj.2016.05.030

Ionescu M (2005) Chemistry and technology of polyols for polyurethanes. Rapra Technology Limited, Shawbury, UK

Jašo V, Glenn G, Klamczynski A, Petrović ZS (2015) Biodegradability study of polylactic acid/ thermoplastic polyurethane blends. Polym Test 47:1–3. https://doi.org/10.1016/j.polymertesting.2015.07.011

Javaid MA, Khera RA, Zia KM et al (2018) Synthesis and characterization of chitosan modified polyurethane bio-nanocomposites with biomedical potential. Int J Biol Macromol 115:375–384. https://doi.org/10.1016/j.ijbiomac.2018.04.013

Kasprzyk P, Błażek K, Datta J (2020) Microcrystalline cellulose management in the production of poly(ether-urethane)s- structure, morphology, and thermal characteristic. Fibers Polym 21:690–700. https://doi.org/10.1007/s12221-020-9482-y

Kasprzyk P, Datta J (2019) Novel bio-based thermoplastic poly(ether-urethane)s. Correlations between the structure, processing and properties. Polymer (Guildf) 160:1–10. https://doi.org/10.1016/j.polymer.2018.11.032

Katarzyna W, Karolina S, Armando G. M, Mariusz M (2016) Characterization of thermal and mechanical properties of lignosulfonate- and hydrolyzed lignosulfonate-based polyurethane foams. Bio-Resources 11:7355–7364. https://doi.org/10.15376/biores.11.3.7355-7364

Khalifa M, Anandhan S, Wuzella G et al (2020) Thermoplastic polyurethane composites reinforced with renewable and sustainable fillers–a review. Polym Technol Mater 00:1–19. https://doi.org/10.1080/25740881.2020.1768544

Kim SM, Park SA, Hwang SY et al (2017) Environmentally-friendly synthesis of carbonate-type macrodiols and preparation of transparent self-healable thermoplastic polyurethanes. Polymers (Basel) 9. https://doi.org/10.3390/polym9120663

Kojio K, Nozaki S, Takahara A, Yamasaki S (2020) Influence of chemical structure of hard segments on physical properties of polyurethane elastomers: a review. J Polym Res 27:64–67. https://doi.org/10.1007/s10965-020-02090-9

Krasowska K, Heimowska A, Rutkowska M (2015) environmental degradability of polyurethanes. Thermoplast Elastomers Synth Appl. https://doi.org/10.5772/60925

Li S, Jose J, Bouzidi L et al (2014) Maximizing the utility of bio-based diisocyanate and chain extenders in crystalline segmented thermoplastic polyester urethanes: effect of polymerization protocol. Polymer (Guildf) 55:6764–6775. https://doi.org/10.1016/j.polymer.2014.11.013

Liu L, Wu Y, Zhu Z (2017) Internal structure and crystallinity investigation of segmented thermoplastic polyurethane elastomer degradation in supercritical methanol. Polym Degrad Stab 140:17–24. https://doi.org/10.1016/j.polymdegradstab.2017.03.003

Magnin A, Hoornaert L, Pollet E et al (2019) Isolation and characterization of different promising fungi for biological waste management of polyurethanes. Microb Biotechnol 12:544–555. https://doi.org/10.1111/1751-7915.13346

Marcano A, Fatyeyeva K, Koun M et al (2019) Recent developments in the field of barrier and permeability properties of segmented polyurethane elastomers. Rev Chem Eng 35:445–474. https://doi.org/10.1515/revce-2017-0033

Mastura MT, Sapuan SM, Mansor MR, Nuraini AA (2018) Materials selection of thermoplastic matrices for 'green' natural fibre composites for automotive anti-roll bar with particular emphasis on the environment. Int J Precis Eng Manuf Green Technol 5:111–119. https://doi.org/10.1007/s40684-018-0012-y

Mi HY, Jing X, Napiwocki BN et al (2017) Biocompatible, degradable thermoplastic polyurethane based on polycaprolactone-block -polytetrahydrofuran- block -polycaprolactone copolymers for soft tissue engineering. J Mater Chem B 5:4137–4151. https://doi.org/10.1039/c7tb00419b

Oh JK, Dae HK, Jong SP, Seong RY et al (2007) Characterization of the polyurethane foam using alginic acid as a polyol. Fiber Poly 8:257–262. https://doi.org/10.1007/BF02877267

Oh SY, Kang MS, Knowles JC, Gong MS (2015) Synthesis of bio-based thermoplastic polyurethane elastomers containing isosorbide and polycarbonate diol and their biocompatible properties. J Biomater Appl 30:327–337. https://doi.org/10.1177/0885328215590054

Ojha U, Kulkarni P, Faust R (2009) Syntheses and characterization of novel biostable polyisobutylene based thermoplastic polyurethanes. Polymer (Guildf) 50:3448–3457. https://doi.org/10.1016/j.polymer.2009.05.025

Parcheta P, Głowińska E, Datta J (2020) Effect of bio-based components on the chemical structure, thermal stability and mechanical properties of green thermoplastic polyurethane elastomers. Eur Polym J 123. https://doi.org/10.1016/j.eurpolymj.2019.109422

Pathak VM, Navneet (2017) Review on the current status of polymer degradation: a microbial approach. Bioresour Bioprocess 4

Pattamaprom C, Wu CH, Chen PH et al (2020) Solvent-free one-shot synthesis of thermoplastic polyurethane based on bio-poly(1,3-propylene succinate) glycol with temperature-sensitive shape memory behavior. ACS Omega 5:4058–4066. https://doi.org/10.1021/acsomega.9b03663

Petrović ZS, Xu Y, Milić J et al (2010) Biodegradation of thermoplastic polyurethanes from vegetable oils. J Polym Environ 18:94–97. https://doi.org/10.1007/s10924-010-0194-z

Prociak A, Pawlik H, Piotr et al (2012) Flexible polyurethane foams modified with natural oil based polyols. J Cel Plastics. 48: 489–499. https://doi.org/10.1177/0021955X12446210

Poussard L, Mariage J, Grignard B et al (2016) Non-isocyanate polyurethanes from carbonated soybean oil using monomeric or oligomeric diamines to achieve thermosets or thermoplastics. Macromolecules 49:2162–2171. https://doi.org/10.1021/acs.macromol.5b02467

Puszka A (2018) Thermal and mechanical behavior of new transparent thermoplastic polyurethane elastomers derived from cycloaliphatic diisocyanate. Polymers (Basel) 10. https://doi.org/10.3390/polym10050537

Rashmi BJ, Loux C, Prashantha K (2017) Bio-based thermoplastic polyurethane and polyamide 11 bioalloys with excellent shape memory behavior. J Appl Polym Sci 134:1–10. https://doi.org/10.1002/app.44794

Rashmi BJ, Rusu D, Prashantha K et al (2013) Development of bio-based thermoplastic polyurethanes formulations using corn-derived chain extender for reactive rotational molding. Express Polym Lett 7:852–862. https://doi.org/10.3144/expresspolymlett.2013.82

Rodriguez J, Chan L, Hernandez-Sanchez F, Cervantes-Uc J (2013) Degradation of polyurethanes for cardiovascular applications. Adv Biomater Sci Biomed Appl. https://doi.org/10.5772/53681

Rogulska M, Maciejewska M, Olszewska E (2020) New thermoplastic poly(carbonate-urethane)s based on diphenylethane derivative chain extender: The effect of diisocyanate structure on some properties. J Therm Anal Calorim 139:1049–1068. https://doi.org/10.1007/s10973-019-08433-z

Rusu D, Boyer SAE, Lacrampe MF, Krawczak P (2011) Bioplastics and vegetal fiber reinforced bioplastics for automotive applications. Handb Bioplastics Biocomposites Eng Appl 397–449. https://doi.org/10.1002/9781118203699.ch15

Rydz J, Sikorska W, Kyulavska M, Christova D (2015) Polyester-based (bio)degradable polymers as environmentally friendly materials for sustainable development. Int J Mol Sci 16:564–596. https://doi.org/10.3390/ijms16010564

Schmidt J, Wei R, Oeser T et al (2017) Degradation of polyester polyurethane by bacterial polyester hydrolases. 4–13. https://doi.org/10.3390/polym9020065

Taha M (2017) A new hybrid approch for conceptual design of sugar palm fiber-reinforced polyurethane composites for automotive anti-roll bar. Universiti Putra Malaysia

Tan L, Su Q, Zhang S, Huang H (2015) Preparing thermoplastic polyurethane/thermoplastic starch with high mechanical and biodegradable properties

Tang D (2011) Biobased thermoplastic polyurethanes synthesized by isocyanate-based and isocyanate-free routes. Technische Universiteit Eindhoven

Tatai L, Moore TG, Adhikari R et al (2007) Thermoplastic biodegradable polyurethanes: the effect of chain extender structure on properties and in-vitro degradation. Biomaterials 28:5407–5417. https://doi.org/10.1016/j.biomaterials.2007.08.035

Tuan Ismail TNM, Ibrahim NA, Sendijarevic V et al (2019) Thermal and mechanical properties of thermoplastic urethanes made from crystalline and amorphous azelate polyols. J Appl Polym Sci 136:1–11. https://doi.org/10.1002/app.47890

Unverferth M, Kreye O, Prohammer A, Meier MAR (2013) Renewable non-isocyanate based thermoplastic polyurethanes via polycondensation of dimethyl carbamate monomers with diols. Macromol Rapid Commun 34:1569–1574. https://doi.org/10.1002/marc.201300503

Wang Z, Yan J, Wang T et al (2019) Fabrication and properties of a bio-based biodegradable thermoplastic polyurethane elastomer. Polymers (Basel) 11:1–13. https://doi.org/10.3390/polym11071121

Whelan D (2017) Thermoplastic elastomers. Brydson's Plast Mater Eighth Ed 653–703. https://doi.org/10.1016/B978-0-323-35824-8.00024-4

Xiang N, Zhang X, Zheng M et al (2020) Microstructure and tensile properties of injection molded thermoplastic polyurethane with different melt temperatures. J Appl Polym Sci 137:1–12. https://doi.org/10.1002/app.48891

Xie F, Zhang T, Bryant P et al (2019) Degradation and stabilization of polyurethane elastomers. Prog Polym Sci 90:211–268. https://doi.org/10.1016/j.progpolymsci.2018.12.003

Yamasaki S, Nishiguchi D, Kojio K, Furukawa M (2007) Effects of polymerization method on structure and properties of thermoplastic polyurethanes. J Polym Sci Part B Polym Phys 45:800–814. https://doi.org/10.1002/polb.21080

Ye S, Xiang X, Wang S et al (2020) Nonisocyanate CO_2-based poly(ester- co-urethane)s with tunable performances: a potential alternative to improve the biodegradability of PBAT. ACS Sustain Chem Eng 8:1923–1932. https://doi.org/10.1021/acssuschemeng.9b06294

Yilgör I, Yilgör E, Wilkes GL (2015) Critical parameters in designing segmented polyurethanes and their effect on morphology and properties: a comprehensive review. Polymer (Guildf) 58:A1–A36. https://doi.org/10.1016/j.polymer.2014.12.014

Zhang Y, Liao J, Fang X et al (2017) Renewable high-performance polyurethane bioplastics derived from lignin-poly(ε-caprolactone). ACS Sustain Chem Eng 5:4276–4284. https://doi.org/10.1021/acssuschemeng.7b00288

Zhao X, Shou T, Liang R et al (2020) Bio-based thermoplastic polyurethane derived from polylactic acid with high-damping performance. Ind Crops Prod 154:112619. https://doi.org/10.1016/j.indcrop.2020.112619

Eco-Friendly Wood Fibre Composites with High Bonding Strength and Water Resistance

G. Rajeshkumar, S. Arvindh Seshadri, M. Bilal Mohammed, K. Srijith, B. Brahatheesh Vikram, and A. Sailesh

Abstract In recent years wood has received an ample attention as the reinforcing filler for producing polymeric composites because these wood fillers have various advantages like biodegradable, renewable, less cost, light weight, less abrasive to equipment and available in abundance. However, challenges are there in replacing the synthetic fibre composites with wood fibre composites because of its poor bonding strength and high moisture absorption and considerable variation in properties with respect to growing conditions. This chapter presents the comprehensive review of bonding strength and water resistance capability of wood fibre composites which will be useful to select appropriate wood fiber composites for various applications.

Keywords Wood fibre · Biodegradable · Bonding strength · Water resistance · Eco-friendly

1 Introduction

The composites are the advanced materials made from two or more materials which offer excellent properties due to the synergistic effect. The fibre/filler reinforced polymer composites are the widely used composite materials in various industrial sectors due to its noteworthy features like biodegradable, renewable, less cost, light weight, less abrasive to equipment and available in abundance (Nagaraja et al. 2020; Rajeshkumar et al. 2020; Kumar et al. 2019). Among various biodegradable fiber/filler composites wood fibre polymer composites (WFPCs) are burgeoning globally in the recent years and it is estimated that the WFPCs market reach US$5.8 billion worth internationally by 2021 with a growth rate of 12.2% per year from 2016 (Chan et al. 2018).

These wood fibers are dimensionally unstable, hygroscopic and susceptible to biological attacks and also these wood fibers have poor interfacial bonding with

G. Rajeshkumar (✉) · S. A. Seshadri · M. B. Mohammed · K. Srijith · B. B. Vikram · A. Sailesh
Department of Mechanical Engineering, PSG Institute of Technology and Applied Research, Coimbatore, Tamilnadu, India
e-mail: grajeshkumar.me@gmail.com

M. Jawaid et al. (eds.), *Eco-Friendly Adhesives for Wood and Natural Fiber Composites*, Composites Science and Technology,
https://doi.org/10.1007/978-981-33-4749-6_5

hydrophobic polymer matrices (Nasir et al. 2015). Therefore, it is very difficult to use these WFPCs in the moist environments and for other structural applications. This can be overcome through modifying the fibre surface and adding suitable coupling agents to polymers. These techniques enhance the bonding strength between the reinforcement and matrix thereby better properties and low water absorption rate could be achieved. In this view, this chapter presents the comprehensive review of bonding strength and water resistance capability of different WFPCs which will be useful to select appropriate WFPCs for various applications.

2 Wood Fibre

Wood fibres, by themselves are a complex, naturally occuring, three-dimensional whose major constituents are cellulose, hemi-cellulose and lignin (Fig. 1). Wood fibres are generally classified into two types: hardwood and softwood, based on their anatomical and botanical characteristics. Softwoods are found in trees that reproduce by seeds (gymnosperms) while hardwoods are produced in flower-reproducing trees (angiosperms). Some of the softwood species are cedars, spruces, pines, firs etc., whereas certain hardwood species include ashes, oaks, beech, birch, eucalyptus, maples etc. (Clemons 2008).

Wood is generally composed of elongated hollow cells known as tracheids at a macroscopic level and properties such as strength and stiffness are due to the thick lignified wall of these tracheids. These cells are arranged parallel to the axis of the tree (Miller 1999). The growth of wood happens at a thin layer situated below the bark known as the *cambium* which further leads to the formation of growth rings. The bark contains dead tissues and its function is to safeguard the inner structure. The

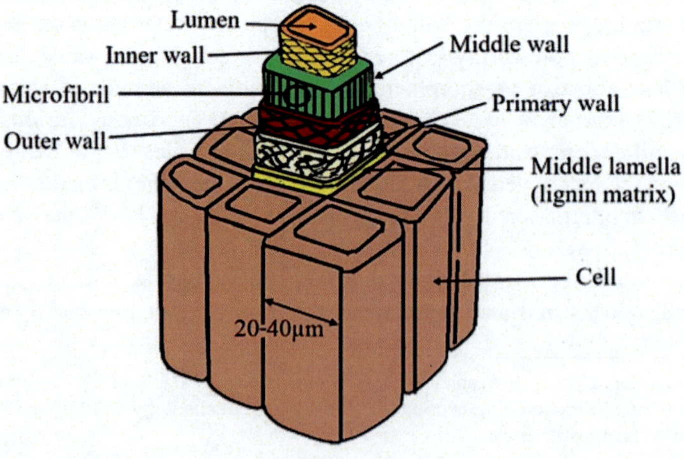

Fig. 1 Structure of wood fibre

structure of wood is complicated and multi-layered. The outer layers of the wood cell wall are called middle lamellar and it is comprised of lignin. The secondary cell wall has two parts namely cellulose microfibrils and a lignin-hemi-cellulose matrix (Chan et al. 2018). The microfibrils are bonded by hydrogen bonds and they are wound in a helix form along the fibre axis. The angle between the fibre axes and the cellulose microfibrils is known as microfibril helix angle (Commonly 5–20%) (Parham and Gray 1984).

As aforementioned, wood fibres have three primary components: cellulose (40–45% by dry weight), hemi-cellulose (25–35%) and liginin (20–30%). In addition to these, there are also some extractives (3–10%) like waxes, resins, gums and proteins (Stokke et al. 2013). Cellulose is crystalline in nature and it provides the structural stability to the wood fibres and also dictates the mechanical properties of wood. Hemi-cellulose is a branched polymer and it has lesser molecular weight than cellulose. However, it also contributes to the structural strength of wood. Lignin is an amorphous, cross-linked polymer network which bonds cellulose and hemi-cellulose together (Pettersen 1984).

Wood can be processed by two methods: pulping and grinding. In pulping, wood fibres are separated from solid wood. The chemical composition and properties of the wood pulp varies with the pulping process utilized (Fardim and Duran 2002). On the other hand, wood fibres are crushed into smaller parts in grinding process. The chemical composition of the fibres remain constant in grinding process but the aspect ratio of the produced wood particles are lower than those obtained from pulping (Schell and Harwood 1994).

The use of wood fibres as reinforcements in polymer matrix composites has been seen as a promising option as it has considerable merits over synthetic fibres. Wood fibres are cheap, available in nature in large amounts, renewable, bio-degradable, and cause minimum abrasion to the processing equipments (Chan et al. 2018).

3 Wood Fibre Polymer Composites

Environmental conservation and sustainable development have become the most important priority for researchers and engineers across the globe. The WFPCs offer a way to achieve this goal and it is produced by reinforcing the wood into the polymer matrix in the form of wood flour, fibres or wood veneers. Predominantly, polymers like polyvinyl-chloride (PVC), polypropylene (PP), polystyrene (PS), polyethylene (PE) and formaldehyde resins are used as matrix materials for wood polymer composites (Perisic et al. 2018). The WFPCs formed combines the characteristics of both polymer and wood fibre and the polymer matrix protects the wood fibre from external agents like moisture, fungal attack etc. (Yue et al. 2019). These composites are light weight, low cost and durable, and have been used in different applications like automobile interiors, furniture, packaging and other nonstructural applications (Chan et al. 2018).

The manufacturing methods used for fabricating WFPCs can be divided into two classes: forming and compounding. In the case of compounding, there are normally two steps involved: (i) melt compounding of wood fibre, matrix, and/ or other substances like coupling agent, compatibilizers etc., and (ii) subsequent processing of the product to obtain the desired output. The compounding of fibre and matrix is generally performed using an extruder and further, the composite specimens for various testing purpose are produced using methods like injection moulding, hot pressing etc. The difficulties involved in the production of WFPCs are low thermal stability of wood fibres, hydrophilic nature of wood fibres, agglomeration of the fibres and tendency to entangle during the process. Certain additives like plasticizers, compatibilizers etc. can be added to the WFPCs during the fabrication stage, to solve some of the above issues (Olakanmi and Strydom 2016).

In WFPCs, the wood fibre is instrumental in providing the required strength and stiffness to the composites whereas the function of polymer is to transfer the load acting on the composite to the reinforcements (Olakanmi and Strydom 2016). The tensile modulus, tensile strength and elongation at break of WFPCs are in the range of 0.8–5.6 GPa, 12–47 MPa and 1.4–5.4% respectively (Chan et al. 2018). The reduction in mechanical and physical properties of WFPCs is due to the poor interfacial bonding between the hydrophilic wood fibre and the hydrophobic polymer matrix. This can be enhanced through surface modification of fibre, alteration of polymer resin, inclusion of additives etc. (Rajeshkumar 2020; Wu et al. 2000).

4 Wood Composites with High Bonding Strength

The bonding strength between the wood fibre and matrix materials are influenced by various factors such as type of matrix materials (adhesives), filler material added to the resin, coupling agents added to the matrix, concentration and molecular weight of these coupling agent, surface treatment of fibre, processing and testing temperatures (Li et al. 2018; Bledzki and Gassan 1999; Summerscales and Grove 2014; Luedtke et al. 2019).

The influence of the coupling agents like maleated polyethylene (MAPE), oxidized polyethylene (OPE) and pure polyethylene (PPE) on the mechanical properties and the interfacial bonding strength of HDPE/wood composites were investigated (Lu et al. 2005). Factors like molecular weight and concentration of the coupling agents were found to have a major effect on the bonding strength, in addition to the type of coupling agent utilized. It was observed that MAPE provided better interfacial bonding strength when compared to OPE and PPE. The maleic anhydride groups in MAPE gets converted to double carboxylic groups easily and these groups were found to be more reactive for esterification than the monocarboxylic groups of OPE and PPE. Furthermore, the succinic structure in MAPE tends to form a stable ester bridge between the matrix and the fibre, thus enhancing the interfacial strength (Lu et al. 2007). The effect of molecular weight of coupling agents was linear at low molecular weights (<10,000), however at larger molecular weights (>30,000), there was not much change in the interfacial adhesion. Finally, the coupling agents with <3 wt.% provided better interfacial bonding and found to be declined for >5 wt.%.

Bio-based composite market has been increasing in the recent years. In line with this the bonding strength of wood fibre composites based on chitosan-based adhesives were studied for different mass ratios of solid content in adhesive and dry wood fibres (Ji et al. 2018). The performance of these composites was also compared with a commercially available urea–formaldehyde/wood composite (control group). It was observed that the bonding strength increased for the mass ratio range of 0–3 wt.% and decreased for the range 3–5 wt.%. The bonding strength was found to be the highest at a mass ratio of 3 wt.%. The high interfacial bonding performance is attributed to factors such as amide linkages between carbonyl groups of wood fibres and amine groups of chitosan adhesive, hydrogen bonding, interlocking and low molecular weight organics.

The effect of processing and testing temperature on the interfacial strength of PLA-wood veneer sandwich composites was studied (Luedtke et al. 2019). The matrices employed were amorphous and semi-crystalline PLA while the reinforcements were maple and beech wood veneers. The specimens were prepared and tested using an automatic bond evaluation machine (ABES) (Gaugler et al. 2019). The specimens were produced at three different pressing temperatures and were tested under four different temperatures. The bond line shear strength of the specimens under different conditions is listed in Table 1.

The higher bond strength at higher processing temperature is attributed to the lower viscosity of PLA at greater temperatures that enables better interlocking between the matrix and fibre due to the greater flow of PLA into the wood fibres. It was also concluded that the bonding strength of these composites were larger than that of wood composites produced with polyolefin or polyurethane as matrices (Gaugler et al. 2019).

In another work, the influence of fibre surface treatment on the interfacial bonding characteristics of wood fibre—PP composites were studied (Wu et al. 2000). The wood fibres in the form of saw dust were treated with acid-silane (vinyl-trimethoxy-silane) solution. The fibre-matrix bonding was enhanced due to the fibre pretreatment and this was confirmed using rheology studies and scanning electron microscope (SEM) analysis. The composites with treated fibres had high complex viscosity and this means that the interfacial bond between the fibre and matrix is strong (Han 1980). Additionally, the fracture surface of wood/PP composites was analyzed using SEM. The microscopic data indicated that the polypropylene particles were strongly attached to the wood fibres and also the fracture was brittle in nature. These two results affirm the existence of strong interfacial bonding.

The fibre-matrix bonding of wood fibre composites can also be improved using a combination of surface treatments methods. The bonding performance of Polyimide (PI)-wood composites was enhanced by the application of silane (3-methacryloxy propyltrimethoxy silane) and plasma treatment onto the black spruce wood fibres. The improvement in interfacial strength was determined using the measurement of inter laminar shear strength (ILSS) through short beam shear test. The ILSS of specimens with 3 min exposure to plasma treatment was 29.7% greater than the ILSS of untreated

Table 1 Bond line tensile shear strength of different PLA-wood veneer composites (Luedtke et al. 2019)

Matrix	Processing temperature (°C)	Testing temperature (°C)	Bonding strength (N/mm^2)	
			Beech	Maple
Amorphous PLA	140	45	7.95	5.78
		60	–	6.42
		80	2.26	2.02
		100	–	1.16
	200	45	9.03	6.89
		60	–	7.84
		80	3.03	2.55
		100	–	1.48
Semi-crystalline PLA	140	45	6.7	5.66
		60	–	5.70
		80	2.26	2.32
		100	–	1.35
	160	45	–	7.02
		60	–	7.76
		80	–	2.47
		100	–	1.64
	200	45	9.56	8.23
		60	–	8.69
		80	3.27	3.14
		100	–	1.82

specimen and also, the ILSS of specimens exposed to silane and plasma treatment (1 min) was more than the untreated specimens by 48.7%. These results revealed the increase in the interfacial strength upon fibre treatment and this is attributed to the plasma treatment that enhances wettability and mechanical interlocking of fibres and matrix and the silane treatment that strengthens the fibre-matrix bonding through an interpenetrating polymer network (Ying et al. 2017).

In order to enhance the bonding in wood fibre composites, laminated wood plastic composites (LWPCs) were fabricated using poplar wood veneer and thermoplastic sheets comprising PP and PE (1:1 weight ratio) with the aid of hot pressing process (Haq and Srivastava 2017). The wood veneer was treated with maleic anhydride (MAH). Two types of composites: the cross three-layered LWPC and parallel multi-layered LWPC were fabricated at pressing temperatures of 140, 160 and 180 °C with wood to plastic ratios of 3:1, 3:2 and 3:3. The bonding performance was evaluated using longitudinal tensile shear strength method and the bonding strength of the composites is computed by using the following Eq. (1).

Table 2 Bonding characteristics of wood fibre composites (Yue et al. 2019)

Pressing temperature (°C)	Wood to plastic ratio	MAH concentration (%)	Bonding characteristics		
			Bonding strength (MPa)	% of wood failure	Mode of failure
140	3:3	0	–	0	Delamination
160	3:1	0	1.09	0	Delamination
160	3:2	0	0.69	0	Delamination
160	3:3	0	0.69	0	Delamination
180	3:3	0	0.95	0	Delamination
160	3:3	4	0.8	100	Tensile failure of wood
160	3:3	8	0.7	100	Tensile failure of wood
180	3:3	4	1.12	100	Tensile failure of wood

$$\text{Bonding strength (Xc)} = \frac{\text{Pmax}}{\text{b}} \times l \qquad (1)$$

where P_{max} is the ultimate bearing capacity (N), and l and b are length and width of the specimen respectively. The bonding characteristics of the laminated wood composites are summarized in Table 2.

Different wood polymer composites were produced with virgin (v) and recycled (r) Polypropylene (PP) reinforced with sawdust from mango wood. In addition to these, specimens of rPP composites were also added with maleic anhydride grafted polypropylene (MAPP). Various experiments like tensile, flexural and melt flow index tests were conducted. The specimens with rPP as matrix exhibited better tensile and flexural properties as compared to those with vPP as matrix. This result was because the wood flours filled the voids in rPP matrix, thus leading to a better interfacial bonding when compared to vPP based composites. Also, the composites that contained MAPP showed superior mechanical properties (Haq and Srivastava 2017). This is because ester bonds are formed between the hydroxyl groups of wood and anhydride groups of MAPP (Kazayawoko et al. 1999). The polyolefin chains get linked with the PP chains during processing and this leads to a stronger fibre-matrix interface.

A novel method of high voltage electrostatic field (HVEF) has also been used to optimize the bonding of urea formaldehyde composites reinforced with Masson pine veneers (He et al. 2019a, b). The composite specimens were subjected to various HVEF treatment times (2, 4, 6 and 8 min) and intensities or voltages (5, 10, 15, 20 kV). The composite preparation (using hot pressing) and HVEF treatment occurred simultaneously. The bonding strength of the specimens was measured with the aid of a universal testing machine. A considerable improvement in the bonding strength was noticed upon rising the treatment time. Maximum enhancement in the bonding strength was obtained at 8 min, 20 kV and the bonding strength was 85% more than that of the untreated specimen at this condition. These positive enhancements in bonding performance were because of the considerably higher resin concentration at the interphase, lower resin penetration and better cross-linking of the matrix material (Frihart 2005).

Similar to the above study, bonding characteristics of HVEF treated Phenol formaldehyde (PF) composites reinforced with laminated veneer lumber (LVL) and Masson pine plywood were investigated (He et al. 2019a, b). The voltage applied during the treatment was 60 kV, treatment time was 20 min and the hot-pressing was done at 145 °C. The bonding strength was measured using universal testing machine and fluorescence microscopy was also conducted to observe the adhesion penetration depth. The bonding strength of untreated specimens was low for both LVL and plywood. On the other hand, the bonding strength of plywood-based composite was greater than LVL composites after the HVEF treatment. The bonding strength of treated plywood and LVL composites were increased by 108.86% and 98.53% respectively on comparing with the untreated specimens. This positive improvement is because of the reduced adhesive penetration depth at the bonding interphase and increased density of bonding interphase.

Another work comprised of the addition of poplar wood fibres to Lithium chloride (LiCl) and 2,2′- (1,4-phenylene) bis (2-oxazoline) (PBO) to polyamide 6 (PA6) resin for producing composites. The primary function of LiCl and PBO was to reduce the melting point of PA6 and increase its molecular weight respectively. The flexural and impact properties of the composites were measured and suggested that the addition of LiCl leads to improved bonding between PA6 and wood fibres. Additionally, the impact strength, flexural strength, modulus of the PBO composites was 43, 30 and 37% more than the specimens without PBO. PBO causes extension of PA6 chains at high temperature and as wood fibres are added, hydrogen bonds are formed between wood and PA6. Further, the N or/and O of PBO also aids the hydroxyl groups, thereby improving the adhesion between PA6 and wood fibres. From the morphology analysis, it was found that as PBO was added, there was a rise in O and N atoms and this led to the reduction of voids in the composite, thus implying enhanced interfacial bonding between PA6 and wood fibres (Xu et al. 2018).

Some studies reported the addition of fillers to improve the interfacial bonding between the wood fibre and the matrix. Different concentrations of activated charcoal were added to urea–formaldehyde resin and medium density fibre boards (MDF) were prepared using the modified resin and mats produced from hardwood fibres with the help of hot-pressing process (Kumar et al. 2013). The average bonding strength was calculated by testing the samples using universal testing machine. The results of the experiment are shown in Table 3.

The bonding strength is related to the cross-linking density of the urea formaldehyde resin and the addition of activated charcoal has led to an increase in the cross-linking density. Also, the resin cured at a faster rate at the core of the mat during the processing of the composite resulted in the better interfacial bonding in the composites.

Table 3 Mean Bonding strength of MDF for different activated charcoal concentrations (Kumar et al. 2013)

Resin	Percentage of activated charcoal added (%)	Internal bonding strength (MPa)
Urea–Formaldehyde	0	0.54
	0.2	0.60
	0.52	0.61
	1.04	0.58

Another work reported the inclusion of carbon fibres to urea formaldehyde resin and preparing MDF utilizing the modified resin and a mat made from rubber wood fibres. Three distinct weight concentrations: 0.1, 0.25, 0.5%, of carbon fibre was added to the matrix. The experimental results disclosed that the internal bonding strength was highest for specimens with 0.1 wt.% of carbon fibre and this was 38.46%, 16.13% and 30.91% greater than for specimens with 0 wt.%, 0.25 wt.% and 0.5 wt.% of carbon fibre respectively. The strong interfacial bonding at 0.1 wt.% of carbon fibre addition was also viewed in the SEM micrographs and it was attributed to the confinement of the fibres to the resin by the filler material. On the contrary, the decline in bonding strength beyond a certain concentration was due to the presence of agglomeration which causes stress concentrations that require lower energy for crack propagation (Khan et al. 2013).

The addition of carbon nano-fibres to Polyacrylamide (PAM)—EFPCs have also caused an increase in the bonding strength (Dang et al. 2018). Various mass fractions of carbon nano-fibres (0, 2.5, 5, 7.5 and 10%) were added to PAM and composites were manufactured using hot-pressing method. The test results are given in Table 4.

The poly (methyl methaacrylate)—wood composites were added with Poly (ethylene terephthalate) (PET) fibres. Additionally, two cross-linking bonding agents: Toulene-2, 4-diiscocynate (TDI) and (3-mercaptopropyl) trimethoxysilane (MPTMS) were included to modify the wood fibres. The wood fibres were treated with TDI and MPTMS separately and the composites were added with 3 types of PET fibres namely virgin PET (VPET), waste PET (WPET) and mixed PET (MPET) with 1:1 weight ratio of VPET and WPET, using hot pressing and investigated its

Table 4 Bonding strength of carbon nano-fibre/PAM/wood composites (Dang et al. 2018)

Matrix	Mass fraction of carbon nano fibre addition (%)	Internal bonding strength (MPa)
Polyacrylamide	0	0.66
	2.5	0.73
	5	0.81
	7.5	0.89
	10	0.87

tensile, impact and flexural properties. Specimens with both MPTMS and TDI exhibited better bonding; however the coverage of fibres was stronger when MPTMS was added. This was also confirmed by the enhancement in the mechanical properties with the inclusion of TDI and MPTMS, which suggests improved fibre-matrix bonding. Also, addition of PET fibres resulted in the increase in adhesion between PMMA and wood fibres. Maximum increase was obtained when MPET was added followed by WPET and VPET (Perisic et al. 2018).

5 Improved Water Resistance Behaviour of Wood Composites

In wood polymer composites, the water absorption is because of mechanical interlocking and the presence of forces such as dispersion, dipole–dipole, hydrogen bonding etc., between the matrix and the hydrophilic surface of the wood reinforcement (Turku and Kärki 2013).

The water resistance behaviour of wood composites can be enhanced by adding nano-fillers to the composite, performing surface treatment of wood fibres, using appropriate fabrication method, controlling the fibre aspect ratio etc. (Li et al. 2020; Gwon et al. 2010; Migneault et al. 2009).

The use of nano clay (NC) and maleic anhydride polypropylene (MAPP) coupling agent to enhance the waster resistance of wood-polypropylene composites was studied. The reinforcement used was waste poplar sawdust and composite specimens with different amounts of nano clay (0 and 3 wt.%) and MAPP (2.5 and 5 wt.%) were prepared (Tabari et al. 2011). The water absorption test was conducted by immersing the composites in distilled water for a day and then computed the amount of water absorbed by using the following Eq. (2).

$$\% \text{ of water absorption} = \left(\frac{W2 - W1}{W1}\right) \times 100 \tag{2}$$

where W_1 and W_2 are weights of the samples before and after immersion respectively.

The results showed that composites that contained 5 wt.% MAPP had lower water absorption that those with 2.5 wt.%. In the absence of NC, water absorption decreased by 35.3% for 5 wt.% MAPP specimens when compared to 2.5 wt.% MAPP specimens. This can be due to the reduction in velocity of diffusion process because of the presence of few gaps at the fibre-matrix interface, hindering of the hydrophilic groups as a result of surface modification and improved bonding between wood fibres and polypropylene. The addition of NC has also decreased the water uptake of wood composites. The water molecules occupy the voids present in the composite and saturate the fibre cell wall. When nanoclay is added, they fill the voids thereby preventing the penetration of water into the composite (Das et al. 2000). In addition, another reason for the reduction in water uptake is that the hydrophilicity of the clay surface immobilizes certain amount of moisture and hinders the diffusion of water into the matrix (Ghasemi and Kord 2009).

Another work was concerned with the influence of construction and demolition waste (CDW) addition on the water resistance properties of WFPCs. A wood-polypropylene (WF-PP) composite was used as reference and three different wood composite specimens were fabricated with recycled high-density polyethylene (rHDPE) as matrix (Hyvärinen et al. 2019). The compositions of the samples are shown in Table 5. The water absorption test was carried out for 28 days and the results are provided in Table 6.

Hybrid epoxy composites containing different proportions of teak and sal wood flour were prepared and their water absorption characteristics were estimated. Five different samples with a total reinforcement content of 33% and varying amounts of sal and teak flour (0–100%) were fabricated. The water absorption test was carried out by immersing the specimens in deionized water for one day. The samples comprising of 50% sal and 50% teak had the least water absorption (3.38%) followed by the sample with 25% teak and 75% sal (4.1%). The sample with 100% teak had the maximum water uptake (5.79%) and this was because teak has high percentage of hemi-cellulose which determines the percentage of water absorbed in

Table 5 Compositions (wt%) of different wood plastic composites with CDW additions (Hyvärinen et al. 2019)

Constituent	WF-PP	WF-rHDPE-CDW1	WF-rHDPE-CDW2	WF-rHDPE-CDW2
Wood fibre	64	44	44	–
Resin	30	30	30	30
Coupling and lubricating agent	6	6	6	6
CDW	–	20	20	64
Wood	–	8	2	6
Plastic film	–	4	6	18
Cardboard	–	2	11	35
Mineral wool	–	6	2	5

Table 6 Water absorption of wood plastic composites (Hyvärinen et al. 2019)

Days	Water absorption (%)			
	WF-PP	WF-rHDPE-CDW1	WF-rHDPE-CDW2	WF-rHDPE-CDW2
0	0	0	0	0
1	3.35	3.54	2.92	3.52
7	9.13	10.59	8.52	8.26
14	14.86	14.37	12.18	11.81
28	19.18	18.17	14.17	15.20

bio-composites. It was concluded that hybridization has led to an enhancement in the water absorption of wood fibre composites (Jain and Gupta 2018).

The effect of manufacturing method on the water absorption of WFPCs was investigated (Hoque et al. 2014). The reinforcement used was Meranti light red sapwood and the composites were prepared using solution casting where epoxy was the matrix and *in-situ* polymerization in which methyl methaacrylate (MMA) was utilized. For the water absorption test, the samples were immersed in distilled water for 2 h and then measurements were taken. The water uptake of pure wood samples was the highest followed by *in-situ* WFPCs, solution casted WFPCs (20% of wood as reinforcement) and solution casted WFPCs (5% of wood reinforcement). This shows that as amount of wood content raises the –OH groups in the samples increases and hence, water uptake also increases (Islam et al. 2012). Water absorption in the case of *in-situ* WFPCs have declined as the MMA occupies the porous structure of the wood fibres. On the contrary, epoxy-based composites exhibited superior water resistance because epoxy is highly resistant to moisture due to its low porosity and hydrophobic nature.

Wood polymer composites with high density polyethylene (HDPE) matrix were manufactured using injection moulding and extrusion process and the influence of fibre size and fabrication method on the water absorption resistance was investigated. The wood fibres with three different L/D ratios were utilized and maleated polyethylene (MAPE) was added as the coupling agents to all specimens. From the water absorption test, it was found that specimens that were injection moulded absorbed 13% less water than those that were extruded. In extruded composites, the fibre ends are exposed to water and this result in higher water absorption whereas in injection moulded composites the wood surfaces are covered by hydrophobic polymer surface. It was also found that the usage of short fibres resulted in enhanced water absorption resistance as compared to long fibres. On an average (for both processes combined), the water absorption of specimens with shorter fibres (L/D = 8.3) was 51% lesser than those with longer fibres (L/D = 21.3). It was suggested that the utilization of longer fibres results in greater probability of voids that act as water sites (Migneault et al. 2009).

The water absorption of poly (methyl methaacrylate) (PMMA) - wood composites in presence of two bonding agents Toulene-2,4-diiscocynate (TDI) and (3-mercaptopropyl) trimethoxysilane (MPTMS) were analyzed. Three samples namely PMMA-wood, PMMA- wood/TDI and PMMA-wood/MPMTS were subjected to water absorption test where the specimens were immersed in distilled water at 20 °C for 28 h. The water resistance of PMMA-wood/MPMTS was maximum followed by PMMA-wood/TDI and PMMA-wood. When bonding agents are added, the surfaces of wood fibres are completely covered by PMMA and this leads to enhanced water resistance. Also, the presence of voids has been significantly reduced by the addition of TDI and MPMTS which in turn reduces water uptake (Perisic et al. 2018).

The effect of fibre surface treatment on the water resistance of rubber wood flour (RWF)—recycled polypropylene (rPP) composites were studied (Khamtree et al. 2020). The wood flour was subjected to alkaline treatment, silane treatment and a combination of these two and specimens were prepared using the modified RWF and rPP by extrusion compression moulding. The silane treatment time was either 2 or 24 h. The specimens were submerged in deionized water for 10 weeks and then the amount of water absorbed was estimated. Surface modification of fibre results in the removal of impurities and hemi-cellulose and due to this, the specimens with treated fibre had less water absorption than samples with untreated RWF (Kaewkuk et al. 2013). Also, the water resistance of samples that had a combination of alkaline-silane treatment was higher than those treated by silane or alkaline methods alone. Minimum voids was witnessed in composites with alkaline treatment of RWF followed by 5% silane treatment for 2 h, as a result of which the water absorption was also the least (5.6%) on comparing with other specimens.

Wood fibres were subjected to treatment with acetate, proionate or benzoate and the effect of these treatments on the water absorption resistance of modified wood-HDPE composites were studied. The specimens were manufactured by extrusion and the water absorption test was carried out by immersing the samples for 61 days. The water absorption was least for composites with benzoylated fibre and the order is as follows: benzoylated fibre WPC (1.37%) < Propionylated fibre WPC (2.46%) < Acetylated fibre WPC (6.15%) < Control fibre WPC (15.19%). On chemical treatment, the hygroscopic nature of wood fibres and the available water sites reduces significantly, thus leading to improved water absorption resistance (Wei et al. 2013).

Another work concentrated on the consequence of hemicellulose and lignin removal on the water absorption properties of HDPE/wood composites (Ou et al. 2014). Different kinds of wood particles namely, wood flour (WF), holocellulose (HC), α-cellulose (αC) and hemicellulose-removed wood flour (HR) were mixed with HDPE and specimens were produced using injection moulding technique. In addition, the specimens were either made with or without the inclusion of maleated polyethylene (MAPE). The water absorption test was conducted by immersing the samples in distilled water for 2 months and then weighing the samples. The experiment was performed five times and average values were taken. At the end of the test, the water absorption resistance of the composites was in the following order: HR > WF > αC > HC. The water absorption of hemicellulose-removed specimens with and without the addition of MAPE was 2.13% and 3.37% respectively. In absence of MAPE, the water absorption of HR/HDPE was lower than WF/HDPE, αC/HDPE and HC/HDPE by 85.75%, 132.65% and 160.53%, respectively. In natural fibres, hemicellulose has the maximum capability to absorb moisture followed by cellulose and lignin. As hemicellulose is removed, the number of accessible hydroxyl groups is reduced and as a consequence, the HR must be properly wetted by the matrix. This leads to a reduction in the penetration of water into the composites (Pelaez-Samaniego et al. 2013).

Similar to the above study, another research work investigated the removal or extraction of hemicellulose at different temperatures on the water absorption characteristics of isotactic polypropylene/wood composites. Hemicellulose was extracted at 140, 155 or 170 °C and specimens were fabricated with and without the addition of coupling agent (MAPP). The control specimen (with and without MAPP) was the one in which hemicellulose was not removed. The composites were immersed in water for 4728 h and then their weights were measured. The specimens in which hemicellulose was extracted at 170 °C (Maximum hemicellulose removed) had the highest water absorption resistance. These composites had absorbed 10.7% (without MAPP) and 7.47% (with MAPP) of water respectively. Some of the reasons for improved water absorption resistance are; better fibre-matrix interfacial bonding at higher extraction temperatures, removal of hemicellulose that decreases the amount of hydroxyl groups and presence of higher proportion of lignin (hydrophobic in nature) on the wood surface after hemicellulose removal (Hosseinaei et al. 2012).

In order to produce eco-friendlier composites, some portions (up to 40%) of urea and formaldehyde were substitute by cottonseed meal (CM) in urea formaldehyde (UF) resin and specimens with this modified matrix and hard maple wood veneers were produced. The samples were immersed in water for 4 h and then their water soaked bonding strength was measured. The water-soaked strength of pure UF resin was 1.35 MPa while it was 2.85, 3.25, 3.27 and 3.39 MPa for specimens with 10%CM, 20%CM, 30%CM and 40%CM respectively. This indicates that the replacement of UF resin with cottonseed meal had led to an improvement in the water resistance of wood polymer composites. This enhancement might be attributed to the hydrophobic property of cottonseed meal and its strong linkage with UF resins (Liu et al. 2018).

Polymer composites with linear low-density polyethylene (LLDPE) as matrix and various wood varieties like oak, pine, acacia and eucalyptus as reinforcements were fabricated and the effect of the different types of wood on the water absorption resistance was studied. During processing, EVOH was added as the compatibilizer. The water absorption was estimated by soaking the specimens in water for 11, 20, 29, 34, 43 and 52 days. After 43 days, the water absorption in all the specimens attained saturated state. The water absorption was minimum for composites with pine as reinforcement, followed by eucalyptus, acacia and oak. The high absorption resistance of pine-based composites was due to the presence of lesser available hydroxyl groups compared to other specimens. The low number of available hydroxyl groups is because of less cellulose content in pine (Shebani et al. 2009).

Wood fibre composites exhibiting superior water absorption resistance were manufactured on a nano-interface assembly with the help of cationic polyethyleneimine-nano-ZNO (PEI-ZnO) and anionic sodium polyacrylate-wood (PAAS-wood). The wood fibres were suspended in PAAS solution while nano-ZnO particles were suspended in PEI solution. The wood fibres were then coated with nano-ZnO at concentrations 1 wt.%, 2 wt.%, 3 wt.%, 4 wt.% or 5 wt.% and the modified fibres were reinforced in polypropylene matrix. The water absorption was estimated by immersing the specimens in water for 6, 12, 24 and 60 h. The water absorption of specimens with 3 wt.% nano-ZnO coating exhibited maximum water absorption resistance. At 6 h of soaking, the water absorption of pure composite was 5.2% whereas, the water absorption of composite with 3 wt.% nano-ZnO coating was 1.83%. Likewise, at 60 h of soaking, the water absorption of pure composite and composite with 3 wt.% nano-ZnO coating were around 22% and 3% respectively. The surface of wood fibre after modification comprised of three distinct layers: PAAS-PEI-nano-ZnO. The assembling of the composites on a nano-interface reduces the surface activity of the fibres and shields the composites against moisture. The shielding can be due to: a) a firm 3-D nano-network that prevents wetting, b) charge-grafting lowers wood surface activity and enhances the mechanical interlocking between the matrix and the surface of wood fibre (Ye et al. 2016).

6 Conclusions and Future Perspective

The wood has received an ample attention as the reinforcing filler for producing polymeric composites for various industrial applications because of its noteworthy features and properties. These fillers have poor bonding with the polymer matrices because of its hydrophilic nature which is responsible for lower strength of composites and higher water absorption rate. The surface modification of filler is one of the best solutions to overcome this problem. Some research works reported that the wood fibre composites have improved bonding strength and reduced water absorption rate of the after the surface modification using different chemicals. Such composites satisfy the requirements in the various industrial sectors. Though huge varieties of wood fibres are available the studies on various characterizations of wood fibre composites are relatively less when compared to the lignocellulose fibres extracted from other parts of the plants and synthetics fibers. This created a research gap for the future researchers to explore the properties of different raw and surface modified wood fibre based composites.

References

Bledzki AK, Gassan J (1999) Composites reinforced with cellulose based fibres. Prog Polym Sci 24(2):221–274

Chan CM, Vandi LJ, Pratt S et al (2018) Composites of wood and biodegradable thermoplastics: a review. Polym Rev 58(3):444–494

Clemons C (2008) Raw materials for wood–polymer composites. In: Niska KO and Sain M (eds) Wood–polymer composites. Woodhead Publishing, United Kingdom, pp 1–22

Dang B, Chen Y, Yang N et al (2018) Effect of carbon fibre addition on the electromagnetic shielding properties of carbon fibre/polyacrylamide/wood based fibreboards. Nano Technol 29(19):195605

Das S, Saha AK, Choudhury PK et al (2000) Effect of steam pretreatment of jute fibre on dimensional stability of jute composite. J Appl Polym Sci 76(11):1652–1661

Fardim P, Duran N (2002) Surface chemistry of Eucalyptus wood pulp fibres: effects of chemical pulping. Holzforschung 56(6):615–622

Frihart CR (2005) Wood adhesion and adhesives. In: Rowell RM (ed) Handbook of wood chemistry and wood composites. CRC Press, Boca Raton, FL, pp 215–278

Gaugler M, Luedtke J, Grigsby WJ et al (2019) A new methodology for rapidly assessing interfacial bonding within fibre-reinforced thermoplastic composites. Int J Adhes Adhes 89:66–71

Ghasemi E, Kord B (2009) Long-term water absorption behaviour of polypropylene/wood flour/organoclay hybrid nanocomposite. Iranian Poly J 18:683–691

Gwon JG, Lee SY, Chun SJ et al (2010) Effects of chemical treatments of hybrid fillers on the physical and thermal properties of wood plastic composites. Compos Part A 41(10):1491–1497

Han CD (1980) Multiphase flow in polymer processing. In: Astarita G, Marrucci G, Nicolais L (eds) Rheology. Springer, Boston, pp 121–128

Haq S, Srivastava R (2017) Wood polypropylene (PP) composites manufactured by mango wood waste with virgin or recycled PP: mechanical, morphology, melt flow index and crystalline behaviour. J Polym Environ 25(3):640–648

He Q, Zhan T, Ju Z et al (2019) Influence of high voltage electrostatic field (HVEF) on bonding characteristics of Masson (Pinus massoniana Lamb.) veneer composites. Eur J Wood Wood Prod 77(1):105–114

He Q, Zhan T, Zhang H et al (2019) Comparison of bonding performance between plywood and laminated veneer lumber induced by high voltage electrostatic field. In: MATEC web of conferences, vol 275. EDP Sciences, p 01013

Hoque ME, Aminudin MAM, Jawaid M et al (2014) Physical, mechanical, and biodegradable properties of meranti wood polymer composites. Mater Des 64:743–749

Hosseinaei O, Wang S, Enayati AA et al (2012) Effects of hemicellulose extraction on properties of wood flour and wood–plastic composites. Compos Part A9 43(4):686–694

Hyvärinen M, Ronkanen M, Kärki T (2019) The effect of the use of construction and demolition waste on the mechanical and moisture properties of a wood-plastic composite. Compos Struct 210:321–326

Islam MS, Hamdan S, Jusoh I (2012) The effect of alkali pretreatment on mechanical and morphological properties of tropical wood polymer composites. Mater Des 33:419–424

Jain NK, Gupta MK (2018) Hybrid teak/sal wood flour reinforced composites: mechanical, thermal and water absorption properties. Mater Res Express 5(12):125306

Ji X, Dong Y, Nguyen TT et al (2018) Environment-friendly wood fibre composite with high bonding strength and water resistance. R Soc Open Sci 5(4):172002

Kaewkuk S, Sutapun W, Jarukumjorn K (2013) Effects of interfacial modification and fibre content on physical properties of sisal fibre/polypropylene composites. Compos Part B 45(1):544–549

Kazayawoko M, Balatinecz JJ, Matuana LM (1999) Surface modification and adhesion mechanisms in woodfibre-polypropylene composites. J Mater Sci 34(24):6189–6199

Khamtree S, Ratanawilai T, Ratanawilai S (2020) The effect of alkaline–silane treatment of rubber wood flour for water absorption and mechanical properties of plastic composites. J Thermoplast Compos Mater 33(5):599–613

Khan TA, Gupta A, Jamari SS et al (2013) Synthesis and characterization of carbon fibres and their application in wood composites. BioResour 8(3):4171–4184

Kumar A, Gupta A, Sharma KV et al (2013) Influence of activated charcoal as filler on the properties of wood composites. Int J Adhes Adhes 46:34–39

Kumar GR, Hariharan V, Saravanakumar SS (2019) Enhancing the free vibration characteristics of epoxy polymers using sustainable Phoenix Sp. fibers and nano-clay for machine tool applications. J Nat Fib. https://doi.org/10.1080/15440478.2019.1636740

Li D, Jiang Y, Lv S et al (2018) Preparation of plasticized poly (lactic acid) and its influence on the properties of composite materials. PLoS ONE 13(3):e0193520

Li Q, Liang Y, Chen F et al (2020) Preparation and performance of modified montmorillonite-reinforced wood-based foamed composites. BioResour 15(2):3566–3584

Liu M, Wang Y, Wu Y et al (2018) "Greener" adhesives composed of urea-formaldehyde resin and cottonseed meal for wood-based composites. J Cleaner Prod 187:361–371

Lu JZ, Wu Q, Negulescu II (2005) Wood-fibre/high-density-polyethylene composites: coupling agent performance. J Appl Polym Sci 96(1):93–102

Lu JZ, Wu Q, Negulescu II (2007) The influence of maleation on polymer adsorption and fixation, wood surface wettability, and interfacial bonding strength in wood-PVC composites 1. Wood Fibre Sci 34(3):434–459

Luedtke J, Gaugler M, Grigsby WJ et al (2019) Understanding the development of interfacial bonding within PLA/wood-based thermoplastic sandwich composites. Ind Crops Prod 127:129–134

Migneault S, Koubaa A, Erchiqui F et al (2009) Effects of processing method and fibre size on the structure and properties of wood–plastic composites. Compos Part A 40(1):80–85

Miller RB (1999) Structure of wood, wood handbook: wood as an engineering material. Madison, WI: USDA Forest Service, Forest Products Laboratory, 1999. General technical report FPL; GTR-113: pp 2.1–2.4, 113

Nagaraja KC, Rajanna S, Prakash GS et al (2020) Improvement of mechanical and thermal properties of hybrid composites through addition of halloysite nanoclay for light weight structural applications. J Ind Text. https://doi.org/10.1177/1528083720936624

Nasir M, Sulaiman O, Hashim R et al (2015) Rubber wood fiber treatment by laccase enzyme and its application in medium density fiber board. J Pure Appl Microbiol 3:2095–2100

Olakanmi EO, Strydom MJ (2016) Critical materials and processing challenges affecting the interface and functional performance of wood polymer composites (WPCs). Mater Chem Phys 171:290–302

Ou R, Xie Y, Wolcott MP et al (2014) Morphology, mechanical properties, and dimensional stability of wood particle/high density polyethylene composites: effect of removal of wood cell wall composition. Mater Des 58:339–345

Parham RA, Gray RL (1984) Formation and structure of wood. In: Rowell E (ed) The chemistry of solid wood. ACS publications, Washington, pp 3–56

Pelaez-Samaniego MR, Yadama V, Lowell E, Amidon TE et al (2013) Hot water extracted wood fibre for production of wood plastic composites (WPCs). Holzforschung 67(2):193–200

Perisic SD, Radovic I, Petrovic M et al (2018) Processing of hybrid wood plastic composite reinforced with short PET fibres. Mater Manuf Processes 33(5):572–579

Pettersen RC (1984) The chemical composition of wood. In: Rowell E (ed) The chemistry of solid wood. ACS publications, Washington, pp 3–56

Rajeshkumar G, Ramakrishnan S, Pugalenthi T et al (2020) Performance of surface modified pineapple leaf fiber and its applications. In: Jawaid M, Asim M, Md. Tahir P et al (eds) Pineapple leaf fibers. Springer, Singapore, pp 309–321

Rajeshkumar G (2020) Effect of sodium hydroxide treatment on dry sliding wear behavior of Phoenix sp. fiber reinforced polymer composites. J Ind Text https://doi.org/10.1177/1528083720918948

Schell DJ, Harwood C (1994) Milling of lignocellulosic biomass. Appl Biochem Biotechnol 45(1):159–168

Shebani AN, Van Reenen AJ, Meincken M (2009) The effect of wood species on the mechanical and thermal properties of wood—LLDPE composites. J Compos Mater 43(11):1305–1318

Stokke DD, Wu Q, Han G (2013) Introduction to wood and natural fibre composites. Wiley, United Kingdom

Summerscales J, Grove S (2014) Manufacturing methods for natural fibre composites. In: Hodzic and Shanks R (eds) Natural fibre composites. Woodhead Publishing, United Kingdom, pp 176–215

Tabari HZ, Nourbakhsh A, Ashori A (2011) Effects of nanoclay and coupling agent on the physico-mechanical, morphological, and thermal properties of wood flour/polypropylene composites. Polym Eng Sci 51(2):272–277

Turku I, Kärki T (2013) Reinforcing wood–plastic composites with macro-and micro-sized cellulosic fillers: comparative analysis. J Reinf Plast Compos 32(22):1746–1756

Wei L, McDonald AG, Freitag C et al (2013) Effects of wood fibre esterification on properties, weatherability and biodurability of wood plastic composites. Polym Degrad Stab 98(7):1348–1361

Wu J, Yu D, Chan CM et al (2000) Effect of fibre pretreatment condition on the interfacial strength and mechanical properties of wood fibre/PP composites. J Appl Polym Sci 76(7):1000–1010

Xu S, Fang Y, Yi S et al (2018) Effects of lithium chloride and chain extender on the properties of wood fibre reinforced polyamide 6 composites. Polym Test 72:132–139

Ye X, Wang H, Zheng K et al (2016) The interface designing and reinforced features of wood fibre/polypropylene composites: wood fibre adopting nano-zinc-oxide-coating via ion assembly. Compos Sci Technol 124:1–9

Ying L, Liu Y, Dan H et al (2017) The effect of coupling agents on the interfacial properties of wood-fibre–reinforced polyimide composites. Surf Interface Anal 49(12):1232–1237

Yue K, Wang L, Xia J et al (2019) Experimental research on mechanical properties of laminated poplar wood veneer/plastic sheet composites. Wood Fibre Sci 51(3):320–331

Use of Innovative High-Density Polyethylene (HDPE) Environmentally Friendly Adhesives for Wood Composites

M. Chandrasekar, K. Senthilkumar, T. Senthil Muthu Kumar, M. R. Ishak, N. Rajini, and Suchart Siengchin

Abstract The plywood laminate-based composites derived from the wood, natural fibres and other agricultural residues provide sustainable solutions to the structural and non-structural applications. However, formaldehyde adhesives used for bonding the lamina together seems to have impact on the production personnel and end user due to their carcinogenic nature. Hence, the alternative resins for bonding plies in wood composite to substitute the conventional formaldehyde resin has been in focus.. This review addresses the published research works on use of high-density polyethylene (HDPE) as the bonding adhesive for wood-based composites.. Various factors were found to influence the bonding characteristics, physical and mechanical properties as follows: (i) curing parameters (ii) coupling agents and treatments such as maleic-anhydride grafted HDPE, vinyltrimethoxysilane, dicumyl peroxide (DCP), silane treatment, thermal treatment, etc., (iii) density of HDPE (iv) HDPE dosage and (v) aging. Studies also highlight that HDPE adhesive satisfies the minimum bonding strength requirements for their use in commercial grade plywood based laminates.

M. Chandrasekar (✉)
School of Aeronautical Sciences, Hindustan Institute of Technology and Science, Padur, Kelambakkam, Chennai 603103, Tamil Nadu, India
e-mail: chandrasekar.25j@gmail.com

K. Senthilkumar · T. Senthil Muthu Kumar
Department of Mechanical Engineering, Kalasalingam Academy of Research and Education, Krishnankoil 626126, Tamil Nadu, India

K. Senthilkumar · T. Senthil Muthu Kumar · N. Rajini
Department of Mechanical and Process Engineering, The Sirindhorn International Thai-German Graduate School of Engineering (TGGS), King Mongkut's University of Technology North Bangkok, 1518 Wongsawang Road, Bangsue, Bangkok 10800, Thailand

M. R. Ishak · S. Siengchin
Department of Aerospace Engineering, Faculty of Engineering, University Putra Malaysia, 43400 UPM Serdang, Selangor, Malaysia

M. R. Ishak
Aerospace Malaysia Research Centre (AMRC), Faculty of Engineering, University Putra Malaysia, Tower block, Level 7, 43400 Serdang, Selangor, Malaysia

M. Jawaid et al. (eds.), *Eco-Friendly Adhesives for Wood and Natural Fiber Composites*, Composites Science and Technology,
https://doi.org/10.1007/978-981-33-4749-6_6

The prospect of commercializing HDPE-adhesives for bonding the plywood based laminates in place of conventional formaldehyde resin is not far.

Keywords HDPE · Adhesive · Wood based composite · Bonding characteristics · Aging

1 Introduction

Wood based composites have been commonly used in structural applications such as windows, door frames, floors, false ceiling panels, cabin deck separation and interior decorations while it is also frequently used in non-structural applications like tables, furniture, and support boards for writing, vegetable chopping, etc. Wood based composites exist in different forms: (1) wood plastic composite which consists of wood flour and polymer matrix with thermosetting and thermoplastic resins, (2) Plywood laminate with thin layers of plywood lamina bonded with the thermosetting and thermoplastic based adhesives and (3) Lumber veneer laminate (LVL) which consists of plywood sheets in the inner layers and veneer board on the top bonded altogether with the adhesives. Adhesives used in bonding LVL and plywood laminates are conventional formaldehyde based thermosetting resins such as phenol–formaldehyde (PF), urea–formaldehyde (UF) and melamine urea–formaldehyde (MUF) mixed with polyvinyl acetate (Chang et al. 2017; Wu et al. 2019). Despite of the successful use of formaldehyde based resin as bonding adhesives, disadvantages such as high production cost, carcinogenic nature of formaldehyde and their impact on environment since it is derived from the fossil based resources (Lustosa et al. 2015; Chang et al. 2018). Thus, research has been focused on the development of formaldehyde-free and environment friendly high-density polyethylene (HDPE) as the adhesive.

1.1 HDPE Film as Adhesive

The idea of using HDPE film for bonding wood plies or veneer was first attempted by Han in 1990 through the flat pressing technique (Han 1990). The characteristics of HDPE such as dimensional stability, non-aqueous, resistance to moisture, lower cost and formaldehyde-free make them a highly suitable alternative for adhesive bonding in plywood and wood veneer laminates.

1.2 Fabrication Method

In the recent studies, LVL or wood boards with the HDPE film as binder was fabricated through the hotpress molding subsequently followed by coldpress. Initially,

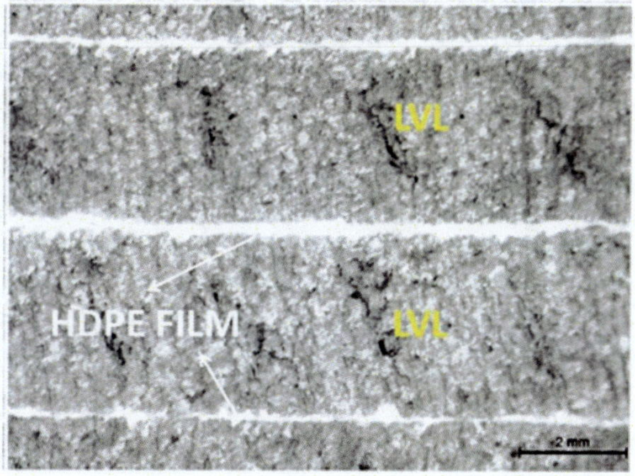

Fig. 1 LVL composite laminate with HDPE film as binder (Lustosa et al. 2015)

HDPE films of thickness ranging from 0.5 to 15 mm were placed in between the plies of the LVL or wood boards. The stacked layers with HDPE film was then heated to curing temperature under pressure for a definite period of time followed by subjecting the cured samples in coldpress for few minutes. The cured samples were found to have good bonding with the plies as shown in Fig. 1 (Lustosa et al. 2015).

2 Factors Influencing Adhesion Properties of HDPE Film with Wood-Based Composites

2.1 Curing Parameters

Tensile shear strength of the Veneer/wood fibre-HDPE/veneer composite panels bonded with HDPE film was found to be between 0.91 and 1.45 MPa at a curing temperature of 70–90 °C between 10 and 20 min. This was above the 0.4 MPa threshold value of GB/T15104-2006 (Sun et al. 2019). However, Lustosa et al. (2015) reported that curing temperature of 100 °C and 120 °C at 10 and 15 min under a pressure of 1 MPa was not sufficient to melt the HDPE and hence, bonding was not achieved as shown in Fig. 2. At 140 °C, HDPE melts and can penetrate the vessels of the wood substrate. Hence, the bonding between the wood substrate at curing temperature greater than 140 °C was due to the mechanical interlocking effect between the HDPE and porous wood substrate. Similar findings were reported by Fang et al. (2013) on the poplar veneer//HDPE laminate. They also showed that tensile shear strength achieved for laminate bonded with HDPE films as adhesive was greater than 0.7 MPa and could match the strength requirements for type II grade plywood of GB/T 9846.3-2004.

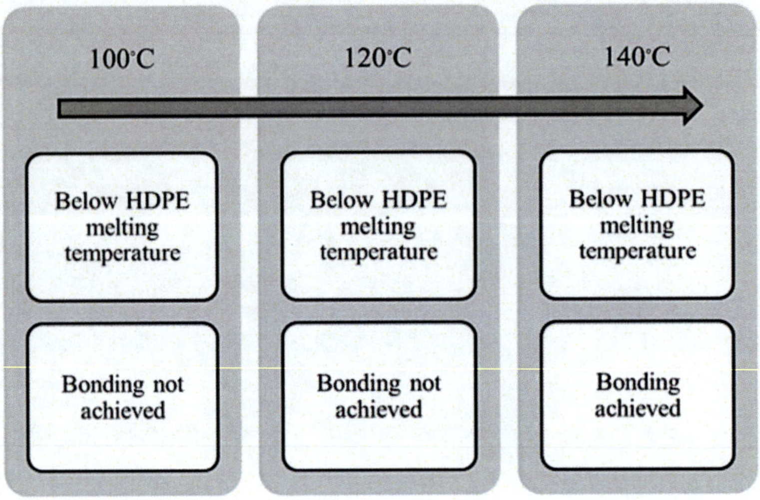

Fig. 2 Influence of curing temperature of HDPE on the bonding characteristics with plywood and veneer

In their work, Liang chang et al. (2017) showed that viscosity of the HDPE film decreased from 4.019 × 10^4/Pa.s at 140 °C to 1.525 × 10^4/Pa.s at 180 °C. Decrease in viscosity with the increasing curing temperature led to better penetration of the HDPE into the wood cells of the laminate ply, thus higher maximum penetration depth (in transverse direction of the plies) and effective penetration depth (in longitudinal direction of the plies) at the interface between the adjacent LVL plies. Tensile shear strength was found to increase in magnitude until 160 °C and continued to decline further. Despite of the increase in penetration depth beyond 160 °C, decrease in tensile shear strength was attributed to the homogeneity of the bond line. Uneven penetration of the HDPE into the wood cells or lumen of the adjacent plies had an impact on the tensile shear strength.

Bekhta and Sedliačik (2019) fabricated alder veneer/HDPE laminate through the hotpress technique. Specimens were cured at a temperature from 140 to 180 °C between 1 to 5 min of the curing time as shown in Fig. 3.

Grīnbergs et al. (2010) fabricated the birch wood veneer/HDPE laminate with curing temperature of 140 ºC for 2 min under pressure ranging from 2 to 8 MPa. Shear strength was found to be approximately 6.8 MPa at a pressure of 2 MPa to 7.61 MPa as the curing pressure was increased to 6 MPa. Shear strength was found to decline by 5.8 MPa with further increase in the hotpress pressure to 8 MPa.

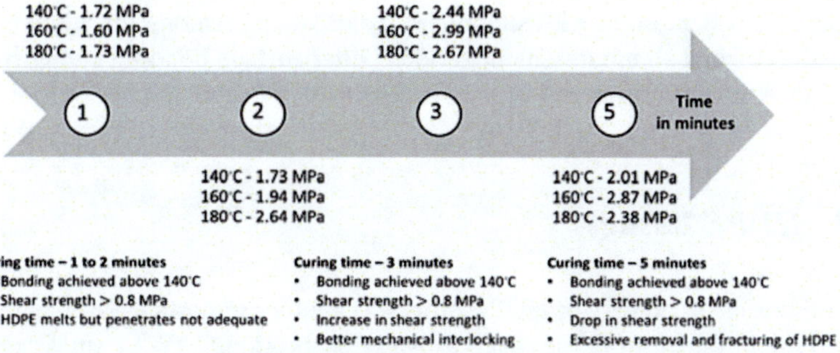

Fig. 3 Influence of hot pressing time on the shear strength of 3 layered alder wood veneer laminate bonded with the HDPE film (Bekhta and Sedliačik 2019)

2.2 Use of Coupling Agents and Treatments

As per Wu et al. (2019), decrease in tensile shear strength beyond curing temperature of 160 °C can be rectified by the use of maleic-anhydride grafted HDPE (MAPE) in place of conventional HDPE. Maleic anhydride in MAPE acts as coupling agent and forms a covalent bond between the carbonyl group of MAPE and hydroxyl group of wood veneer which led to increase in tensile shear strength at curing temperatures beyond 160 °C.

Fang et al. (2014) treated the wood veneer with the 2 wt% of vinyltrimethoxysilane and 0.05 wt% dicumyl peroxide (DCP). Their results indicate that silane treated poplar veneer laminate showed tensile shear strength of 1.7 to 1.9 MPa when 1–3 layers of HDPE film was used. The reason for significant improvement in the tensile shear strength for the silane treated poplar veneer with HDPE film is due to the following reasons: a) formation of covalent bonds between the silanol groups and hydroxyl group of wood veneer and b) oxy radicals of DCP reacts with the HDPE at curing temperature and releases vinyl radicals which further forms covalent bond with the poplar veneer and HDPE film. In another study, Fang et al. (2017) used two methods to improve interfacial bonding between poplar veneer and HDPE film: (a) 2wt% silane and 0.05wt% DCP and (b) conditioning of veneer boards in the oven at 200 °C for 30 min. Both the silane treatment and thermal treatment yielded considerable increase in tensile shear strength, modulus of elasticity and modulus of rigidity. Furthermore, water contact angle on the poplar wood surface also increased form 28° to 119° and 97° for the thermal and silane treatments respectively. Thermal treatment of poplar veneer reduces their hygroscopic nature and cause reduction in polarity which in turn enhances the bonding between the veneer and hydrophobic HDPE film. Similar trend on water contact angle was reported by Liu et al. (2019) on poplar wood veneer laminate treated with r-aminopropyl triethoxy silane and trimethoxy vinyl silane. A significant finding from their study is that water contact angle decreased for the untreated wood veneer while it remain unchanged for silane treated wood veneer after 30 s due to the fact that hydroxyl groups of wood veneer formed a covalent bond with silanol groups of silane. Among the silane treatments, trimethoxy vinyl silane treatment yielded better surface bonding strength and water resistance than the former.

2.3 HDPE Dosage

HDPE dosage or density of HDPE films was another important parameter that influenced the physical properties such as water absorption % (WA), thickness swelling (TS) and mechanical properties such as modulus of rigidity, modulus of elasticity and tensile shear strength. According to Fang et al. (2013), both WA and TS were lower for LV with HDPE film than LVL with conventional UF resin. In addition to the non-aqueous and hydrophobic characteristic of the HDPE, it also penetrates into the micro-pores of the wood veneer blocking the moisture diffusion into the laminate along the bonded interface. According to Lustosa et al. (2015), tensile shear strength increased from 2.05 to 2.51 MPa, flexural modulus from 88.5 to 98.8 MPa, 9.641 to 10.520 GPa as the HDPE dosage was increased from 150 to 250 g/m^2. Similar trend on the tensile shear strength was reported by Chang et al. (2017) in poplar veneer wood laminates where there was significant increment in bonding strength with the addition of HDPE film layers.

2.4 Fibre Weight to HDPE Ratio

Guo et al. (2018) fabricated sandwich panels with aspen wood fibre board (WFB) as core bonded on the top and bottom surface with poplar veneers through the HDPE film. WFB:HDPE was varied between 60:40, 70:30 and 80:20 in terms of weight %. They observed decline in bonding strength with the greater WF content in the order: 60:40 > 70:30 > 80:20 while use of MAPE (which involves small addition of maleic anhydride) resulted in similar bonding strength at all wt%. A similar observation was made on WF/poplar veneer laminate bonded with HDPE film by Liu et al. (2019). They reported that the bonding strength declined with the higher WF: lower HDPE ratio and at 80:20 ratios, maximum delamination between the WF and poplar veneer was observed.

2.5 Aging

Water resistance of the wood veneer laminate bonded with HDPE film were studied by exposing the specimens to the following conditions: (a) 4 h immersion in boiling water, dried for 16 h followed by subsequent boiling for 4 h and (b) 72 h of boiling water. It can be observed from their study that single lap joint shear strength decreased from 7.62 MPa for dry specimens to 2.34 and 2.10 MPa for the exposed specimens. However, the shear strength was still greater than the threshold value of 1.8 MPa for the Class-1 plywood and 1 MPa for the Class-3 plywood and hence, they recommended HDPE as adhesive suitable for the outdoor applications (Kajaks et al. 2012). Tang et al. (2012) exposed red pine veneer/HDPE laminate to 2 thermal cycles by immersing the specimens in boiling water for 4 h, followed by drying at 63 ± 3 °C for 20 h. The exposed specimens did not delaminate after exposure and possessed tensile shear strength of 1.09 MPa which is above the threshold value of 0.4 MPa as per the plywood GB/T15104-2006 requirements.

3 Future Perspective

HDPE bonded wood laminates were found to have met the minimum tensile shear strength required as per the plywood GB/T15104-2006 requirements before and after the weathering trials. Studies also showed that bonding between the plywood laminas in the wood based laminate can be achieved by optimizing the HDPE film thickness, curing temperature and pressure. Due to these features, HDPE can be an efficient and environmental friendly alternative for the traditional phenol formaldehyde resin in their use as adhesive for bonding the wood based laminates. Since the morphological features vary from one wood to the other, determination of the film thickness for optimum bonding and assessment of the interfacial bonding characteristics on a microscopic level can be helpful in identifying the effectiveness of bonding between the particular species of wood laminas. On the other hand, adhesively bonded wood laminates have application in the indoor and outdoor structures where they are susceptible to degradation from the environmental factors such as humidity, moisture and temperature. Thus, future studies can be extended in assessing the degradation characteristics and ability of the HDPE based wood laminates to retain strength under various aging conditions.

4 Conclusion

This article highlights the use of HDPE as adhesive for bonding the plywood laminas in the wood based composites. Following is the observation from the reported works.

- Mechanical properties of the wood based composites were found to increase by increasing the curing temperature. Bonding strength, viscosity, and penetration of HDPE into the woods were also influenced by the curing temperature.
- The usage of coupling agents such as silane treatments and maleic anhydride grafting of HDPE have shown promising results in enhancing the properties of the wood based composites. For example, the water contact angle of the poplar

wood surface was improved from 28° to 119° and 97°. It was ascribed to the thermal and silane treatments, respectively.

- Bonding strength declined with the reduction in HDPE dosage.
- There was a reduction in the tensile shear strength due to aging. However, the strength was still above the threshold values required for the commercial grade plywood laminates. Hence, it was suggested that the wood composites bonded with HDPE adhesives could be employed in the outdoor applications.

References

Bekhta P, Sedliačik J (2019) Environmentally-friendly high-density polyethylene-bonded plywood panels. Polymers (Basel) 11:1166

Chang L, Guo W, Tang Q (2017) Assessing the tensile shear strength and interfacial bonding mechanism of poplar plywood with high-density polyethylene films as adhesive. BioResources 12:571–585

Chang L, Tang Q, Gao L et al (2018) Fabrication and characterization of HDPE resins as adhesives in plywood. Eur J Wood Wood Prod 76:325–335

Fang L, Chang L, Guo W et al (2013) Preparation and characterization of wood-plastic plywood bonded with high density polyethylene film. Eur J Wood Wood Prod 71:739–746

Fang L, Chang L, Guo W et al (2014) Influence of silane surface modification of veneer on interfacial adhesion of wood–plastic plywood. Appl Surf Sci 288:682–689

Fang L, Xiong X, Wang X et al (2017) Effects of surface modification methods on mechanical and interfacial properties of high-density polyethylene-bonded wood veneer composites. J Wood Sci 63:65–73

Grīnbergs U, Kajaks J, Reihmane S (2010) Usage of ecologically perspective adhesives for wood bonding

Guo L, Wang W, Wang Q, Yan N (2018) Decorating wood flour/HDPE composites with wood veneers. Polym Compos 39:1144–1151

Han G-S (1990) Preparation and physical properties of moldable wood-plastic composites. Dr Diss Grad Sch Agric Kyoto University

Kajaks J, Reihmane S, Grinbergs U, Kalnins K (2012) Use of innovative environmentally friendly adhesives for wood veneer bonding. Proc Est Acad Sci 61:207

Liu Y, Guo L, Wang W et al (2019) Modifying wood veneer with silane coupling agent for decorating wood fiber/high-density polyethylene composite. Constr Build Mater 224:691–699

Lustosa ECB, de Del Menezzi CHS, de Melo RR (2015) Production and properties of a new wood laminated veneer/high-density polyethylene composite board. Mater Res 18:994–999

Sun Y, Guo L, Liu Y et al (2019) Glue wood veneer to wood-fiber–high-density-polyethylene composite. Int J Adhes Adhes 95:102444

Tang L, Zhang Z, Qi J et al (2012) A new formaldehyde-free adhesive for plywood made by in-situ chlorinating grafting of MAH onto HDPE. Eur J Wood Wood Prod 70:377–379

Wu Y, Xu J, Deng L et al (2019) Modified high-density polyethylene film as the adhesive for veneer overlaying of wood-based panel. BioResources 14:7840–7851

Utilization of Renewable Biomass and Waste Materials for Production of Environmentally-Friendly, Bio-based Composites

Carlo Santulli

Abstract The introduction of renewable biomass into a polymer matrix is an option competing with other possibilities, such as energy recovery and/or re-use in the carbonized state, or production of chemicals, such as, in the case of ligno-cellulosic waste, concentrates on the production of simple sugars, then possibly leading to the development of biopolymers. These competitive applications have also some interest and market, however with a considerable energy, water and materials consumption, due also to the not always high yielding. Other possibilities for renewable biomass are therefore being used as fillers to increase mechanical performance of polymers or to allow e.g., the absorption of toxic chemicals. This review concentrates on the use of biomass as close as possible to the "as received" state, therefore avoiding whenever suitable any thermal treatment. More specifically, it focuses on its introduction into the three categories of oil-based (or bio-based replacement) of engineered polymers, into industrial biopolymers, such as poly(lactic acid) (PLA) and self-developed biopolymers, such as thermoplastic starch (TPS).

Keywords Renewable biomass · Biocomposites · Thermoplastic starches

1 Introduction

1.1 Definition of Renewable Biomass

The production of more sustainable materials suggests as a possible option the use of renewable biomass, in particular derived from the waste fraction that is normally disposed by energy recovery or landfilled, since re-using them in new materials would result in the reduction of waste production, which is the best strategy proposed by recent environmental regulations, such as EC98/2008 Directive. This will have also

C. Santulli (✉)
School of Science and Technology, Geology Section, Università di Camerino, via Gentile III da Varano 7, 62032 Camerino, Italy
e-mail: carlo.santulli@unicam.it

M. Jawaid et al. (eds.), *Eco-Friendly Adhesives for Wood and Natural Fiber Composites*, Composites Science and Technology,
https://doi.org/10.1007/978-981-33-4749-6_7

the additional benefit of allowing a smaller fraction of materials to derive from non-renewable sources, such as hydrocarbons and various chemicals obtained from cracking of extracted oil.

Although in general terms it could appear straightforward to identify what "biomass" is, being correlated to forest management, a harmonized definition has been reached in recent years, according to which biomass includes all aboveground biomass compartments of living trees and plants, namely" the aboveground part of the stump, the stem from stump to top, dead and living branches, and foliage" (Avitabile and Camia 2018). In Fig. 1, an estimate is offered of the harmonized presence of biomass, according to EFDAC (European Forest Data Center), derived by the 26 participating National Forest Inventories (NFIs) organizations EFDAC biomass dataset.

This term is competitive with "agrowaste", which is more restrictive, in that it has been often considered in connection with some potential in carrying out production

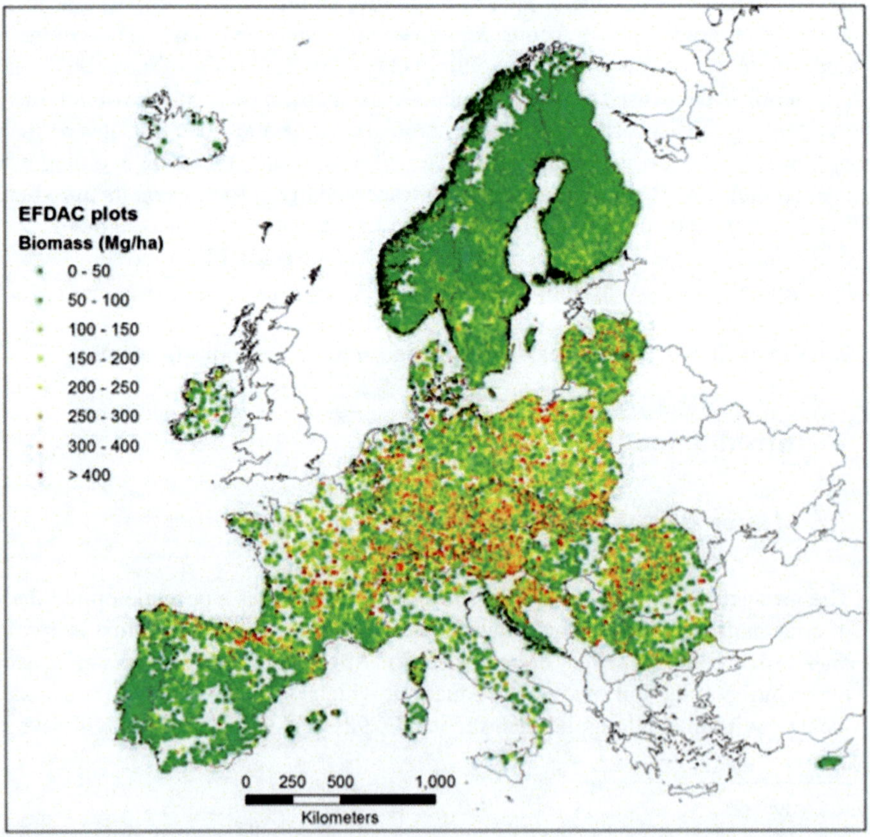

Fig. 1 Spatial distribution of the harmonized EFDAC biomass plots for the 26 countries included in the EFDAC dataset (Avitabile and Camia 2018)

of chemicals for the synthesis of bio-based materials, a possibility which is going to be treated further down, yet in general terms puts emphasis on the derivation from the agricultural sector. It can also be suggested that a part of biomass can also be a residue from agrowaste, but in common language mentioning "biomass" has become an indication of the possibility to dispose of the material by combustion and energy recovery (Zhang and Long 2010).

In itself, this does not indicate that all the various waste materials originated from biomass would possibly offer a sufficient calorific power into combustion. However, regardless including the prefix "bio", it is not sure that the whole of biomass would have sufficient biological action i.e., bio-deterioration followed by bio-degradation in the presence e.g., of micro-organisms. In practical terms, other definition of biomass are still operational, which arrive even to include a part of construction and demolition waste. This for example includes in some countries a significant amount of wood, although not derived directly from forests, yet passed from a phase of waste disposal, for which an energy recovery path is well traced (Fig. 2), from which it may not be easy to subtract material for other uses (Nuss et al. 2013). The elected destination for the obtained materials is often related with the use as combustible material, therefore for the production of briquettes, pellets, etc., In this sense, concern for particle emissions for this kind of use has been often been raised (Johansson et al. 2013). On the other side, this leaves unresolved the issue of how to use ashes generated

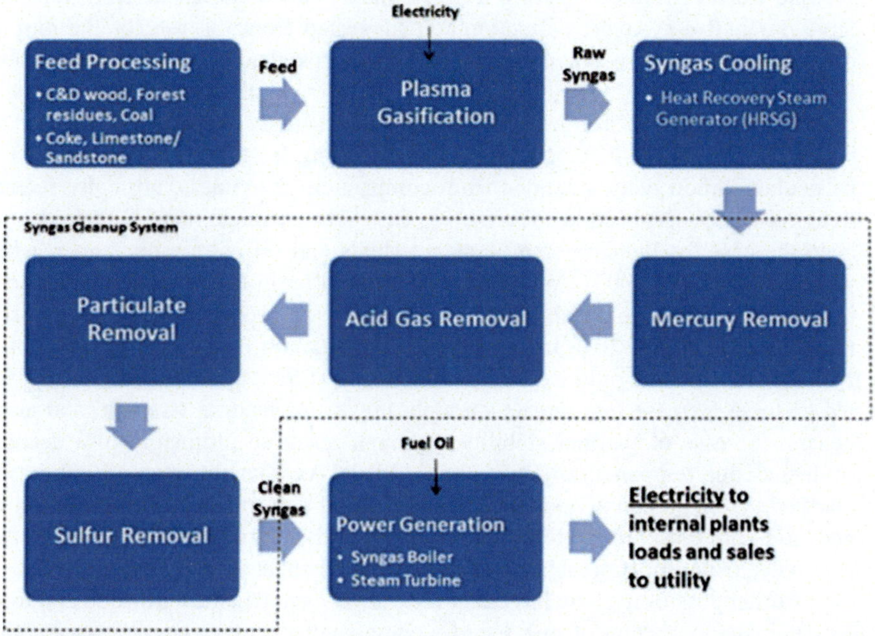

Fig. 2 "Energy recovery" path for biomass from construction and demolition waste (CDW) (Nuss et al. 2013)

from combustion: a frequently explored option concerned their introduction into cementitious matrices (Ukrainczyk et al. 2016; Martirena and Monzó, 2018). The biomass can also play the role to absorb some chemicals, such as it is the case for 2,4.6-trichlorophenol inserted in kaolinite clay matrix, used to decontaminate effluents from water treatment plants: in the specific case (Olu-Owolabi et al. 2017), *Carica papaya* and pine-cone seeds were used.

However, the specification as "renewable" suggests also the fact that this biomass derives from plants, hence increases the likeliness of it being shifted from energy recovery at end-of-life to more sustainable scenarios. These would include recycling in other materials, therefore possibly resulting in a mitigation of the impact of global warming, especially in the case that the use of renewable biomass is considered with limited processing, therefore possibly without any prior combustion process. The renewability characteristics allow consequently the biomass playing different possible roles, among which also the manufacturing of materials and products can also be considered (Dhillon and von Wuehlisch 2013).

Since the main option still considered for biomass is still energy recovery from combustion and of course generation of ashes, in some cases its introduction in materials was performed only after its transformation into carbon. In particular, this is interesting because it may involve the transformation into active carbons or in any case in materials that will increase the porosity of the hosting materials, may be also according to a hierarchical structure (Pan et al. 2014). This is particularly suitable when carried out starting from secondary raw materials derived from very large productive systems: an example is suggested from the use for the purpose of cotton stalks, where active carbons obtained proved adapted for dye absorption (Fathy et al. 2010). A number of proposals have been raised also for the waste of other crops, diffused in a large portion of the world, namely rice husk biomass, among which the fabrication of biocomposites is also considered though (Pode 2016). In particular, carbon black obtained from combustion of biomass, normally referred to as "biochar", has been introduced e.g., in a petrochemical matrix, such as it has been the case for bamboo stem, coconut shells and oil palm empty fiber, which were first heated to 700 °C, and then used as the filler in epoxy resin: this led to an improvement in thermal stability and also in flexural properties of the composites (Abdul Khalil et al. 2010). In contrast, the introduction in a vinylester resin of 10 wt.% of carbon black produced from combustion at 600 °C of jatropha seeds shells led to some increase into tensile strength, not so for flexural strength, and also a general increase of thermal stability, although resulting ultimately in a decrease of degradation temperature (Aprilia et al. 2014). Another study on oil palm ashes clarified that in an epoxy resin tensile and flexural performance of the resin do not decrease until the introduction of ashes result in their agglomeration, which is the case when their amount reaches 30% by weight (Rizal et al. 2019).

A further possibility, which is also widely investigated in the agro-food and forest sector, is the extraction of one specific component for other possible production, possibly including materials for further use. In particular, cellulose can be extracted from food waste rich in polysaccharides with the idea to produce bio-hydrogen (Yasin et al. 2013), or lignin reprocessed for the production of materials, chemicals and fuels, in the still new yet already diffused concept of "biorefineries" (Stafford et al. 2020). A possibility to apply lignin, more relevant for the use in material, is the possibility to use it as a curing agent for thermosetting polymers, such as epoxy (Abdul Khalil et al. 2011). Crosslinking agents for epoxies are also been proposed to be extracted from biomass in the form of unspecified protein hydrolizates (Mekonnen et al. 2013). Among other chemicals obtainable from agrowaste is ethanol, which can also possibly lead to the production of bio-based polymers, structurally similar to the ones obtained from oil, such as bio-polyolefins (polyethylene, polypropylene, etc.) (Bhatia et al. 2012).

1.2 Other Possibilities for Reprocessing Biomass Waste

There are more complex ways, also in chemical terms, to deal with the matter, for example the production of biopolymers from glucose-based raw materials obtained from biomass. In particular, some types of agricultural and forestry waste showed a significant yielding in providing sufficient amounts of polysaccharides, in particular gluten, starch and proteins, to be potentially able to carry out the production of biopolymers (Santulli et al. 2015). A possibility in this regard is offered by fungal enzymes, where in particular the lignocellulosic biomass can be reprocessed to lead to glucose and xylose (Fig. 3) and then possibly re-polymerize the polysaccharide based polymers, such as poly(lactic acid) from glucose, or poly(hydroxybutirate) (PHB) from specific bacterial strains from xylose (Garcez Lopes et al. 2009).

A comprehensive review of studies performed from algal biomass is for example reported in Noreen et al. 2016, where emphasis is put on the production of bio-based polyesters, not only for structural applications, also e.g., for drug release and as biomedical scaffolds. Also, lipids from food waste and algal biomass hydrolizates proved suitable for application as surfactants and plasticizers by the transesterification with glycerol (Pleissner et al. 2015). In other cases, developed materials are intended for absorption of toxic chemicals, such as heavy metals, both by fungal biomass (Rashid et al. 2014) or algal biomass (He and Chen 2014). In this case, biomass, especially when porous or sponge-like, can be used as the ideal environment for the development of fungal mass, in view of the removal also of chlorinated compounds (Iqbal et al. 2005). Other possibilities are in absorbing dyes, such as Crystal Violet using peanut hulls (Tahir et al. 2017) and methylene blue using orange peel strips (Jain et al. 2010) from water.

To try to summarize all the aforementioned approaches in a concept, the above studies concern the use of biomass, in some cases also for production of materials for specific uses. However, the materials obtained most frequently do involve neither the requirement of durability, nor a particularly interesting structural performance as the replacement of conventional polymer reinforced composites. Rather, they are manufactured with the idea to control biodegradability to occur in a rather short time. In this way, the specific type of biomass waste used, and its production system can also be unknown, or else these solutions are suitable whenever there is no real

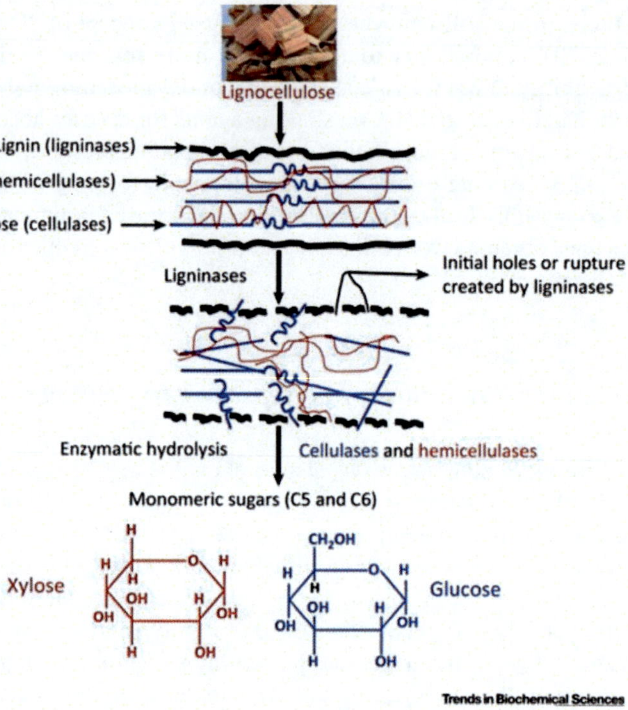

Fig. 3 Enzymatic degradation path of ligno-cellulose biomass down to simple sugars (Gupta et al. 2016)

separation between the different waste inflows that contribute to biomass waste, such as it is normally the case in fields like forest management.

The objective of the present chapter is focusing exclusively on the use of as-received biomass waste, ligno-cellulosic, protein-based or ceramic, from agro-food and forestry industries, with minimal modifications for its introduction into biopolymers, hence not involving combustion before prospected use. The idea of the production of these biocomposites would be disposing of waste into a new material, instead of using it in energy recovery or for specific application. In the case of introduction of ligno-cellulosic biomass waste, this may present some similarities with what has been done over the last three decades using plant fibers, to fabricate materials that have been defined as "natural fiber composites" (NFCs) or "plant fiber composites" (PFCs). On these, a number of recent and comprehensive reviews also available, in view of the use of natural fibers as the replacement of glass fibers or wood fibers, such as proposed by Madsen and Gamstedt (2013), and Ahmad et al. (2015). Other review papers have also been published, which otherwise concentrates on more specific aspects of natural fiber composites, for example on their mechanical properties (Elanchezhian et al. 2018), dynamical properties (Saba et al. 2016), or on liquid molding (Francucci and Rodriguez 2016).

However, the use of waste biomass discussed in this chapter does not involve the manufacturing of textile products, yet the use of the material as random fibers, or particles, in order to minimize the preparation of waste for introduction into biocomposites. It is quite obvious that this mode of application presents some critical aspects, the main being the possibility to behave as filler rather as a proper reinforcement in the biocomposite and in some cases the risk that only a limited amount of it is introduced in the material. On the other side, using it as much as possible as-received reduces mechanical energy consumption and the use of chemicals. Moreover, it preserves the characteristics of waste, which may include for example lightweight, proneness to water absorption, etc.

1.3 Biomass Waste for Introduction into Polymers

The sections of the chapter are organized according to the type of matrix used: a first possibility can be to use a non-biodegradable matrix (Sect. 2), such as the ones based on petrochemical polymers. This usually only serve to a first orientation in the possible re-use of the waste biomass in new materials, and it is often aimed at evolving into studies with the use of more sustainable matrices.

Among the latter, the chapter distinguishes between those that are of industrial origin, such as poly(lactic acid) (PLA), to which Sect. 3 is devoted, or expressly formulated, such as the category of thermoplastics starches (TPS), dealt with in Sect. 4. This will result in a more sustainable end-of-life, given the fact that often the introduction of waste into traditional oil-based polymers does not result in their easy recyclability, if not directly hindering it. Plus, the scope for the materials will not be disposal as combustible and not either chemicals removal, but the production of materials for further and possibly durable use. A particular emphasis will also be given to those composites that allow offering value, also in economical terms, and visibility to the agricultural waste introduced, resulting therefore in a real upcycling process of the material itself. The cases in which specific agro-food waste is used, with particular interest, are related in Sect. 5: in this situation, waste biomass is not only introduced in the material, yet also communicated in the way of offering particular situations, aimed at making use for the composite into the same productive system that generated that particular type of waste. This generates other complexities, such as it for application in food system, linked to hygienic if not anti-bacterial characteristics, which on the other side represent challenges for further use of this type of waste in more added-value materials.

2 Studies on Non-biodegradable Matrices

The introduction of biomass waste in a thermosetting matrix, in most cases epoxy, would result into a better appreciation of the potential for biomass waste material for its use in composites. It can be suggested that the final aim might be evolving towards the use of biopolymers. On the other side, the performance of these is not matching, at least at the moment, the one of conventional thermosetting resins, and biodegradability may for some applications (furniture, automotive, boatbuilding, etc.) be detrimental rather that advisable. The consequence is that applying epoxy or other thermosetting resins as the matrix for these composites does not only serve as a reference, but can potentially indicate also a mode for further use, which reduces the environmental impact in any case. In particular, it is worth noting that more recently bio-based epoxy resins, as well as curing agents, have also been developed, in which part of the carbon derives from plant sources (Baroncini et al. 2016).

A number of studies performed in this sense are reported in Table 1.

Table 1 Some studies on introduction of biomass in non biodegradable matrix composites

Biomass waste	Matrix	Main results	References
Sea shells	Bio-based epoxy	Up to 30 wt.% sea shells produces increase of flexural modulus	Fombuena et al. (2014)
Banana, sugarcane, wheat husk, coir, groundnut	Styrenated-methacrylate-epoxy	Problems due to random orientation of the biomass	Patel et al. (2015)
Date palm seeds	Epoxy with polyamidic hardener	Homogeneous distribution, yet not full impregnation	Ruggiero et al. (2016)
Waste hemp fibers	Diamine-benzoxazine	With 30 wt.% waste increased mechanical and impact performance and reduced curing temperature	Dayo et al. (2017)
Sunflower husk, hazelnut shell, walnut shell	Conventional epoxy	Complex in prediction of mechanical properties depending on the filler characteristics, plasticization resulting from sunflower husk	Barczewski et al. (2019)
Bagasse, oil palm, coconut, cornhusk, groundnut shell and rice husk fibers	Vinylester	Decrease of tensile strength, increase of hardness up to 10 wt.% fibers	Ogah and James (2018)

In contrast, the introduction into thermoplastic non-biodegradable polymers, such as polypropylene (PP), revealed that their thermal stability is increased by the presence of biomass: this has been studied using *Eucalyptuys globulus*, Norway spruce, energy grass, *Brassica rapa*, pine cones, and grape seeds (Parparita et al. 2014). This can be attributed from the combined effect of the physical barrier created by PP matrix, which prevents at low temperatures the transport of volatile products from biomass out of the composites, and to the consumption of the radicals resulting from PP degradation through reactions with structures containing oxygen atoms from lignin degradation. In this case, widely available biomass, such as rice husk, has also been used in combination with recycled thermoplastic polymers, such as high density polyethylene (HDPE), basically for a considerable reduction in price and the possibility to enhance the profile of application of recycled polymer. Materials obtained from this kind of operations are often referred to as biomass-filled plastic composites. In some cases they have been subjected to more moderate thermal treatment than total combustion, hence e.g., torrefaction, which was applied on polyamide composites filled with sunflower hulls and flax shives, leading to a reduced variability of the mechanical performance of the final composite (Vold et al. 2015). In accordance to the higher mechanical performance required with respect to what expected by a recycled polymer, have also been investigated e.g., on the interaction between creep and low cycle fatigue during their service (Guo et al. 2018).

3 Studies on Industrial Biopolymers

Among the industrial biopolymers, poly(lactic acid) (PLA) proved particularly suitable for the introduction of fillers with limited treatment. Of course, passing from thermosetting resins to biopolymers, it can be easily suggested that some criticalities in the reduction of performance, especially above a given level of introduced waste, can be revealed. This occurs in particular because PLA presents a thermal behavior that is e.g., substantially limited by the occurrence of a not very high glass transition temperature. This can be solved e.g., by restricting the movement of the chains at T_g by adding a crosslinker, or lowering it below ambient temperature by the addition of a plasticizer (Jin et al. 2019). As the result, it is logical to suppose that the influence of adding waste biomass in a non-carbonized state over the thermal properties of PLA may not be negligible. In practical terms, a number of different types of food waste have been attempted for introduction into PLA, with the idea to establish the maximum amount of polymer than can effectively bond the different types of food waste, dried and mashed with simple cutting devices. The data are reported in Fig. 4.

The flexibility of PLA has been proven in multifold situations, for example in the comparison carried out by Spiridon et al. (2016), using different fibers, from celery root (*Apium graveolens*), grape pomace (*Vitis vinifera* Chambourcin), poplar (*Populus Alba* sp.) seed hair and *Asclepias syriaca*. This, depending on the fibers, created a difference in prevalent damage phenomena between pull-out and debonding, the decrease of transparency and water contact angle: in the case of grape

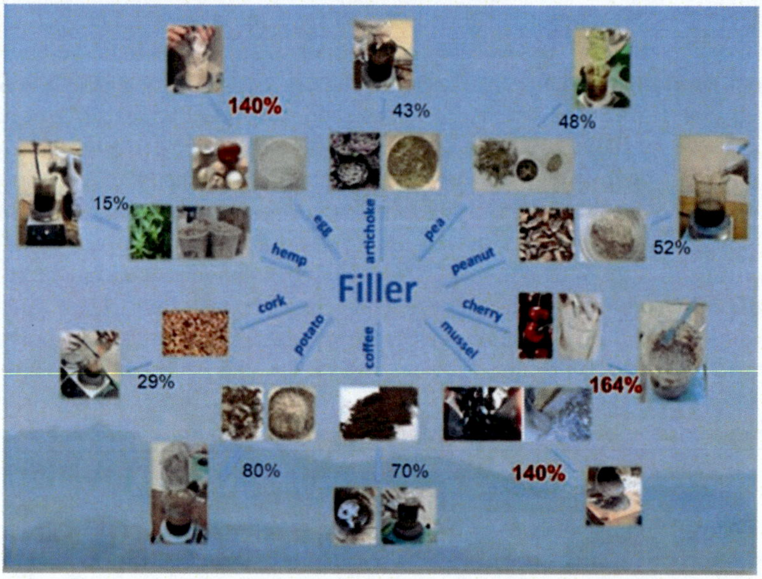

Fig. 4 Maximum amounts for introduction of different types of food waste and other biomass waste, such as hemp shives, taking poly(lactic acid) amounts as equal to 100 (Cecchi et al. 2019)

pomace, some anti-microbial activity against *Escherichia coli* and *Staphylococcus aureus* after 24 h incubation time was also encountered.

A possibility that has been explored for the introduction of biomass waste into biopolymers is the extraction of highly crystalline fractions in the form of cellulose nanowhiskers (CNW-S). In the case of this operation, the main concern remains, apart from the intensive chemical treatment needed, also the possibility of a low yielding at the end of the process. This has been performed on oil palm biomass, resulting in a higher tensile strength of the nanocomposites after adding 3 vol.% of CNW-S in a poly(lactic acid) (PLA) matrix, with also an increase of glass transition temperature (T_g), cold crystallization temperature (T_{cc}) and melting temperature (T_m), as revealed by DCS analysis (Haafiz et al. 2016).

In terms of cost reduction, though, the availability of local biomass can represent another interesting option for introduction as filler in PLA. An example of this procedure is reported by Salak et al. (2015), using Kudzu (*Pueraria lobata*) biomass, obtained in the form of char after thermal pyrolysis.

Interesting results were also obtained on the improvement of the tensile performance of poly(caprolactone) (PCL) by the addition of biomass from agricultural waste, such as date palm sheath fibers (Dhakal et al. 2018). This was to be attributed though in particular to the filler geometry, since in another case, dealing with harder and less elongated ones, such as olive stones, date seeds, and even wheat bran, tensile strength was reduced instead and especially in the case of wheat bran, water uptake was even doubled. On the other side, no thermal decomposition was revealed, which is promising in terms of material processing (Hejna et al. 2020).

4 Studies on Self-produced Polymers

Instead of using industrially produced biopolymers, it is also possible to develop starch-based polymers for example by modifying starch particles with various processes. In particular, this involves the production of a thermoplastic starch (TPS), being based on the plasticization of starch with e.g., glycerol, which results, depending on their respective content into the lowering of glass transition temperature (T_g) at temperatures below ambient temperature, so that in common use the material would be in a rubbery state. It has been proved that a glycerol/starch over 2 is normally sufficient for the purpose, of course yielding a material with lower stiffness yet more ductile with respect to the ones with lower amount of glycerol (Mali et al. 2005). TPS can be further crosslinked, on the example of industrial biopolymers, such as PLA, by the addition of acids, such as acetic acid and lactic acid (Majzoobi and Beparva 2014), or citric acid (Reddy and Yang 2010). Another possibility for crosslinking and altering the characteristics of starch granules, so to make them more suitable for the possible production of biopolymers, able eventually to allocate waste biomass, is their plasma treatment: on this aspect, a comprehensive review is available (Zhu 2017).

The combination of starch-plasticizer-acid has facilitated the accommodation of biomass waste of different kinds, for example in a self-produced milk-starch matrix treated with glycerol (Troiano et al. 2018), which was able to effectively include peanut hull fragments, improving hardness of the composite: the effect is clearly visible in the micrograph in Fig. 5. Also the combined presence of different waste types of various origins, namely hemp shives and waste clay from pottery production, has been considered for introduction in self-produced TPS (Santulli et al. 2019). In

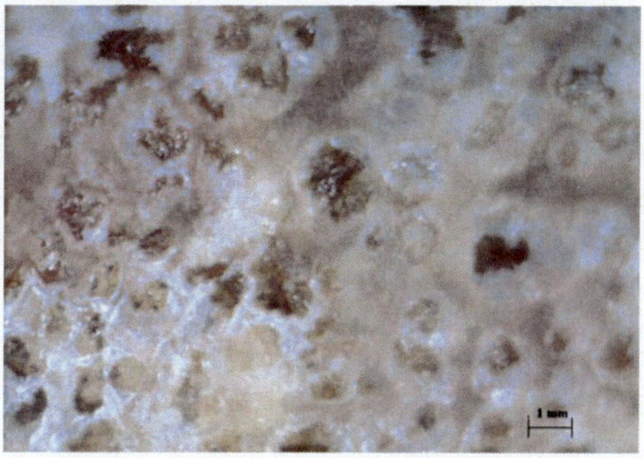

Fig. 5 Introduction of peanut hull fragments into a starch-milk matrix (Troiano et al. 2018)

particular, an option frequently considered is the extraction of cellulose whiskers, especially for fibers not available or suitable for the production of yarns or fabrics, for their geometrical variability: this has been summarized into a specific review by Pérez-Pacheco et al. (2016). The number of ligno-cellulosic materials used is very large, however, up to waste from garden pruning, such as for example Opuntia fibers from garden waste (Scognamiglio et al. 2019) or the respective mucilage, which offered a considerable plasticization potential added to the one of TPS (Scognamiglio et al. 2020). A study on malic acid modification of starch particles allowed comparing the results from the introduction of microcrystalline cellulose with two types of cellulose-based biomass, namely *Asclepias syriaca* L. and *Populus alba* L. seed hair (Bodirlau et al. 2014). The outcome of this study indicated a slight increase in water resistance and tensile strength by cellulose-based biomass, although conversely the stiffness was somehow reduced.

5 Conclusions and Future Perspectives

The introduction of waste biomass for the production of biocomposites represents a very important challenge, which may have interest for the creation of more sustainable materials, although this can only be discussed from case to case due to a number of specific aspects. First of all, waste biomass is supposed to be subtracted from the environmental balance to materials that are disposed by landfilling, but are normally not compostable and often biodegradable with a very low rate over time, or that could be disposed by combustion and energy recovery, therefore leaving carbon-rich residues. In some cases, these materials have been used after combustion in combination with polymers, and this offers interesting possibilities, such as exploiting the inherent presence of porosity in carbon. However, a less energy consuming approach would probably appear using the biomass as much as possibly as-received. Here, a number of distinctions need to be made between the types of biomass used (botanical species and type of waste used e.g., seeds, leaves, branches, etc.) and the family of polymers in which this is introduced. It is of course possible to tend only to the incorporation into a traditional oil-based polymer resin, such as epoxy, this is interesting in terms of assessing the potential of the biomass for use. However, most environmentally friendly options are in its inclusion in biopolymers of industrial development, such as poly(lactic acid) and polycaprolactone, or in, mainly starch-based, self-developed polymer matrices.

The future perspective is likely to involve either a more effective disposal of biomass into materials, without affecting biodegradation at end-of-life, either due to further developments in sustainable biopolymers or to progresses into the synthesis of oil-based polymers (or bio-based substitute of these) able to be disintegrated and mineralized by specific bacterial strains.

References

Abdul Khalil HPS, Firoozian P, Bakare IO, Akil HM, Noor AM (2010) Exploring biomass based carbon black as filler in epoxy composites: flexural and thermal properties. Mater Des 31:3419–3425

Abdul Khalil HPS, Marliana MM, Issam AM, Bakare IO (2011) Exploring isolated lignin material from oil palm biomass waste in green composites. Mater Des 32:2604–2610

Ahmad F, Choi HS, Park MK (2015) A review: natural fiber composites selection in view of mechanical, light weight, and economic properties. Macromol Mater Eng 300:10–24

Aprilia MAS, Abdul Khalil HPS, Bhat AH, Dungani R, Hossain MS (2014) Exploring material properties of vinyl ester biocomposites filled carbonized jatropha seed shell. BioResources 9(3):4888–4898

Avitabile V, Camia A (2018) An assessment of forest biomass maps in Europe using harmonized national statistics and inventory plots. for Ecol Manage 409:489–498

Barczewski M, Sałasińska K, Szulc J (2019) Application of sunflower husk, hazelnut shell and walnut shell as waste agricultural fillers for epoxy-based composites: a study into mechanical behavior related to structural and rheological properties. Polym Testing 75:1–11

Baroncini EA, Yadav SK, Palmese GR, Stanzione JF III (2016) Recent advances in bio-based epoxy resins and bio-based epoxy curing agents. J Appl Polym Sci 133(45):44103

Bhatia L, Johri S, Ahmad R (2012) An economic and ecological perspective of ethanol production from renewable agro waste: a review. AMB Expr 2:65. https://doi.org/10.1186/2191-0855-2-65

Bodirlau R, Teaca CA, Spiridon I (2014) Green composites comprising thermoplastic corn starch and various cellulose-based fillers. BioResources 9(1):39–53

Cecchi T, Giuliani A, Iacopini F, Santulli C, Sarasini F, Tirillò J (2019) Unprecedented high percentage of food waste powder filler in poly lactic acid green composites: synthesis, characterization, and volatile profile. Environ Sci Pollut Res 26(7):7263–7271

Dayo AQ, Xu Y-L, Zegaoui A, Nizamani AA, Wang J, Zhang L, Liu W-B, Shah AH (2017) Reinforcement of waste hemp fibres in aromatic diamine-based benzoxazine thermosets for the enhancement of mechanical and thermomechanical properties. Plast. Rubber Compos 46(10):442–449

Dhakal H, Bourmaud A, Berzin F, Almansour F, Zhang Z, Shah DU, Beaugrand J (2018) Mechanical properties of leaf sheath date palm fibre waste biomass reinforced polycaprolactone (PCL) biocomposites. Ind Crops Prod 126:394–402

Dhillon RS, von Wuehlisch G (2013) Mitigation of global warming through renewable biomass. Biomass Bioenerg 48:75–89

Elanchezhian C, Vijaya Ramnath B, Ramakrishnan G, Rajendrakumar M, Naveenkumar V, Saravanakumar MK (2018) Review on mechanical properties of natural fiber composites. Mater Today Proc 5:1785–1790

Fathy NA, Girgis BS, Khalil LB, Farah JY (2010) Utilization of cotton stalks-biomass waste in the production of carbon adsorbents by KOH activation for removal of dye-contaminated water. Carbon Letters 11(3):224–234

Fombuena V, Bernardi L, Fenollar O, Boronat T, Balart R (2014) Characterization of green composites from biobased epoxy matrices and bio-fillers derived from seashell wastes. Mater Des 57:168–174

Francucci G, Rodriguez E (2016) Processing of plant fiber composites by liquid molding techniques: An overview. Polym Compos 37:718–733

Garcez Lopes MS, Rocha RCS, Zanotto SP, Cabrera Gomez G, Silva L (2009) Screening of bacteria to produce polyhydroxyalkanoates from xylose. World J Microbiol Biotechnol 25(10):1751–1756

Guo Y, Li D, Zhu S, Chen Y (2018) Interaction of low cycle fatigue and creep in biomass-filled plastic composites. BioResources 13(2):3250–3258

Gupta VK, Kubicek CP, Berrin J-G, Wilson DW, Couturier M, Berlin A, Filho EXF, Ezeji T (2016) Fungal enzymes for bio-products from sustainable and waste biomass. Trends Biochem Sci 41(7):633–645

Haafiz MK, Hassan A, Abdul Khalil HPS, Nurul Fazita MR, Islam MS, Inuwa IM, Marliana MM, Hussin MH (2016) Exploring the effect of cellulose nanowhiskers isolated from oil palm biomass on polylactic acid properties. Int J Biol Macromol 85:370–378

He J, Chen JP (2014) A comprehensive review on biosorption of heavy metals by algal biomass: materials, performances, chemistry, and modeling simulation tools. Biores Technol 160:67–78

Hejna A, Sulyman M, Przybysz M, Saeb MR, Klein M, Formela K (2020) On the correlation of lignocellulosic filler composition with the performance properties of poly(ε-Caprolactone) based biocomposites. Waste and Biomass Valorization 11:1467–1479

Iqbal M, Saeed A, Edyvean RGJ, O'Sullivan B, Styring P (2005) Production of fungal biomass immobilized loofa sponge (FBILS)-discs for the removal of heavy metal ions and chlorinated compounds from aqueous solution. Biotech Lett 27:1319–1323

Jain N, Basniwal Suman RK, Srivastava AK, Jain VK (2010) Reusable nanomaterial and plant biomass composites for the removal of Methylene Blue from water. Environ Technol 31(7):755–760

Jin F-L, Hu R-R, Park S-J (2019) Improvement of thermal behaviors of biodegradable poly(lactic acid) polymer: a review. Compos B 164:287–296

Johansson LS, Tullin C, Leckner B, Sjövall P (2003) Particle emissions from biomass combustion in small combustors. Biomass Bioenerg 25(4):435–446

Madsen B, Gamstedt EK (2013) Wood versus plant fibers: similarities and differences in composite applications. Adv Mater Sci Eng 2013, Article ID 564346. https://doi.org/10.1155/2013/564346

Majzoobi M, Beparva P (2014) Effects of acetic acid and lactic acid on physicochemical characteristics of native and cross-linked wheat starches. Food Chem 147:312–317

Mali S, Grossmann MVE, Garcia MA, Martino MN, Zaritzky NE (2005) Mechanical and thermal properties of yam starch films. Food Hydrocolloids 19:157–164

Martirena F, Monzó J (2018) Vegetable ashes as supplementary cementitious materials. Cem Concr Res 114:57–64

Mekonnen T, Mussone P, Alemaskin K, Sopkow L, Wolodko J, Choi P, Bressler D (2013) Biocomposites from hydrolyzed waste proteinaceous biomass: mechanical, thermal and moisture absorption performances. J Mater Chem A 1:13186–13196

Noreen A, Mahmood Zia K, Zuber M, Ali M, Mujahid M (2016) A critical review of algal biomass: a versatile platform of bio-based polyesters from renewable resources. Int J Biol Macromol 86:937–949

Nuss P, Gardner KH, Jambeck JR (2013) Comparative Life Cycle Assessment (LCA) of Construction and Demolition (C&D) derived biomass and U.S. Northeast forest residuals gasification for electricity production. Environ Sci Technol 47:3463−3471

Ogah AO, James TU (2018) Mechanical behavior of agricultural waste fibers reinforced vinyl ester bio-composites. Asian J Phys Chem Sci 5(1), Article no AJOPACS.35841

Olu-Owolabi BI, Alabi AH, Diagboya PN, Unuabonah EI (2017) Düring R-AM, Adsorptive removal of 2,4,6-trichlorophenol in aqueous solution using calcined kaolinite-biomass composites. J Environ Manage 192(1):94–99

Pan J, Pan J, Cheng X, Yan X, Lu Q, Zhang C (2014) Synthesis of hierarchical porous silicon oxycarbide ceramics from preceramic polymer and wood biomass composites. J Eur Ceram Soc 34:249–256

Parparita E, Nistor MT, Popescu M-C, Vasile C (2014) TG/FTIR/MS study on thermal decomposition of polypropylene/biomass composites. Polym Degrad Stab 109:13–20

Patel JV, Patel JP, Bhatt RD, Parsania PH (2015) Mechanical and electrical properties of jute-biomass-styrenated methacrylate epoxy resin sandwich composites. J Sci Ind Res 74:577–581

Pérez-Pacheco E, Canto-Pinto JC, Moo-Huchin VM, Estrada-Mota IA, Estrada-León RJ, Chel-Guerrero L (2016) Thermoplastic starch (TPS) cellulosic fibers composites: mechanical properties and water vapor barrier: a review. In: Composites from renewable and sustainable materials, Chapter 5, Intechopen 2016. https://doi.org/10.5772/65397

Pleissner D, Yan Lau K, Zhang C, Ki Lin CS (2015) Plasticizer and surfactant formation from food-waste and algal biomass-derived lipids. Chemsuschem 8:1686–1691

Pode R (2016) Potential applications of rice husk ash waste from rice husk biomass power plant. Renew Sustain Energy Rev 53:1468–1485

Rashid A, Nawaz Bhatti H, Iqbal M, Noreena S (2014) Fungal biomass composite with bentonite efficiency for nickel and zinc adsorption: a mechanistic study. J Eur Ceram Soc 34(2):249–256

Reddy N, Yang Y (2010) Citric acid cross-linking of starch films. Food Chem 118:702–711

Rizal S, Fizree HM, Saurabh CK, Gopakumar DA, Aprilia NAS, Hermawan D, Banerjee A, Fazita MRM, Haafiz MKM, Khalil Abdul HPS (2019) Value-added utilization of agro-waste derived oil palm ash in epoxy composites. Journal of Renewable Materials 7(12):1269–1278

Ruggiero A, Valasek P, Müller M (2016) Exploitation of waste date seeds of *Phoenix dactylifera* in form of polymeric particle biocomposite: Investigation on adhesion, cohesion and wear. Compos B 104:9–16

Saba N, Jawaid M, Alothman OY, Paridah MT (2016) A review on dynamic mechanical properties of natural fibre reinforced polymer composites. Constr Build Mater 106:149–159

Salak F, Uemura S, Sugimoto K (2015) Thermal pretreatment of Kudzu biomass (*Pueraria lobata*) as filler in cost-effective PLA biocomposite fabrication process. Polym Eng Sci 55(2):341–348

Santulli C, Puglia D, Kenny JM (2015), Biopolymers directly developed from biomasses from agrowaste sources. In: Thakur VM, Thakur MK (eds) Handbook of sustainable polymers: structure and chemistry , Chapter 13, Pan Stanford, ISBN 9789814613569

Santulli C, Vita A, Scardecchia S, Forcellese A (2019) A material proposed for re-use of hemp shives as a waste from fiber production. Mater Today Proc. https://doi.org/10.1016/j.matpr.2019.11.063

Scognamiglio F, Mirabile Gattia D, Roselli G, Persia F, De Angelis U, Santulli C (2019) Thermoplastic starch films added with dry nopal (Opuntia Ficus Indica) fibres, Fibers 7, paper no 99

Scognamiglio F, Mirabile Gattia D, Roselli G, Persia F, De Angelis U, Santulli C (2020) Thermoplastic Starch (TPS) films added with mucilage from opuntia ficus indica: mechanical, microstructural and thermal characterization. Materials 13(4), January, 1000

Spiridon I, Darie-Nita RN, Hitruc GE, Ludwiczak J, Cianga Spiridon IA, Niculaua M (2016) New opportunities to valorize biomass wastes into green materials. J Clean Prod 133:235–242

Stafford W, De Lange W, Nahman A, Chunilall V, Lekh P, Andrew J, Johakimu J, Sithole B Trotter D (2020) Forestry biorefineries. Renew Energy 154:461-475

Tahir N, Nawaz Bhatti H, Iqbal M, Noreena S (2017) Biopolymers composites with peanut hull waste biomass and application for Crystal Violet adsorption. Int J Biol Macromol 94A:210–220

Troiano M, Santulli C, Roselli G, Di Girolami G, Cinaglia P, Gkrilla A (2018) DIY bioplastics from peanut hulls waste in a starch-milk based matrix. FME Transactions 46(4):503–512

Ukrainczyk N, Vrbos N, Koenders EAB (2016) Reuse of woody biomass ash waste in cementitious materials. Chem Biochem Eng Q 30(2). https://doi.org/10.15255/CABEQ.2015.2231

Vold JL, Ulven CA, Chisholm BJ (2015) Torrefied biomass filled polyamide biocomposites: mechanical and physical property analysis. J Mater Sci 50:725–732

Yasin NHM, Mumtaz T, Hassan MA, Abd Rahman NA (2013) Food waste and food processing waste for biohydrogen production: a review. J Environ Manage 130:375–385

Zhang G, Long W(2010), A key review on emergy analysis and assessment of biomass resources for a sustainable future. Energy Policy 38(6):2948-2955

Zhu F (2017) Plasma modification of starch. Food Chem 232:476–486

Current and Future Challenges of Bio-Based Adhesives for Wood Composite Industries

Haisong Wang, Xiaoyu Gong, and Jingwei Gong

Abstract Traditional trialdehyde adhesives are the most prominent adhesives in wood industry, as well as they are also the main reason for the release of formaldehyde during the use of wood glue products such as wood-based panels. There is a growing demand for adhesives prepared with environment-friendly from biomass materials to transition toward a more sustainable society. The development and application of new promising adhesives not only involve resource reasonable utilization and environmental protection, but also have great significance to the profound development of adhesive industry. In this chapter, the characteristics, preparation technology, modification methods and current research progress of several bio-based adhesives (lignin, tannin, soybean protein and starch) are reviewed, the application status and the existing problems of bio-based wood adhesive are analyzed, and the development direction of bio-based wood adhesives is prospected.

Keywords Bio-based adhesives · Wood composite · Lignin-based adhesives · Tannin-based adhesives · Soybean protein adhesives · Starch-based adhesives

1 Market and Performance Requirements of Adhesives for Wood Composite Industries

1.1 Adhesives for Wood-Based Panels

As one of the largest green economies in recent years, the progress of wood composite industries in the national sustainable development would bring forward a new impetus, particularly the wood-based panel development in industry. Given the environmentally friendly development of forestry, the wood composite is a beneficial method to improve the utilization of wood resources effectively. In 2014, China and

H. Wang (✉) · X. Gong · J. Gong
School of Light Industry and Chemical Engineering, Dalian Polytechnic University, Liaoning 116034, China
e-mail: wanghs@dlpu.edu.cn

M. Jawaid et al. (eds.), *Eco-Friendly Adhesives for Wood and Natural Fiber Composites*, Composites Science and Technology,
https://doi.org/10.1007/978-981-33-4749-6_8

Indonesia were recognized as the world's most important suppliers of wood-based panels, accounting for more than 55% of total exports. In 2018, the total output of China's wood-based panels reached up 299 million m^3, among which plywood accounted for 60%, and China remained the largest producer of wood-based panels in the world. It also shows that the global wood-based panel market is expected to grow at a compound annual growth rate (CAGR) of more than 3%, exceeding 35 billion dollars by 2023. Due to the continued economic growth in Asia, the recovery of the North American market, and the expansion of bioenergy targets, the production shows great potential to be further explored in the future. With the rapid development of the international wood composite industries, the emergence of adhesives has attracted enormous attention in saving raw materials and simplifying the processing technologies of wood products.

Based on incomplete data, the production of wood adhesives in the world accounts for 50–60% of the total production of adhesives, meanwhile the production of wood-based panel adhesives accounts for more than 80% of the production of wood adhesives. Thereby, it is of advantage to some extent to develop the wood-based adhesives from the perspective of a steady growth of the world economy.

More high requisitions for curing speed, curing temperature and viscidity have been brought forward.

Currently, a variety of adhesives have been developed. Trialdehyde adhesives, accounting for more than 90%, have been widely used in the market. However, environmental pollution, chemical raw materials prices rising and the urgent demand of energy conservation and emission reduction have become a pressing need to explore adhesives with high performance, functionalization green cleaning and non-toxic broadening wood related research to the functional materials. Therefore, bio-based adhesives have become new requirements for wood-based panel manufacturing industries. At present, the main bio-based adhesives for wood-based panel include lignin-based adhesives, tannin-based adhesives, soybean protein adhesives and starch-based adhesives. From the view of a sustainable improvement for synthesis technology, bio-based adhesives are widely used for its remarkable and promising applications in the field of bioscience.

1.2 Wood Adhesives for Decorative

Plywood in the market of decoration industry is not as good as that in the wood composite industry, but its development potential is not underestimated. Currently, the number of populations has increased significantly, and land resources are more and more tight, leading to the buildings have begun to multi-story and high-rise development, so people have higher and higher requirements for building decoration materials with the lightness and high strength characteristics. In today's society where new materials are constantly emerging, the reason why wood materials still play an extremely important role in building decoration is that it has incomparable advantages over other materials, such as irreplaceable natural properties, typical green materials and excellent physical and mechanical properties. Moreover, in order to take the sustainable development path, building decoration materials has been considered a trend, with green, energy saving, low-carbon and environmentally friendly and the market of building decoration materials is absolutely inseparable from these standards in the future. Only in this way can people's living standard be effectively improved and it is not difficult to foresee that the decoration industry in the world is also a "sustainable resources, evergreen format" industry, and green wood furniture and wood decoration will inevitably become a trend of decoration professions, and the proportion of wood adhesive for decoration will also increase.

Adhesives used in furniture and the secondary processing of wood-based panels have certain requirements on bond strength, endurance and cold resistance. Now the products used in the market are mainly polyvinyl acetate emulsion (PVAc), acrylate emulsion, hot melt adhesive and other adhesives. PVAc could be widely used in the adhesion of porous materials, especially wood composite products because it has fascinating advantages, such as, non-toxicity, safety, and no environmental pollution. However, drawbacks such as poor water resistance, heat resistance and frost resistance still remain. The advantages of acrylate emulsion lie in its wide sources, pollution-free, and simplicity of synthesis. However, how to improve water resistance is attractive but remains challenging. Some scholars have applied the polymer emulsifier to the synthesis of acrylate copolymer emulsion adhesive to improve the water resistance. Hot melt adhesive as a synthetic adhesive, which has accounted for more than 20% of the total production of synthetic adhesives in developed countries, has emerged for the purpose of artificial board surface decoration and wood sealing due to its excellent properties, including pollution-free, large bonding surface, fast bonding speed, suitable for continuous production, easy to storage and transportation, and has contributed to successful development of environmentally friendly adhesives. Natural, ecological, green, healthy and recyclable manufacturing are still the main trends of decorative adhesives in the current world.

2 Recent Developments of Bio-Based Adhesives for Wood Composite Industries

In recent years, petroleum-based products have innumerous uses in various industries for its cheap availability, but it eventually decreases the availability of fossil fuel which will raise the prices of petroleum-based raw materials. To address the issues regarding environmental pollution, we suggest to use natural adhesives for industrial applications. Bio-based adhesives, as a kind of environmentally friendly adhesive with renewable, have given rise to a whole new and green range of uses.

The development of formaldehyde free environment-friendly adhesives by using cheap renewable resources is the research direction of global wood composite industry adhesives. Overall, the materials of adhesives from biomass alternatives have attracted attention and converting lignin into bio-based adhesives may pave a promising way to substitute petroleum-derived products.

Great potential has been gained in research for past few years, however, bio-based adhesives still have some major disadvantages like high viscosity, low bonding strength and poor water resistance. In this chapter, the research status of lignin-based adhesives, tannin-based adhesives, soybean protein adhesives and starch-based adhesives are discussed.

2.1 Lignin-Based Adhesives

Lignin is the second largest renewable natural resource after cellulose and the only non-petroleum resource of renewable aryl compounds in nature. The annual yield of 6×10^{14} t worldwide makes lignin the largest natural polyhydroxy aromatic compound. Besides, lignin is also the primary byproduct of pulp and paper industries. Of these, cooking waste liquor is comprised of up to 50 million tons of technical lignin worldwide annually. Structural unit consists of phenylpropane linked together through carbon–carbon bonds and ether bonds forming a three-dimensional network polymer and has been also made into other materials such as emulsifiers, dispersants and polyvalent chelating agents as well as other fields. Meanwhile, as compared to the phenolic resin, lignin has relatively good self-adhesive properties (Fig. 1) and can become the substitute raw material of the phenolic resin adhesives prepared to be extensively applied to the adhesive industries. To some extent, the amount of formaldehyde can be decreased. However, the lignin cannot meet the demands modern society for high performance materials due to high steric hindrance, complex structure, small number of active groups and large resistance of aromatic ring structure. In order to increase the reactivity of lignin, activation modification is often used to increase the content of existing active groups or introduce new active groups, and the molecular weight is reduced to a certain extent, which can be used to fabricate different types of adhesives with high performances.

Because the side chain of lignin macromolecules has hydroxyl groups, ether bonds, double bonds and other groups, the α position of the side chain is an active center. Therefore, phenol and its derivatives can be used to modify lignin, which introduces phenol group in order to improve the content of phenolic hydroxyl. the reactivity of lignin could be improved effectively by the introduction of phenolic hydroxyl group, because of the small size and high activity of phenolic hydroxyl group. A ternary gradual copolymerization approach was reported for the preparation of steam explosion lignin phenol formaldehyde adhesives with excellent performance (Zhao et al. 2016), meanwhile the maximum substitution rate of phenol reached 70 wt% and phenolic hydroxyl content of lignin increased by 130%. Raj et al. (2020) reported that 100% phenolic lignin was utilized for the synthesis of LF resin and evaluated the effect of isolation method on the physicochemical properties of lignin through three different approaches-Alkaline hydrolysis, mild acidolysis and organosolv. The results showed that lignin has unique features and unique properties, such

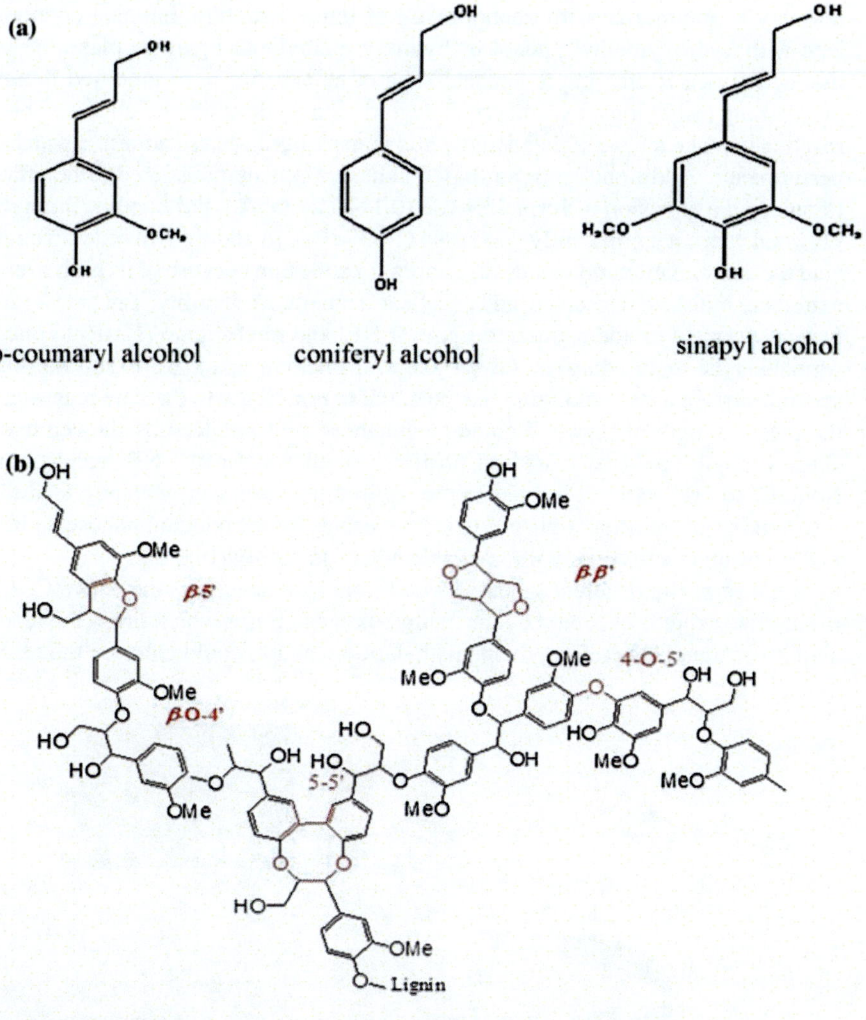

Fig. 1 **a** Three monomers of lignin; **b** Representative connection type of lignin (Naseem et al. 2016)

as a number of G type units of alkali and organic solvent lignin, free ortho-position (C_5) and appropriate molecular weight. These features make lignin suitable in the preparation of adhesives.

With the condition of alkali, the free phenolic hydroxyl group on the benzene ring of lignin reacts with formaldehyde to produce hydroxy methylated lignin by introducing hydroxymethyl groups. Although the active site of lignin might not be increased by hydroxy methylated modification, the introduced hydroxymethyl could not only copolymerize with phenol or C_5 of lignin structure, but also copolymerized with hydroxymethyl phenol or hydroxy methylated lignin to make the reaction activity easier for improvement. Yang et al. (2015a; b, c) modified lignin by bio-refining technology, hydrothermal degradation, hydroxymethylation and phenol reaction to make it have higher activity and prepare lignin-based adhesive with better performance. Additionally, lignin-based adhesive was demonstrated to be efficient to reduce the emission of formaldehyde (0.11–0.24 mg/L), but bonding strength of plywood was not significantly decreased (1.35 MPa). In addition, in order to circumvent the delamination, plywood was chosen for adhesion because of its high strength, extremely moisture resistance, and excellent temperature stability. Yang et al. (2019) first put forward to add nanosized lignin (LNP) and microlignin (LMP) to phenol–formaldehyde resin adhesive. Given plentiful phenylpropane units and the higher specific surface area, nanolignin has potential to contribute to the cure reaction, and the shear strength of plywood could be enhanced by LNP. Results showed that the shear strength could be increased positively with 5 wt% of LNP, which ranging from 8.7 to 10.9 MPa. It is safer to the environment and can generally lead to the crosslinking of traditional phenol wood adhesives by introducing nanoscale lignin, and resulting in adhesives with favorable adhesion ability (Fig. 2).

Because of the different kinds of wood fiber raw materials, the active C_3/C_5 of the aromatic ring is occupied by the methyl oxygen groups which limits the reaction activity of lignin. Therefore, the demethylation reaction could enhance the reaction

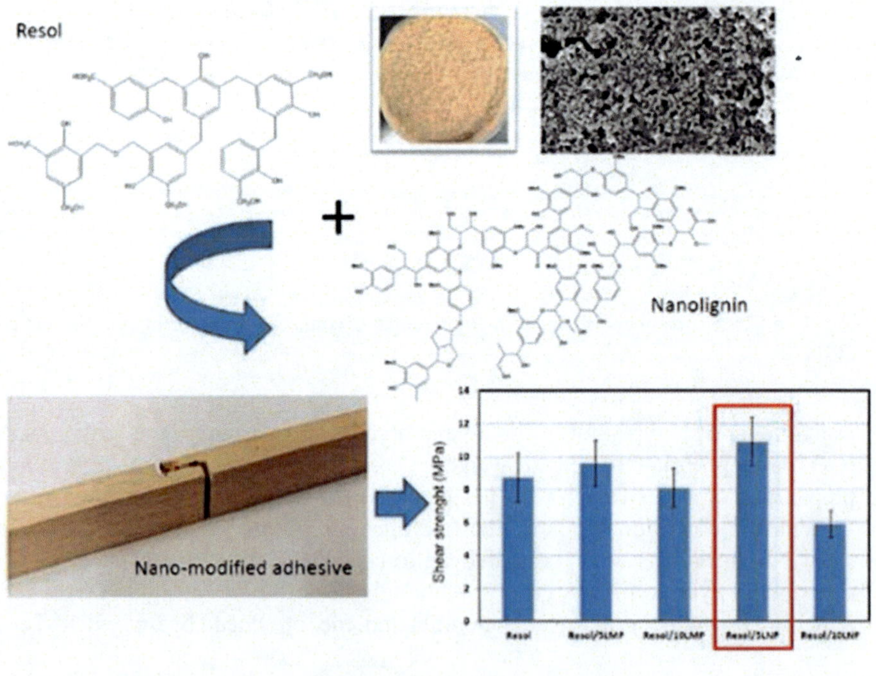

Fig. 2 Reaction diagram of lignin-based adhesives (Yang et al. 2019)

activity of lignin, since the methoxy may be partially or completely converted into phenolic hydroxyl. At present, sulfide demethylation is the most common method for modification of lignin. According to the basic conditions, the process of dismethylation of lignin was used by sulfur, and the results showed that the content of phenolic hydroxyl increased from 3 to 5.6%, the content of methoxide decreased from 12 to 7.6%, and the molecular weight also decreased significantly. In conclusion, the lignin reactivity was significantly improved after modification. Song et al. (2016) used an in-situ generated Lewis acid to demethylate wheat straw alkali lignin. The results showed that the relative content of the phenolic hydroxyl group increased from 5.2 to 16.0%, while the relative value of methoxy group decreased significantly from 0.82 to 0.17% after the demethylation process. Therefore, the demethylated wheat straw alkali lignin was then used to prepare lignin-based phenol formaldehyde adhesives with replacing 60 wt % of phenol, which could open up a new area of applications in the phenol formaldehyde adhesives. Wang et al. (2019) reported that alkali lignin was demethylated with HI and HBr to improve the content of hydroxyl and its reactivity for the phenolic resins preparation. Prepared by copolymerization, lignin-containing resins conformed to the bonding strength standard of external plywood. The demethoxide process not only increased the hydroxyl content but also imparted faster cure times and lower formaldehyde emission of lignin-containing resins. Venkatesagowda and Dekker (2019) studied The dehydrogenation of Kraft lignin (KL) by wood-rot fungi. The O-methyl/methoxyl groups of lignin were removed by demethylation of the action of O-demethylases liberating methanol, and a demethylated KL enriched in vicinal-hydroxyl groups was produced with potential to serve as lignin-based phenol–formaldehyde polymers. Their study clearly revealed the mechanism of lignin demethylation, and it was possible to manifest that the O-methyl/ methoxyl groups had reported similar demethylase activities as model lignin compounds via enzyme activity usually shown by O-demethylases, thus providing critical evidence for the presence of specific fungal enzymes that catalyze such reactions.

In recent years, ionic liquid comprised of anions and cations as a new solvent system can completely dissolve lignin fiber raw materials, greatly promoting the basic researches of lignin conversion and application. In the process of dissolution, lignin can be activated and modified in the state of dissolution of ionic liquid due to its high coordination capacity. Furthermore, lignin could be an excellent candidate for chemical modification material because of its great functionality, low molecular weight and high activity for the development of bio-based materials. Different from other modification methods, it is more advantageous to use lignin modified by ionic liquid to prepare phenolic resin adhesive instead of phenol. The ionic liquid interacts with lignin mainly through hydrogen bond, which is stronger than the interaction between lignin molecules, resulting in lignin dissolves in the ionic liquid. The result suggests that the addition of reverse phase solvent can lead to lignin regeneration (water, ethanol, etc.) (Ji et al. 2012). Younesi-Kordkheili and Pizzi (2017) investigated the physical and mechanical properties of ionic liquid modified lignin-phenol–formaldehyde (LPF) resin, soda bagasse lignin modified by 1-ethyl-3-methylimidazolium acetate ionic liquid plywood, and then, various contents of modified lignin were added as an alternative for phenol in phenolic-formaldehyde resin (PF). The research results showed that the viscosity and solid content of the synthesized resins increase, while the gel time and density by addition of treated lignin decrease. With the increase of lignin content percentage varying from 0 to 20 wt%., the mechanical properties of the sheet could be significantly improved.

Microwave reaction technology provides unique advantages such as fast heating rate, high heat transformation efficiency and uniform heating, to reduce the activation energy needed for the reaction and vividly improve the reaction rate. Microwave-assisted biomass hydrothermal liquefaction is a thermal decomposition process in the presence of water, where heat can be transferred from the biomass interior to the surface by microwave irradiation. This process of preparing liquid fuels and high value-added chemicals has attracted worldwide attention due to its low cost and short reaction time. These characteristics indicate that microwave-assisted hydrothermal liquefaction may be beneficial to the preparation of lignin-PF products. Li et al. (2017) extracted lignin from bagasse with 93% acetic acid solution and rapidly degraded by microwave in hot compressed water with oxalic acid as catalyst to produce phenolic formaldehyde (PF) adhesives and the molecular weight, viscosity, and adhesive strength of the adhesives were found to be lower than those of pure PF adhesives.

2.2 *Tannin-Based Adhesives*

The word tannin has been used extensively to define two different types of compounds which are primarily phenolic in nature: hydrolysable tannins and condensed tannins. The former correlated with mixtures of simple phenols, like pyrogallol and ellagic acid, and esters of a sugar, mainly glucose, with gallic and digallic acids. Contrary to it, condensed tannins, accounts for more than 90% of the world's total commercial tannins (200,000 tons per year) and are chemically and economically more suitable for the preparation of adhesives and resins. Tannin is one of the most successful forest resources to be used as wood adhesive consisting of condensed tannin as the main raw material and curing agent, which has fast curing speed, low price and good sizing performance. It has been proven that selecting appropriate curing agent in preparation of wood adhesive is useful for enhancing the time of cold curing and release of free formaldehyde. Tannin molecules are generally composed of 5–11 sulfonanol units. The two phenolic rings (A and B) in the molecule are held together by a heterocyclic ring. As a polyphenolic compound, tannins have a high nucleophilic property of their sulfonanol unit A ring and can react with formaldehyde and finally form insoluble polymers under the action of alkali catalyst, whose principle is the same as that of phenol and formaldehyde reaction to prepare phenolic resin and also has similar adhesive properties and aging resistance to phenolic resin adhesives. Although considerable efforts are still being devoted to extending other applications to tannin adhesives, high molecular weight, high viscosity, low crosslinking degree, short applicable period and poor physical and mechanical properties are also its main setbacks.

Pizzi team (2006) has had in-depth research on the synthesis and preparation of classic tannin adhesives and been widely used in many countries. The technologies of interest here are the new ones according to the use of hardeners which are either non-emitting or manifestly non-toxic. Two lines of investigation were pursued and their function well as hardeners for a variety of tannin-based adhesives while affording considerable side advantages to the adhesive and to the bonded wood joint. Tannin-modified phenolic resin has some unique features such as high molecular weight, high viscosity, high space resistance, and short production circle. Therefore, it is still very difficult to use concentrated tannins without pretreatment. To address this issue, Liu et al. (2020) carried out the depolymerization of condensed tannins in a mixed aqueous solution of sodium hydroxide and urea. Macromolecular condensed tannins were pretreated with sodium hydroxide/urea solution and depolymerized into small molecule tannins solution. The results showed that the water solution with NaOH/urea = 0.2/0.175 (mol/mol, every 100 g of treatment solution) had a better depolymerization effect on Acacia mangium tannin. With the aim of have a good contribution in innovation of green chemistry, notably, this cheap and environmentally friendly catalyst has been demonstrated to be of great for the synthesis process as a feedstock can be used as a feedstock for the synthesis process; Degradation products do not need to be recovered and purified, which avoids the dehydration and purification of waste liquid and makes it safer for the environment (Fig. 3).

In the wood composite industry, the combination of tannin adhesives and other adhesives may have unexpected effects. Eghtedarnejad and Reza (2015) investigated the possibility of combination of tannin adhesive and natural and renewable adhesive

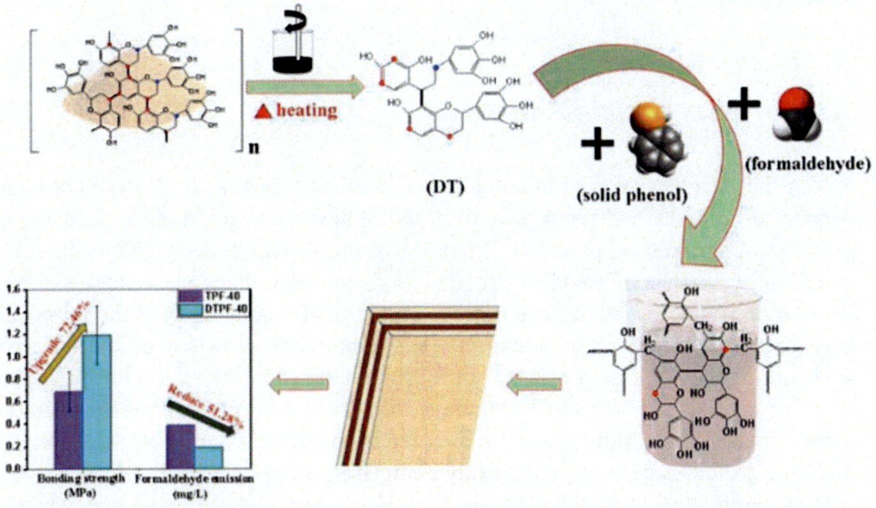

Fig. 3 Degradation mechanism diagram (Liu et al. 2020)

of syrysh and the prepared adhesives were measured to be 35 kg/cm^2 at 210 °C. When the tannin content was 6 + 4%, the level of adhesive resistance was the highest, at 0.89 Mpa. The amount of water absorption and thickness swelling were decreased and reached standard level by increasing the percentage of the dry weight of the syrysh adhesive. For the sake of finding partial substitution for bisphenol A-based commercial epoxy resin (CE), a bio-based epoxy resin was prepared by Shnawa (2020) from eucalyptus tannins via a typical glycidylation reaction with epichlorohydrin. Tannins used here are extracted from the barker of eucalyptus tree and consist mainly of condensed tannins. The tannin-based epoxy resin (TE) can be used as curing accelerator, which has an unsatisfactory effect on curing performance of CE resin, such as curing enthalpy. And up to 20 wt/wt% of CE could be replaced by TE without any significant reduction in the curing behavior of CE. For TE used at 20% loading in the CE, the maximum curing peak is at 116.7 °C and curing enthalpy equals to 127.4 J g^{-1} closed to that of pure CE. In future, the study of chemically modified tannins as unconventional organic fillers prepared by bio-based composites will be a promising perspective. The exploration of mechanical and physical properties of the composites produced has been a subject of environmental interest because of its attractive advantages. Chen et al. (2020) developed a bio-based adhesive from soybean meal and larch tannin that was designed for exterior use. The addition of Phenol hydroxymethylated tannin oligomer (PHTO) in soybean meal-based adhesive decreased the moisture absorption rate of the resultant adhesives, and effectively increased the wet shear strength of the resultant plywood. The wet shear strength and formaldehyde emission of the resultant plywood were 1.04 MPa and 0.08 mg/L, respectively, which both satisfiedthe bonding strength requirements of outdoor plywood and E_0 level. The improvement of adhesion performance was mainly due to the cross-linking structure among PHTO and proteins, as well as the cross-linking of PHTO itself.

2.3 Soybean Protein Adhesives

Protein-based adhesive is a kind of adhesive with protein or its hydrolyzed products, denatured products and plant protein-based adhesives as the main processing by-products of agricultural products. It has abundant sources and good reproducibility, and mainly consists of soybean protein. Soybean with abundant, renewable, inexpensive and easy to obtain have emerged as critical component for the purpose of exploring new composite applications. In the preparation of soy-based adhesives, soybean products are modified by appropriate means to make them have the properties of natural polymer synthetic resins. Finally, a kind of adhesives prepared by adding a series of additives can be directly used in the wood composite industry. Soybean protein is a compound mainly comprised of soybean globulin and soybean whey protein, 90% of which is soybean globulin. Soy globulin has a special spherical structure, so that most of the hydrophobic groups are wrapped in it. Meanwhile, the impact of exposed hydrophilic groups of soybean protein adhesive on water resistance is considerable.

The modification of soybean protein adhesives is generally modified to improve the mechanical properties of soybean protein adhesives, and it uses biochemical and physical factors to alter amino acid residues and polypeptide chains. Due to the destruction of hydrogen bonds and other chemical bonds, the secondary, tertiary, and quater-level structures are destroyed, which leads to the original irregular bending, folding, and spiral shapes are gradually extended to form a loose linear peptide bond structure, so as to improve the viscosity of the modified adhesive of soybean protein and increase the contact area for adhesive. Meanwhile, the formed soybean protein adhesive should be special structure with the hydrophobic structure inward and hydrophilic outward. The hydrophobic group inside the modified soybean protein will be exposed, which prevents the soybean protein from forming hydrogen bond with water, and then hinder the damage of water to the binding body of the soybean protein adhesive, thus improving the water-resistant property of the soybean protein adhesive. With a small amount of uniform distributed chemical bonds as the backbone core, the resultant force formed together with a large number of hydrogen bonds could be the glue relay source of modified soybean adhesive, which not only guarantees the dry bond strength but also improves the water resistance.

Physical modification to high-level structure and intermolecular aggregation mode of proteins are beneficial under the conditions of high temperature, ultraviolet light, and ultrasonic waves. In the water phase medium, the strong pressure, shear force and high temperature of the protein molecules, the hydrogen bond fracture between the protein molecules, the amino acid is exposed, the soybean protein is trans-modified, the original biological activity is lost, and the solubility of the protein and the strength of the glue are significantly improved. Yang et al. (2010) investigated effects of ultrasonic wave on properties of modified soybean protein adhesives for duplex paper and obtained the optimum conditions by orthogonal experiment as follows. The results showed that treatment temperature had the greatest impact on adhesive strength of soy protein adhesive, followed by urea concentration, protein concentration, treatment time, and ultrasonic time. Apart from ultrasonic, high-pressure homogenization is also a common method of physical modification. Soybean meal (SM), a byproduct of bean oil production, had been alternative formaldehyde-based wood to develop bio-adhesive. However, SM adhesives suffer from drawbacks such as uneven particle size distribution, and hydrophilic hydrogen bonds which result in SM adhesives' low adhesion stability and poor water resistance. To overcome these problems, a stable high-performance bio-based adhesive was developed using a simple, clean and physical high-pressure homogenization (HPH) technology in conjunction with eco-friendly cross-linking by Zhang et al. (2020). Results showed that HPH treatment reduced the SM's particle size by 62%, improved the particle size distribution uniformity, and significantly enhanced bonding stability of the adhesives. Furthermore, many active functional groups within the soy protein were exposed simultaneously during HPH, which increased their reactivity with the cross-linker. Moreover, the method has been successfully applied to industrial production on a pilot scale and has a broad practical application prospect.

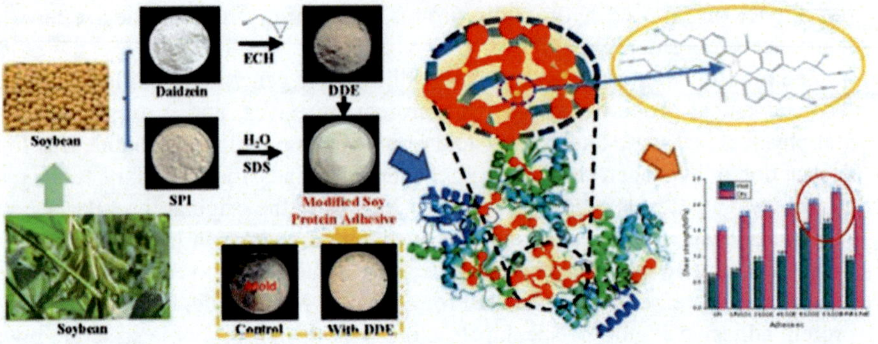

Fig. 4 Possible diagram of soy protein adhesive with high strength and mildew resistance (Xu et al. 2020)

In view of the factors of affecting the chemical modification, the reactivity of functional groups and modifier will be interesting. We mainly modified the side chain groups of soybean protein. Eslah et al. (2016) investigated the potential of chemically modified soybean flour (SF) as a strong water-resistant adhesive for plywood composites and revealed that SF with NaOH, AA, and PEI had the best adhesion performance. In addition, cross-linking agent with high reactivity was added to soybean protein adhesive (As shown in Fig. 4), and its bonding strength and water resistance were significantly improved. Soluble copper, chromium, zinc salts, aliphatic epoxides and epoxides can be used as active curing agents for basic soybean protein adhesives. Under the same technological conditions, the soybean protein adhesives obtained in the presence of epoxides have higher bonding strength. With the aim to modify soy protein-based adhesive, Pang et al. (2020) reported that a waterborne epoxy emulsion (WEU) was prepared by grafting hydrophilic groups onto bisphenol-A (E44) followed for the oil-in-water emulsion by a phase-transformation process. It is an effective way to improve the water resistance and bonding performance of soybean protein adhesives that introducing epoxy crosslinking agent. The results showed that triethylamine neutral agent could increase the wet shear strength of soybean protein adhesive by 20% compared with N, N-dimethyl ethanolamine. In comparison with pure soybean protein adhesive, the prepared adhesive reached 1.14Mpa, 192.5% higher. These properties meet the requirements for interior-use plywood ($\geq$0.7 mpa). The prepared soybean protein/WEU adhesive provides a new approach for the development of high-performance engineered wood products.

2.4 *Starch-Based Adhesives*

Starch molecules formed by glucose units that remove a water molecule and hold it together via glycosidic bonds. Therefore, development of starch-based adhesive with numerous the advantages such as easy workup, cost-effectiveness, non-toxic, biodegradability and environmentally friendliness. However, because there are too many hydrophilic hydroxyl groups in the main chain of starch molecules, the adhesive strength is not ideal and it is difficult to meet the requirements of wood bonding strength. The water resistance of starch-based adhesive can be increased by chemical modification, which modification mechanism is that in the process of synthesizing starch-based adhesive, active groups such as aldehyde group, carboxyl group, amide group and isocyanate group are introduced into the liquid system of starch adhesive through oxidation, condensation, cross-linking and other reactions to change the structure of starch adhesive, so as to enhance the activity of the adhesive. After the introduction of these groups, during the curing process of the adhesive, the condensation reaction of these active groups will occur due to the change of temperature and pressure, and stable methylene bonds, urea bonds, ammonia ester bonds and other water-resistant chemical bonds will be further generated. After the cross-linking, the reticular skeleton structure will be formed, which can effectively prevent water molecules from wedging in and avoid the deterioration of adhesive performance caused by oxygen bond damage. The modified starch-based adhesive is improved by the combination of force of the newly inserted water-resistant chemical bond and a large amount of hydrogen bonds. Moreover, the structure of starch was strengthened, which improved the water resistance dramatically.

The application of starch for the development of adhesive is mainly limited to its formation naturally, which is not suitable. Therefore, it is inevitable to use various technologies and modification processes to modify starch in order to improve the performance and bonding strength of the derived adhesive. Zhou used epichlorohydrin to cross-link the oxidized starch, and then used butyl acrylate to graft copolymerize the cross-linked modified starch emulsion, to prepare the cross-linked-graft modified starch adhesive, whose properties have been greatly improved compared with the unmodified starch adhesive and with all the indicators of plywood hot-pressed dehydrated urea–formaldehyde resin, which was shown that the starch-based adhesive can adapt well to the existing production process. The most popular starch adhesive synthesis process is shown in Fig. 5. Du et al. reported that with starch as main raw material, H2O2 as oxidant, $FeSO_4$ as catalyst, borax as crosslinker, and sodium thiosulfate as stabilizer, the synthesis conditions of starch adhesive were preferred by orthogonal experiment, and then the optimal process conditions of adhesive for biochar compound fertilizer granulation were preferred. The results showed that excessive NaOH would decrease the initial viscosity of adhesive during high temperature gelatinization and in low temperature gelatinization, increasement of NaOH content can improve the initial viscosity of adhesive. At a volume of 2 or 4 mL NaOH solution was added into the oxidation stage for improving initial viscosity of prepared adhesive, thereby starch oxidation reaction was completed. When 6 mL NaOH was added in the oxidation stage, the initial viscosity of the prepared adhesive was relatively low due to excessive oxidation. The 30 wt% concentration of citric acid–starch-based adhesive with *paraserienthes falcataria* veneer was prepared by Kusumah et al. (2020) and citric acid–starch were applied in the manufacturing of plywood. The variations of compositions ratio (citric acid: starch) were 90:10, 80:20, 70:30, 60:40, 50:50 respectively. The results showed that the lamination rate

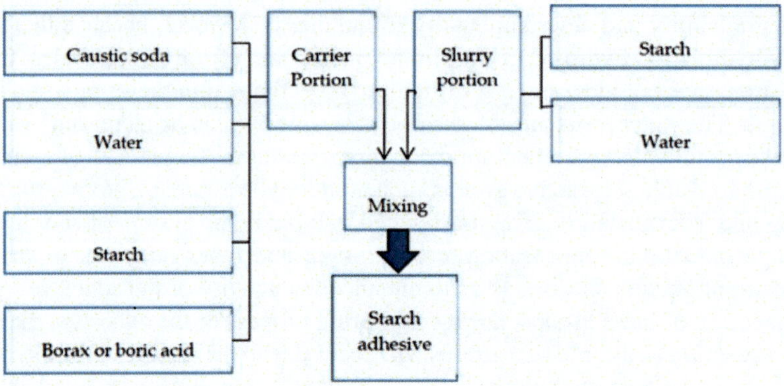

Fig. 5 The Stein Hall process (Nasiri et al. 2020)

and shear strength of plywood were the lowest when the starch content was 50 wt%. The shear strength of plywood for general utilization satisfied Standard National Indonesia (SNI).

2.5 Other Bio-Based Adhesives

The woody biomass liquefaction adhesives. Lignocellulosic biomass liquefaction refers to the thermochemical process of converting lignocellulosic biomass into liquid in the presence of some organic compounds. The liquid products of lignocellulosic biomass liquefaction in the presence of phenol can be treated as raw materials for the production of adhesives. Gao et al. prepared water-resistant wood adhesives with phenolated soy flours. Liquefaction of soybean meal in the presence of phenol degradation, not only destroyed the spherical molecules that were linked together by hydrogen bonds inside the soy protein, tightened up the spherical structure and released the spherical protein in polar group, also of the peptide chain rupture and peptides and the reaction of phenol, the activity of liquefaction reaction with formaldehyde in the flour group increase in the number, which could be used in the preparation of weather-resistant plywood. The production of adhesives by liquefaction of woody biomass has a promising prospect, but the process is not mature at present. Lignocellulosic biomass liquefaction is carried out in the presence of certain organics, which can decrease the reaction temperature to a certain extent and achieve the purpose of energy saving, but the use of liquefier increases the complexity of the process.

The pyrolytic bio-oil adhesive. The pyrolytic bio-oil adhesive uses pyrolytic bio-oil and phenol as raw materials for the synthesis of PF resin adhesives. Pyrolytic bio-oil is made from agricultural and forestry residues, which are cheap and easy to be obtained. Under the condition of isolated or limited oxygen, the organic polymer molecules in the materials can be quickly broken into short chain molecules to obtain liquid products. The pyrolysis bio-oil of plant raw materials being rich in lignin and tannin is mainly composed of phenolic substances, which can partially replace the polymerization of phenol and formaldehyde in phenolic resin to form viscose. At present, the research on the production and applications of pyrolytic bio-oil mostly focuses on the application of pyrolytic bio-oil in energy, and the main objective of the present study is to prepare a series of adhesives for pyrolytic bio-oil.

However, the components of lignocellulosic biomass liquefaction products and the process of pyrolysis bio-oil are complex, the physical and chemical properties are relatively unstable and phase separation and precipitation may occur in long-term storage. Moreover, there are distinctive differences between the properties of different liquefaction products and pyrolytic bio-oil, so in terms of performance, the bio-oil-phenolic resin adhesive prepared by partially phenol cannot completely replace pure phenolic resin adhesive.

In addition, the phenolic hydroxyl on lignin in wood is oxidized to phenolic oxygen free radical via catalyzing of laccase, and the free radical generated is coupled to produce high polymer mass and amorphous dehydrogenation polymer. By using this reaction, the binding force of wood itself can be improved, and it is demonstrated that the use of a simple and economical approach for preparation of fiberboard and particleboard without any synthetic adhesives. The products obtained from biological engineering can be used as adhesive materials to completely get rid of the tension of chemical raw materials, which has the characteristics of self-breeding, self-fermentation, self-synthesis and self-use when manufactured.

Acetonitrile extracting lignin-based adhesive. Cui et al. (2018) reported that a lignin rigid oligomer was obtained by directly extracting lignin from pulp manufacturers with acetonitrile, which is highly polar with significantly increased carboxylic acid and reduced methoxy compared with the as-received lignin and makes the acetonitrile extracting lignin (AEL) more reactive for the preparation of adhesives. For example, Gong et al. (2020) prepared a lignin-based wood adhesive with no formaldehyde in which the AEL was used as raw material, phenolated and mixed with aqueous of polyvinylpyrrolidone. The shear strength of the adhesive could reach up to 1.70 MPa and it was still as high as 1.02 MPa after soaking for 24 h in water, meeting the National Primary Standard of plywood and according with Chinese national standard.

3 Future Challenges of Bio-Based Adhesives for Wood Composite Industries

At present, the industrial application of tannin and lignin-based adhesives is rare. Due to the high molecular weight of tannin, the prepared adhesive has high viscosity and the short application period, which limits its development and application to some extent. Lignin has also large molecular weight, complex structure, low reaction activity, and requires longer hot-pressing time and higher temperature for plate making. Therefore, more lignin was used in the modification of resin and less lignin-based adhesive was prepared directly. And lignin-based adhesive also needs special processing after hot pressing. The product is black with lower physical and mechanical properties and lower water resistance. Therefore, how to reduce the molecular weight of tannin and lignin, improve the reactivity and increase the proportion of its application in viscose will be the main development trend.

Currently, soybean protein adhesives are relatively widely used in wood composite industries. Therefore, isolating soybean flour or soybean protein can be used as raw materials. The former is cheap, but the water resistance and bonding strength of the prepared adhesives are not as good as the latter. Nevertheless, the challenges such as exploring the processing technology of soybean protein separation, improving the water resistance and bonding strength of soybean protein adhesive and reducing cost, still remain, restricting its popularization application.

Starch-based adhesive has the advantages of renewable, environmental protection and low cost, but it also has disadvantages such as poor water resistance and corrosion resistance, which limits its development and application in wood composite industries. Therefore, the preparation of starch-based adhesive usually requires modification of starch first and the modified starch-based adhesive is non-toxic and environmental friendly, with good film-forming performance and adhesion performance. In addition to the mentioned above problems, the glue viscosity, low adhesive strength, short application period and dark color defects of bio-based adhesives also seriously restrict its development, becoming the main research topic in the future.

At the same time, because bio-based adhesives are bonded to other objects mainly through intermolecular forces, the strong bonding strength between wood materials and adhesives depends on many factors, including the wettability, the roughness the permeability, moisture content, the presence of wood surface extract, and the moisture absorption, the chemical composition and the pH of the wood. The bio-based raw materials can be improved by chemical methods and become wood adhesive to meet the needs of industrial production. In addition, the combination of bio-based adhesives and petrochemical adhesives, replacing some of the petrochemical adhesives, will be a good development direction. Bio-based adhesives are a kind of renewable resources with environmental protection characteristics that petrochemical adhesives cannot match, which can effectively realize sustainable development, but compared with petrochemical adhesives, corrosion resistance and drying rate need to be further improved. At present, the research and development of all kinds of biological adhesives in the world are still at the infancy stage and increasing the research intensity of biological adhesives and improving its performance are of positive significance for developing a new adhesive market and protecting the environment.

References

Chen M, Zhang Y, Li Y, Shi SQ, Li J, Gao Q, Guo H (2020) Soybean meal-based wood adhesive enhanced by phenol hydroxymethylated tannin oligomer for exterior use. Polymers (Basel) 12(4)

Cui M, Nguyen NA, Bonnesen PV, Uhrig D, Keum JK, Naskar AK (2018) Rigid oligomer from lignin in designing of tough, self-healing elastomers. ACS Macro Lett 7(11):1328–1332

Eghtedarnejad N, Reza MH (2015) Building wooden panels glued with a combination of natural adhesive of tannin/Eremurus root (syrysh). Eur J, Wood Prod

Eslah F, Jonoobi M, Faezipour M, Afsharpour M, Enayati AA (2016) Preparation and development of a chemically modified bio-adhesive derived from soybean flour protein. Int J Adhes Adhes 71:48–54

Gong X, Liu T, Yu S, Meng Y, Lu J, Cheng Y, Wang H (2020) The preparation and performance of a novel lignin-based adhesive without formaldehyde. Ind Crops Prod 153

Ji W, Ding Z, Liu J, Song Q, Xia X, Gao H, Wang H, Gu W (2012) Mechanism of lignin dissolution and regeneration in ionic liquid. Energy Fuels 26(10):6393–6403

Kusumah SS, Jayadi Wibowo DT, Pramasari DA, Widyaningrum BA, Darmawan T, Ismadi Dwianto W, Umemura K (2020) Investigation of eco-friendly plywood bonded with citric acid—starch based adhesive. In: IOP Conference Series: Earth and Environmental Science, p 460

Li Y, Li B, Du F, Wang Y, Pan L, Chen D (2017) Microwave-assisted hydrothermal liquefaction of lignin for the preparation of phenolic formaldehyde adhesive. J Appl Polym Sci 134(10)

Liu J, Wang L, Li J, Li C, Zhang S, Gao Q, Zhang W, Li J (2020) Degradation mechanism of Acacia mangium tannin in NaOH/urea aqueous solution and application of degradation products in phenolic adhesives. Int J Adhes Adhes 98

Naseem A, Tabasum S, Zia KM, Zuber M, Ali M, Noreen A (2016) Lignin-derivatives based polymers, blends and composites: a review. Int J Biol Macromol 93(Pt A):296–313

Nasiri A, Wearing J, Dubé MA (2020) Using lignin to modify starch-based adhesive performance. ChemEng 4(1)

Pang H, Zhao S, Wang Z, Zhang W, Zhang S, Li J (2020) Development of soy protein-based adhesive with high water resistance and bonding strength by waterborne epoxy crosslinking strategy. Int J Adhes Adhes 100

Pizzi A (2006) Recent developments in eco-efficient bio-based adhesives for wood bonding: opportunities and issues. J Adhes Sci Technol 20(8):829–846

Raj A, Devendra LP, Sukumaran RK (2020) Comparative evaluation of laccase mediated oxidized and unoxidized lignin of sugarcane bagasse for the synthesis of lignin-based formaldehyde resin. Ind Crops Prod 150

Shnawa HA (2020) Curing and thermal properties of tannin-based epoxy and its blends with commercial epoxy resin. Polym Bull

Song Y, Wang Z, Yan N, Zhang R, Li J (2016) Demethylation of wheat straw alkali lignin for application in phenol formaldehyde adhesives. Polymers (Basel) 8(6)

Venkatesagowda B, Dekker RFH (2019) Enzymatic demethylation of Kraft lignin for lignin-based phenol-formaldehyde resin applications. Biomass Conv Biorefinery 10(2):203–225

Wang H, Eberhardt TL, Wang C, Gao S, Pan H (2019) Demethylation of alkali lignin with halogen acids and its application to phenolic resins. Polym (Basel) 11(11)

Xu C, Xu Y, Chen M, Zhang Y, Li J, Gao Q, Shi SQ (2020) Soy protein adhesive with bio-based epoxidized daidzein for high strength and mildew resistance. Chem Eng J 390

Yang G, Sui N, Yang B (2010) Effects of ultrasonic on properties of modified soybean protein-based adhesives for duplex paper. Adv Mater Res 152–153:1866–1872

Yang S, Wu J-Q, Zhang Y, Yuan T-Q, Sun R-C (2015a) Preparation of lignin-phenol-formaldehyde resin adhesive based on active sites of technical lignin. J Biobased Mater Bioenergy 9(2):266–272

Yang S, Yuan TQ, Li MF, Sun RC (2015b) Hydrothermal degradation of lignin: products analysis for phenol formaldehyde adhesive synthesis. Int J Biol Macromol 72:54–62

Yang S, Zhang Y, Yuan T-Q, Sun R-C (2015c) Lignin-phenol-formaldehyde resin adhesives prepared with biorefinery technical lignins. J Appl Polym Sci 132(36):n/a-n/a

Yang W, Rallini M, Natali M, Kenny J, Ma P, Dong W, Torre L, Puglia D (2019) Preparation and properties of adhesives based on phenolic resin containing lignin micro and nanoparticles: a comparative study. Mater Des 161:55–63

Younesi-Kordkheili H, Pizzi A (2017) Properties of plywood panels bonded with ionic liquid-modified lignin–phenol–formaldehyde resin. The J Adhes 94(2):143–154

Zhang Y, Shi R, Xu Y, Chen M, Zhang J, Gao Q, Li J (2020) Developing a stable high-performance soybean meal-based adhesive using a simple high-pressure homogenization technology. J Cleaner Prod 256

Zhao M, Jing J, Zhu Y, Yang X, Wang X, Wang Z (2016) Preparation and performance of lignin–phenol–formaldehyde adhesives. Int J Adhes Adhes 64:163–167

Various Types of Natural Fibers Reinforced Poly-Lactic Acid Composites

Moussaoui Nafissa, Benhamadouche Lamia, Mohammad Jawaid, and Mohammad Asim

Abstract Advance composites are considered as the substitute of solid waste wood. By utilizing these waste materials with different type of polymers enhance the properties of material and also make efficient to use as product. Advance composites are also fulfilling the requirement of environment concerns. Polylactic acid (PLA) is one of the biopolymer, have been studied extensively, the basic studies such as structural and extraction process have been discussed. Many different type natural fibres and biopolymers are being used to prepare biodegradable materials. The properties of natural fibre have been studied on the basis of the chemical compositions that decide the properties of single fibre. There are several test have been conducted such as mechanical, thermal, physical, and weathering test on natural fibre reinforced PLA composites. In this chapter, PLA based natural fibre composites have been studied and their properties have been elaborated.

Keywords Polylactic acid · Natural fibre · Biocomposites · Thermal properties · Mechanical properties · Weathering properties

1 Introduction

The evolution of civilization has generally been influenced by man's need to find out new materials, enabling him to introduce the necessary properties to the increasingly more sophisticated tools of destruction. Companies that have gained experience in exploiting new materials, however, have enjoyed distinct advantages over their neighbors (Oksman et al. 2003). This has been widely illustrated in terms of key materials in which humankind has passed from the Stone Age to that of Bronze, the Iron Age,

M. Nafissa · B. Lamia
Department of Mechanical Engineering, M′sila University, 28000 M′sila, Algeria

M. Jawaid · M. Asim (✉)
Institute of Tropical Forestry and Forest Products (INTROP), Universiti Putra Malaysia (UPM), 43400 Serdang, Selangor, Malaysia
e-mail: khanfatehvi@gmail.com

M. Jawaid et al. (eds.), *Eco-Friendly Adhesives for Wood and Natural Fiber Composites*, Composites Science and Technology,
https://doi.org/10.1007/978-981-33-4749-6_9

and steel. Bio composites are among the most popular materials of contemporary life or modern times which have transformed our lifestyle.

Nowadays, conception with a view to reconstruction or in a slightly broader perspective, eco-design is increasingly a philosophy that applies to products and materials. With the recent increase in environmental awareness, the use of natural fibers has become a major focus and important (Gassan and Bledzki 1999; Kalia et al. 2009). Natural fibers own many benefit, such as lightness, small density, availability and relatively lofty mechanical properties, like that as specific module and specific resistance (Asim et al. 2018; Gassan and Bledzki 1999; Omrani et al. 2016). Recently, natural cellulosic fibers are increasingly attractive to scientists as an alternative reinforcement in polymer composites (Begum and Islam 2013; Mohammed et al. 2015; Sepe et al. 2018; Thakur et al. 2014; Wambua et al. 2003). They are obtainable in large quantities and provide a new generation of reinforcements in polymeric materials. These eco-friendly fibers have been used as replacements for fiberglass, carbon, Kevlar, and other plastic polymer fibers in a variety of applications (Atiqah et al. 2014; Begum and Islam 2013; Wambua et al. 2003). Natural fiber composites aid in the conservation of non-renewable resources, which are the primary source of most materials used in today's applications.

Reinforcing building materials (concrete, mortar, and composite) with fibers is an increasingly popular technique. The introduction of nanocellulose and micro cellulose as reinforcements in polymer matrices (biodegradable or not) used in composites (such as glass fibers and carbons) (Chen et al. 2009; Frone et al. 2011; Liu et al. 2010; Lu et al. 2008; Sanjay et al. 2018).

Vegetable fibers can be associated with matrices of different natures. The use of plant fibers in a thermoplastic or thermosetting matrix (polypropylene polyethylene, epoxy or phenol–formaldehyde resins, polystyrene, and styrene) was the subject of many studies in the last few years (Frone et al. 2011; Liu et al. 2010; Lu et al. 2008; Nakagaito and Yano 2008). It is a technique already used and industrialized especially in the automobile (interior panels, floors of trucks, etc.), the construction (coating, profiles of doors and windows, closure) where is still domestic equipment (furniture of the garden) (Edhirej et al. 2017). They can replace glass fibers because of their mechanical performance.

The use of vegetable fibers (straw, flax, abaca, jute, banana, date palm, bamboo ...), as a reinforcement of material, has aroused a keen interest during the last decade (Edhirej et al. 2017; Thakur 2013). Strengthening materials with these fibers reduces disposal problems because they are easier to recycle or burn than mineral fiber materials. When biodegradable materials are reinforced with plant fibers, these materials remain biodegradable and can be used in situations where they are subjected to loads. In recent decades, a great deal of works has been done to investigate the possibility of using natural fibers as a reinforcing element of thermoplastic and thermosetting materials, due to their environmental benefits and low cost (Thakur 2013). They offer further perspectives in waste collection because they are of biological origin and can hence give rise to vastly functional composite materials used in association with thermoplastic polymers (Wollerdorfer and Bader 1998).

The substitution of non-decomposable matrices by decomposable materials makes it possible to obtain entirely degradable and renewable composites (Atiqah et al. 2014; Begum and Islam 2013; Mohammed et al. 2015; Sepe et al. 2018; Thakur et al. 2014). Materials derived from renewable resources are receiving increasing interest from the academic and industrial worlds. They are very varied in nature and often have interesting characteristics. In the bioplastics market, there is a wide variety of polymers from renewable resources, including polyesters such as polylactide (PLA). The latter is without a doubt one of the most prospective competitor for future advancement as it is not only biodegradable but also made from renewable resources like corn starch (Garlotta 2001; Hartmann 1998; Jonoobi et al. 2010; Oksman et al. 2003). Poly-lactic acid have a high potential to replacement petroleum-based plastic due to its elevated mechanical properties and its process ability in comparison of other bio polymers, making it suitable with a variety applications such as automotive, medical and agriculture (Dorgan et al. 2005; Haafiz et al. 2013; Lim et al. 2008; Najafi et al. 2012; Oliveira et al. 2016). Poly lactic acid is a versatile polymer, completely degradable, produced from pure substances regenerative that ferment in acid. The ratio of the two acid molecules (D-lactic and L-lactic), that are composed of an amorphous to semi-crystalline or crystalline material, could alter the characteristic of PLA (Shen et al. 2009).

2 Natural Fibers

Natural cellulosic fibers have been utilized for 3000 years in composites, for example in antique Egypt, where clay and straw were blended to build walls. In the latest decade, natural fiber-reinforced polymer have attracted the attention from both academia and industry because of the advantages they over offer synthetic reinforcing fibers such as carbon and glass. The development of natural fibre composites has therefore been the subject of research over the last ten years. These natural cellulosic fibers are economical, very eco-friendly with specific properties, for example, they are non-abrasive and biodegradable (Li et al. 2007; Saheb and Jog 1999).

2.1 Natural Fibers Types'

Natural fibers, which are often cellulose fibers, are sorted according to their origin; they come from different areas of the plant: leaves (palm, banana, Alfa, pineapple, etc.), Seeds (Kapok, cotton, milkweed, etc.), Bast or stem (Kenaf, hemp, jute, ramie, bamboo, etc.), Fruit (coconut fiber, etc.) (Van Rijswijk et al. 2001).

As with all natural products, the mechanical and physical properties of natural fibers change significantly. These properties are governed by structure, chemical composition and depend on fiber types, age, plant organs and growing conditions.

Table 1 Classification and examples of natural fibers (Bismarck et al. 2005)

Origin	Provenance	Examples
Vegetal	Seeds	Cotton, kapok, milkweed
	Fruits	Coconut
	Bastor stem	Hemp, jute, ramie, kenaf
	Leaves	Sisal, henequen, abaca, pineapple
	Wood	Wheat, corn, barley, rye, oats, rice
	Stalk	Bamboo, bagasse, Alfa, reed
	Canes and reeds	
Animal	Wool/hair	Wool, hair, cashmere
	Silkworms	Mulberry silk, tussah silk
Mineral		Asbestos, Wollastonite

Table 2 Commercially important fiber sources(Salit et al. 2015)

Fibre source	Species	World production (10^3 tonnes)	Origin
Wood	(>10,000 species)	1,750,000	Stem
Bamboo	(>1250 species)	10,000	Stem
Cotton lint	Gossypiumsp	18,450	Stem
Jute	Corchorussp	2,300	Stem
Kenaf	Hibiscus cannabinus	970	Stem
Flax	Linumusitatissimum	830	Stem
Sisal	Agave sisilana	378	Stem
Roselle	Hibiscus sabdariffa	250	Stem
Hemp	Cannabis sativa	214	Stem
Coir	Cocos nucifera	100	Stem
Ramie	Boehmerianivea	100	Stem
Abaca	Musa textiles	70	Stem
Sunn hemp	Crorolariajuncea	70	Stem

The main component of all plant fibers is cellulose; it varies from one fiber to another (Tables 1 and 2).

2.2 *Chemical Composition of Natural Fibers*

Cellulosic fiber is a composite in itself. The reinforcement is constituted by layers of partially crystalline cellulose microfibrils. The latter is coated with an amorphous polysaccharide matrix (hemicelluloses and pectin) which is associated with hydrogen

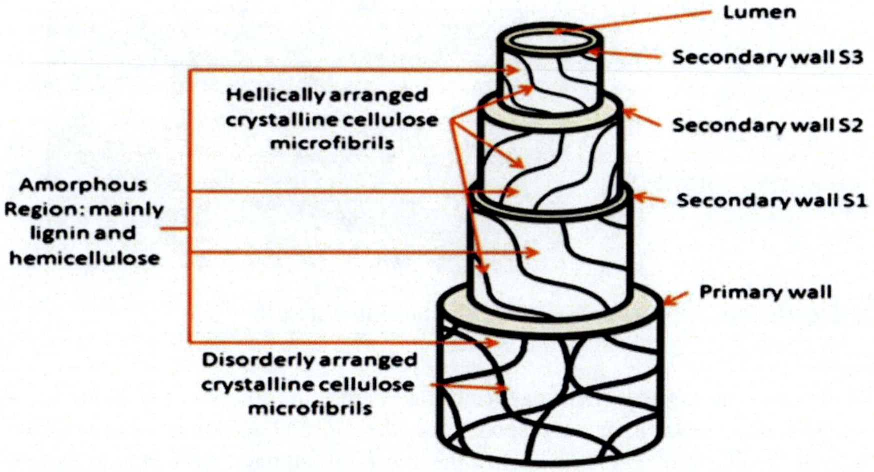

Fig. 1 Model for describing the structure of a plant fiber cell (Salit et al. 2015)

bonding and covalent lignin (Fig. 1) shows that the cellulose microfibrils are helically rolled up along the axis of the core fiber.

The orientation of the micro-fibril about the axis of the cell plays a large role on the mechanical properties of the fiber walls, more angle of the micro-fibril increases, the Young module decreases, while the extensibility of the walls increases (Dicker et al. 2014). The moreover, hemicellulose has a key role in the humidity absorption, bio decomposition, and thermal degradation of natural fibers. However, Lignin is responsible for the rigidity and hardness of wood and of plants. The proportions of theses constituents vary greatly according to the fibers and cellulose is the major constituent element. In general, natural fibre contains other organic components, in addition to cellulose, hemicelluloses and lignin. It contains about 2–6% starch, 2% deoxidized saccharine, 2–4% materials grasses, and 0.8–6% protein (Saheb and Jog 1999). The quantity of each constituent of a fiber depends in particular on the quality of the soil where the plant was grown, its level of maturity, the quality of the retting process, the conditions for measuring properties (humidity, temperature), etc. All these parameters explain the great differences between the values found in the literature (Beg 2007).

Cellulose is a natural polymer whose molecules, formed by long chains, consist of D-anhydro-glucopyranoses bound by β-(1,4)-glycosides bonds in positions C1 and C4 (Fig. 2). The degree of polymerization (DP) is about 10,000, it varies according to the plant species. The hydroxyl groups play a large role in the crystalline (Beg 2007; John and Anandjiwala 2008).

Unlike other fiber components that have an amorphous structure, cellulose has a large crystalline structure. Crystalline cellulose is one of the polymers with the highest modulus of elasticity, about 136 GPa compared to 75GPa of fiberglass (Beg 2007; John and Thomas 2010).

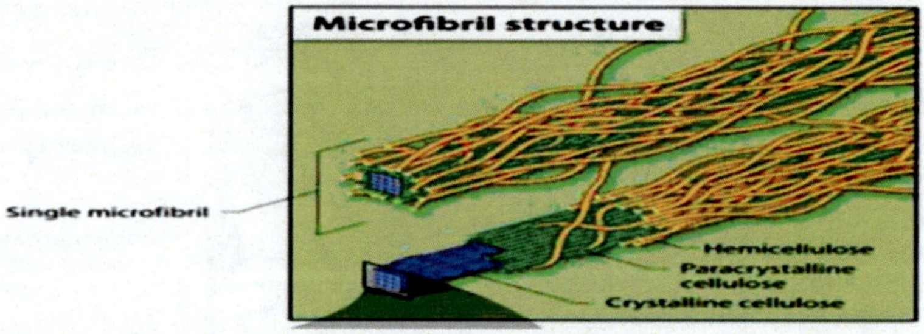

Fig. 2 Structure of the cellulose microfibrillar (Siqueira et al. 2010)

Lignin's an extremely heterogeneous macromolecule composed of aromatic polymers. It represents, after polysaccharides, the more abundant natural polymer in nature. It helps protect plants against the attack of pathogenic organisms and structural rigidity of cell walls. Lignin is made up of three-dimensional phenolic polymers, amorphous with three different penylopropane-type units: p-coumaryl, coniferyl, and sinapyl alcohols (Fig. 3). In the process of lignification, plant phenoloxidases as laccases are implicated and permit of the different elementary units to polymerize. After synthesis, lignin combines with various polysaccharides to form a matrix that forms the plant wall (Klyosov 2007).

Hemicellulose consists of a selection of polysaccharides (Dxylopyranose, D-glocopyranose, D-galactopyranose, etc.) which occur in short and highly branched chains. The degree of polymerization in this polymer is 20–300 times lower than

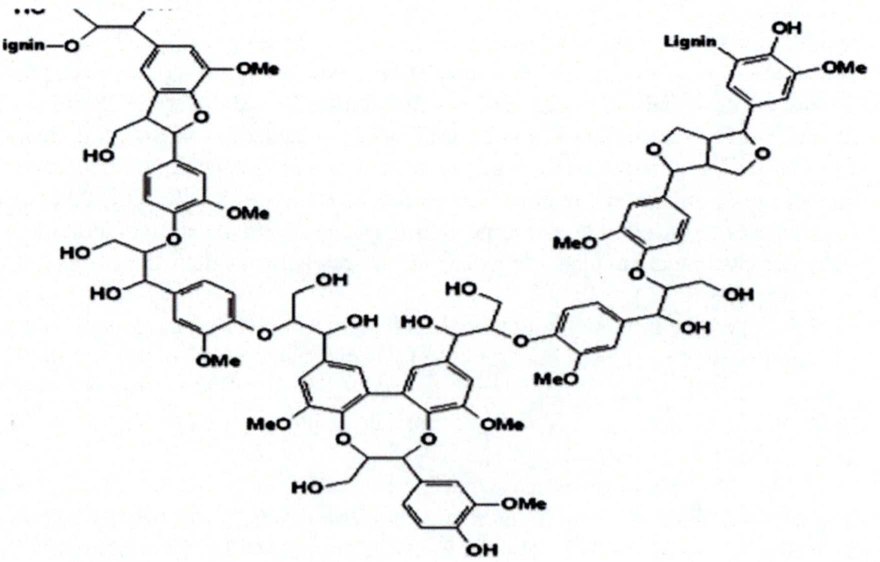

Fig. 3 Schematic representation of lignin (Klyosov 2007)

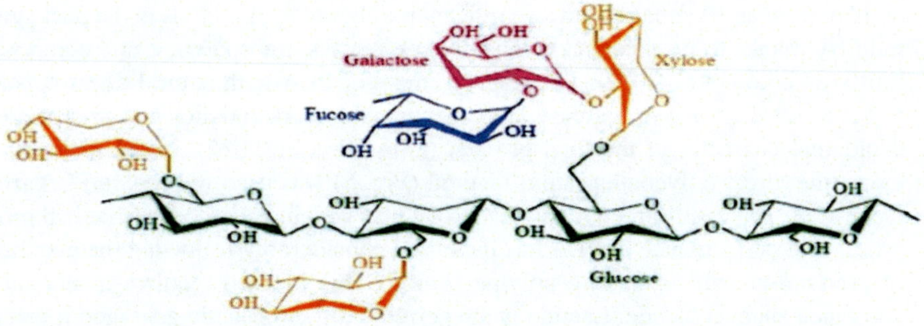

Fig. 4 Structure of a type of xyloglucan (Thygesen 2006)

in cellulose (Thygesen 2006). Hemicelluloses are different from celluloses by their multiple varieties of saccharides constituting the polymer that is different from one plant to another. Unlike cellulose, which is a linear polymer, hemicelluloses have a considerable degree of chain branching. Also, hemicellulose is rarely crystalline and is responsible for the sensitivity of the fibers to water through the interface (hydroxyl groups). This increases the risk of degradation due to external agents (bacteria, acids, etc.). One of the most studied hemicellulose classes corresponds to xyloglucans (Fig. 4).

A pectin is a group of complex heteropolysaccharides that typically represent about 35% of the primary walls in dicotyledons and 2–10% in monocotyledons. The structure of the pectin depends on the plant species and the proportions of the different types vary according to their position in the cell walls (Sakakibara 1980). Pectin's also the more hydrophilic compounds in plant fibers because of the existence of carboxylic acid groups.

3 Polylactic Acid

Polylactic acid (PLA) is aliphatic polyester; generally, it is based on α-hydroxy acids. These include polymandelic acid or polyglycolic acid. These acids are compostable and biodegradable. PLA is a thermoplastic polymer, with high mechanical characteristics (strength and modulus). Annually, it may be produced from renewable resources (Garlotta 2001).

Consequently, it has generated a lot of interest in potentially replacing petroleum-based polymers. Special qualities of PLA have been improved for biomedical uses even before its introduction as a packaging and base material. The commercial launch of biobased PLA in 2003 paved the way for more common applications. Its biocompatibility and bioresorbability made it an appropriate choice for applications including blood vessels, drug delivery systems, sutures etc.

In particular, PLA has found a significant number of applications in the packaging industry thanks to its excellent mechanical properties, transparency, and compostability (Saeidlou et al. 2012). The reinforcement of PLA with natural fibers appears to be an ideal choice to increase their mechanical characteristics and preserve the ecological character of the final product (Girones et al. 2012). In the majority of cases, the direct polycondensation method (Fig. 5) was used to prepare PLA from lactic acid. The resulting PLA had low molecular weight and poor mechanical properties. The performance of PLA has increased considerably following the evolution of production with open ring polymerization. This method requires a secondary substance such as lactide. Lactide is the periodic dimer of lactic acid, and it can be in the form of L-lactide, L, D-lactide (meso-lactide), and D-lactide stereo complex (Fig. 6). Currently, PLA synthesis rarely starts with chemically synthesized lactic acid. The lactic acid that is used is resulting from the transformation of carbohydrates like starch and cellulose. The majority of it comes from corn and cassava harvests. Fermentation created on microorganisms gives principally L-lactic acid (Ebnesajjad 2012).

For PLA to be produced on large-scale production lines in applications such as blow molding, injection molding, extrusion, and thermoforming, the polymer must have appropriate thermal stability to avoid degradation and retain its molecular weight and qualities. PLA is thermally degraded at temperatures exceeding 200 °C (392 °F) (Iwata 2015) by hydrolysis, lactide regeneration, dissociation of the main chain by oxidation, and by intra- or intermolecular transesterification processes.

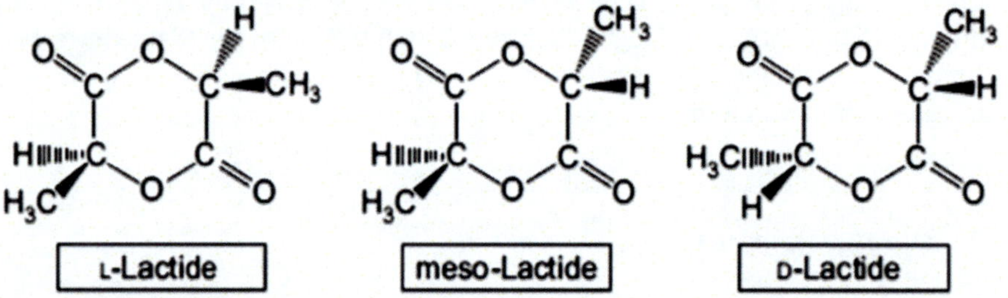

Fig. 5 D-Lactide stereo complex

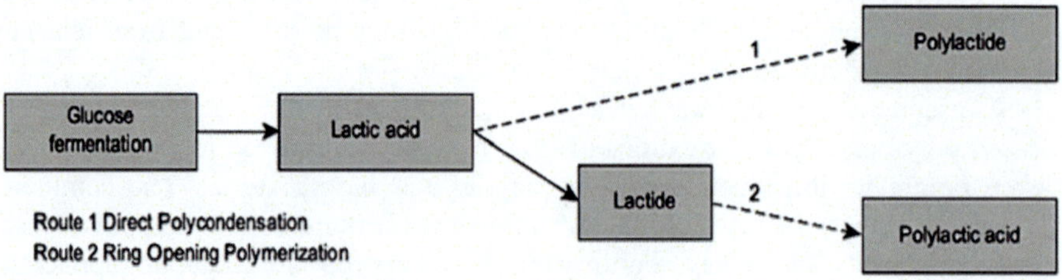

Fig. 6 General route of PLA production

The main reasons for PLA degradation are low molecular weight impurities, catalyst concentration, temperature, and time (Iwata 2015). Oligomers and catalysts increase the degree of degradation and reduce the degradation temperature of PLA. They can also cause changes in viscosity and rheology, fumes in the treatment process, and low mechanical characteristics.

Poly (lactic acid) homopolymers have a glass transition and melting temperature of approximately 175 °C and 55 °C respectively. They need processing temperatures between 185 and 190 °C (Spinu et al. 1996). At this stage, chain splitting and decompression reactions are known to occur, leading to a decrease in molecular weight and thermal degradation. Therefore, PLA homopolymers have a very short treatment period. The most common method to improve the transformability of PLA is based on the reduction of the melting point by random incorporation of small doses of lactic enantiomers of opposite form in the polymer (i.e. the addition of a small amount of D-lactide to L-lactide to get PDLLA). Unfortunately, the reduction in melting point leads to a significant decrease in crystallinity and crystallization rates (Spinu et al. 1996).

3.1 *Synthesis of Polylactic Acid*

The elemental component of PLA is lactic acid, which was first isolated in 1780 from sour milk by the Swedish chemist Scheele and produced for the first time in commerce in 1881 (Hartmann 1998). Food-related uses represent the main exploitation of lactic acid in the United States and account for approximately 85% of the product's market it is present in many processed foods as a buffer, acid flavoring agent, acidulant, and bacterial inhibitor. Lactic acid is formed either by chemical synthesis or by carbohydrate fermentation, which is the most predominant method (Benninga 1990; Frant 1981; Kharas et al. 1994).

Lactic acid (2-hydroxy propionic acid) is the most elemental of the hydroxy acids that have an asymmetric carbon atom and can be in two visually active compositions.

The L(+) isomer is produced in mammals including humans, while the D(−) and L(+) enantiomers are both derived from bacterial systems. Most of the lactic acid marketed worldwide is obtained by carbohydrates bacterial fermentation, from homolactic organisms such as various optimized or modified strains of the genus Lactobacilli, which constitute only lactic acid (Hartmann 1998). The Organisms producing mainly the L(+) isomer are L. bavaricus, Lactobacilli amylophilus, L. maltaromicus, L. casei, and L. salivarius. While isomer D or mixtures of both are given by strains such as L. delbrueckii, L. jensenii, or L. acidophilus (Hartmann 1998). The bacteria considered homofermentative produce lactic acid via Embden-Meyerhof and convert up to 1.8 mol of lactic acid per mole of hexose (i.e. more than 90% produce glucose-based lactic acid). From feedstocks, these strains produce high carbon conversions for animals under standard fermentation conditions, such as temperatures around 40 °C, relatively small to neutral pH, and low oxygen concentrations (Enomoto et al. 1994).

The different types of carbohydrates that can be used in fermentation processing are due to the specific strain of Lactobacillus. In principle, the majority of simple sugars extracted from agricultural by-products can be used. These sugars contain (1) sucrose from cane or beet sugar; (2) glucose, maltose, and dextrose from corn or potato starch; and (3) lactose from cheese whey. In addition to carbohydrates, organisms also require protein and other complex nutrients such as B vitamins, amino acids, and nucleotides, which must be obtained from corn liquor, yeast extract, cottonseed meal, or soy flour. These needs are highly dependent on the different species and it is common to create strains around existing nutrients, which can lead to considerable additional costs (Benninga 1990; Enomoto et al. 1994; Kharas et al. 1994).

Fiber reinforcement is a possibility to improve thermal stability. Polylactide polymers are rigid and brittle materials, so plasticizers are needed to improve stretching and impact resistance properties. The polylactide is completely biodegradable. Degradation is by hydrolysis to lactic acid, which is metabolized by microorganisms to water and carbon monoxide. By composting with other biomasses, biodegradation occurs in 15 days, and the material has completely disappeared in three or four weeks (Oksman et al. 2003).

4 The Properties of Natural Fiber/PLA Composites

4.1 Thermal Properties

Knowledge of the thermal properties of composites natural fiber/PLA is essential to develop the best methods for transforming these materials into useful products and to predict performance throughout the life of the products. It also provides essential information for troubleshooting when the material is not functioning as intended or when a product or process element needs to be modified, such as raw materials. The determination of principal characteristics of polymer systems, such as thermogravimetric analysis (TGA), glass-to-rubber transition temperature (Tg), and melting point (Tm), is essential using differential scanning calorimetry (DSC) experiments. In the latter case, Tg is usually determined as the inflection point of the specific heat increment at the glass–rubber transition. Dynamic mechanical analyses (DMA) can also be utilized to evaluate the Tg. However, a relaxation mechanism is demonstrated in the Tg temperature range by DMA analyses. Today, PLA is considered the most credible alternative to replace conventional polymers in composites, their thermal properties investigated for the potential use in load-bearing applications such as body-in-white and body structures in the automotive sector. The Tg of PLA depends on the molecular weight, optical purity (enantiomer), and thermal history of the polymer (Sarasua et al. 1998). The degree of crystalline is based on many parameters such as molar mass, temperature and annealing time, as well as on many key properties of a semi-crystalline polymer, including stiffness, fragility, toughness, etc. (Ahmed et al. 2009).

For example, Pan et al. (2007) observed that the addition of Kenaf (KF) considerably increased the crystallization rate and the tensile and storage modulus. The PLLA crystallization can be completed from the melt at 5 °C/min with the addition of 10% by weight of Kenaf during the cooling process. Also found that the size of the spherulite decreases in isothermal crystallization and the nucleation density increases considerably with the presence of Kenaf. Furthermore, with the addition of 30% by weight of Kenaf, the half times of isothermal crystallization were lowered to 46.5% and 28.1% of pure PLLA at 120 °C and 140 °C, respectively. Tawakkal et al. (2014) noted that thermo gravimetric analysis of the incorporation of 5% and 10% (w/w) thymol in pure PLA and PLA/kenaf composites shows that there is no significant change in decomposition temperature, while analysis by differential scanning calorimeter indicate a decrease in all key thermal transitions (Tg, Tcc, Tm and Hm). The test results for thermal properties suggest that thymol in these composite functions as a plasticizing agent. Tokoro et al. (2008) have incorporated three varieties of bamboo fibers into a PLA matrix to improve its impact strength and heat resistance. Their studies have led to the creation of good quality composite materials. The presence of bamboofibers considerably increases the impact resistance and thermal properties. Qin et al. (2011) confirmed that PLA/RSF (rice straw fiber) composites with PBA increased thermal stability. DSC data showed that RSF acted as a nucleating agent and that PBA made crystallization of PLA more difficult and incomplete. The available literature shows that thermo-mechanical analysis (TMA) should be performed in the future to improve the understanding of the influence of the addition of natural fibers in PLA composites.

4.2 Mechanical Properties

Recent research has indicated that PLA is a very suitable polymer matrix for natural fiber-filled composites. For example, Tokoro et al. (2008) have reinforced a PLA matrix with 3 kinds of bamboo fibers to increase its impact. Their studies are caused in good quality composite materials, with the presence of bamboo fibers significantly increasing impact performance. Ochi (2008) studied kenaf/PLA composites with diverse ratios of fibers. He showed that Young's modulus, tensile, and bending strength increased linearly up to 50% wt of the fiber content in the composite. The study found that unidirectional biodegradable composites made from emulsion-type PLA resin and kenaf fiber at 70% by weight had high bending and tensile strengths of respectively 254 MPa and 223 MPa. Shibata et al. (2004) have developed materials with PLA/lyocell fabric by compression molding. The tensile strength and modulus of the lyocell/PLA composites have been upgraded by increasing the fiber content. The impact resistance was significantly greater than that of pure PLA. Pan et al. (2007) as well developed PLA/kenaf composites by injection molding with kenaf fiber contents between 0 and 30%. At 30%, a 30% improvement in tensile strength was noted.

The extensive use of biocomposites in the manufacture of vehicle components would greatly benefit the industry, but to date, their low mechanical characteristics, associated mostly with thermal stability and the interactions between the matrix and the fibers, limit their usage. While studies mention the properties and potential improvements of biocomposites, the general mechanical behavior from a structural point of view is not fully understood and the mechanical performance is under target. Biocomposites are not yet adopted in mass-produced products in the automotive transport industry because their mechanical properties and performance in different environments and their effect under different loading conditions are not sufficiently taken into account. Biocomposites are not yet adopted in mass production in the automotive industry because their mechanical properties and performance in different environments and their effect under different loading conditions are not sufficiently taken into account (Nassiopoulos and Njuguna 2015).

The mechanical properties of fiber-reinforced composites are influenced by the length of fibers, the degree of filling, the aspect ratio of fibers or fiber bundles, the morphology of the fibers and the orientation of the fibers or bundles of fibers, and the degree of polymerization of the cellulose in the fiber. The mechanical characteristics of PLA-based composites also probably depend on the specific properties of the polylactide (e.g. residual lactide content, molar mass) as well as the processing conditions (e.g. catalyst, pre-drying, and processing temperatures) (Mohanty et al. 2005). Lanzillotta et al. (2002) found that the tensile strength of injection-molded PLA was not upgraded by adding 20–40% w/w of flax fiber. The authors justified this to poor adhesion between flax and PLA. The significant reduction in fiber length observed by these researchers may also partly justify the absence of an increase in the tensile strength of composites. Though, with increasing flax content, Young's modulus of these biocomposites increases linearly. After studying the chemical modification of fibers (for example, etherification, esterification) and reactive extrusion to modify PLA during treatment, the authors also concluded that improvements to the composite properties derived from the composite did not justify the associated costs.

The research group of the Structural Mechanics Institute of the German Aerospace Centre has developed and controlled PLA biocomposites reinforced with natural fiber mats. A composite Young's modulus of about 10 GPa and tensile strength of about 90 MPa were the results of their work (Riedel and Nickel 1999). The same authors cited in a later report that the bending modulus for PLA biocomposites was comparable to that for glass fiber reinforced polymers (Nickel and Riedel 2001).

The main objective of incorporating fibers into a polymer matrix is to significantly improve the modulus of the material, but this effect depends on the type and content of fiberin the matrix. It has been shown that the modulus of elasticity measured for PLA reinforced with 30% flax or cotton increases by 1.5–2.5 times that of crude PLA, depending on the type of fiber modification (Thakur and Thakur 2016).

4.3 Weathering Properties

It was found that the mechanical characteristics of PLA can be similar to those of traditional petroleum-based polymers. For this reason, PLA will be considered as an alternative polymer that can be exploited in other fields, particularly in the automotive sector. In this context, it is necessary to study the external properties of PLA, i.e. its resistance to weathering. In particular, the influences of UV and moisture exposure on the mechanical and other characteristics of PLA become critical (Kaynak and Kaygusuz 2016).

However, in the literature, there are not many studies (Copinet et al. 2004; Ikada 1997; Janorkar et al. 2007; Ndazi and Karlsson 2011; Shinzawa et al. 2012) on the consequences of weathering and their degradation processes. In general, the principal degradation process of ultraviolet (UV) irradiation observed is the random "main chain separation" in the chemical bonds of the PLA backbone by the absorption of a photon (Ikada 1997; Janorkar et al. 2007). The molecular weight of PLA is significantly reduced via "photolysis" of the C–O and C–C bonds, and/or the main "photo-oxidation" to produce carboxylic acid and ketones. The most important moisture degradation process described in the bibliography (Copinet et al. 2004; Ndazi and Karlsson 2011) after a decrease in molecular weight is the "hydrolysis" of the C–O bonds in the ester bonds of the PLA structure, which can also produce carboxylic acid and ketones. We have found other several works based on the study of the natural weathering and accelerated aging properties of composites PLA reinforced such as keratin (Spiridon et al. 2013), montmorillonite (Jandas et al. 2013; Zaidi et al. 2010), hempfiber(Islam et al. 2010), epoxy natural rubber and rice starch (Yew et al. 2009), cellulose whiskers (Grigsby et al. 2013) and condensed tannin (Bolio-López et al. 2013). In general, these studies have shown that the added fillers most often increase the level of deterioration without having an important influence on the processes.

References

Ahmed J, Zhang J-X, Song Z, Varshney S (2009) Thermal properties of polylactides. J Therm Anal Calorim 95:957–964

Asim M, Jawaid M, Abdan K, Ishak M (2018) The effect of silane treated fiber loading on mechanical properties of pineapple leaf/kenaf fiber filler phenolic composites. J Polym Environ 26:1520–1527

Atiqah A, Maleque M, Jawaid M, Iqbal M (2014) Development of kenaf-glass reinforced unsaturated polyester hybrid composite for structural applications. Compos B Eng 56:68–73

Beg MDH (2007) The improvement of interfacial bonding, weathering, and recycling of wood fiber reinforced polypropylene composites. The University of Waikato

Begum K, Islam M (2013) Natural fiber as a substitute to synthetic fiber in polymer composites: a review. Res J Eng Sci 2278:9472

Benninga H (1990) A history of lactic acid making: a chapter in the history of biotechnology. Springer Science & Business Media

Bismarck A, Mishra S, Lampke T (2005) Plant fibers as reinforcement for green composites. Natl Fibers Biopolym Biocompos 10:9780203508206

Bolio-López G, Veleva L, Valadez-González A, Quintana-Owen P (2013) Weathering and biodegradation of polylactic acid composite reinforced with cellulose whiskers. Revista Mexicana De Ingeniería Química 12:143–153

Chen Y, Liu C, Chang PR, Cao X, Anderson DP (2009) Bionanocomposites based on pea starch and cellulose nanowhiskers hydrolyzed from pea hull fiber: effect of hydrolysis time. Carbohyd Polym 76:607–615

Copinet A, Bertrand C, Govindin S, Coma V, Couturier Y (2004) Effects of ultraviolet light (315 nm), temperature and relative humidity on the degradation of polylactic acid plastic films. Chemosphere 55:763–773

Dicker MP, Duckworth PF, Baker AB, Francois G, Hazzard MK, Weaver PM (2014) Green composites: A review of material attributes and complementary applications. Compos Appl Sci Manuf 56:280–289

Dorgan JR, Janzen J, Clayton MP, Hait SB, Knauss DM (2005) Melt rheology of variable L-content poly (lactic acid). J Rheol 49:607–619

Ebnesajjad S (2012) Handbook of biopolymers and biodegradable plastics: properties, processing and applications. William Andrew

Edhirej A, Sapuan SM, Jawaid M, Zahari NI (2017) Cassava: its polymer, fiber, composite, and application. Polym Compos 38:555–570

Enomoto K, Ajioka M, Yamaguchi A (1994) Polyhydroxycarboxylic acid and preparation process thereof. Google Patents

Frant M (1981) Kirk-Othmer encyclopaedia of chemical technology, ion selective electrodes. Wiley, New York, Place Published, pp 720

Frone A, Berlioz S, Chailan JF, Panaitescu D, Donescu D (2011) Cellulose fiber-reinforced polylactic acid. Polym Compos 32:976–985

Garlotta D (2001) A literature review of poly (lactic acid). J Polym Environ 9:63–84

Gassan J, Bledzki AK (1999) Possibilities for improving the mechanical properties of jute/epoxy composites by alkali treatment of fibers. Compos Sci Technol 59:1303–1309

Girones J, Lopez J, Mutjé P, Carvalho AJFd, Curvelo AAdS, Vilaseca F (2012) Natural fiber-reinforced thermoplastic starch composites obtained by melt processing. Compos Sci Technol 72:858–863

Grigsby W, Bridson J, Lomas C, Elliot J-A (2013) Esterification of condensed tannins and their impact on the properties of poly (lactic acid). Polymers 5:344–360

Haafiz MM, Hassan A, Zakaria Z, Inuwa IM, Islam MS, Jawaid M (2013) Properties of polylactic acid composites reinforced with oil palm biomass microcrystalline cellulose. Carbohyd Polym 98:139–145

Hartmann M (1998) High molecular weight polylactic acid polymers. In: Biopolymers from renewable resources. Springer, Place Published, pp 367–411

Ikada E (1997) Photo-and bio-degradable polyesters. Photodegradation behaviors of aliphatic polyesters. J Photopolym Sci Technol 10:265–270

Islam MS, Pickering KL, Foreman NJ (2010) Influence of accelerated aging on the physico-mechanical properties of alkali-treated industrial hemp fiber reinforced poly (lactic acid) (PLA) composites. Polym Degrad Stab 95:59–65

Iwata T (2015) Biodegradable and bio-based polymers: future prospects of eco-friendly plastics. Angew Chem Int Ed 54:3210–3215

Jandas P, Mohanty S, Nayak S (2013) Sustainability, compostability, and specific microbial activity on agricultural mulch films prepared from poly (lactic acid). Ind Eng Chem Res 52:17714–17724

Janorkar AV, Metters AT, Hirt DE (2007) Degradation of poly (L-lactide) films under ultraviolet-induced photografting and sterilization conditions. J Appl Polym Sci 106:1042–1047

John MJ, Anandjiwala RD (2008) Recent developments in chemical modification and characterization of natural fiber-reinforced composites. Polym Compos 29:187–207

John MJ, Thomas S (2010) Cellulosic fibril–rubber nanocomposites. In: Rubber nanocomposites: preparation, properties, and applications, p 197

Jonoobi M, Harun J, Mathew AP, Oksman K (2010) Mechanical properties of cellulose nanofiber (CNF) reinforced polylactic acid (PLA) prepared by twin-screw extrusion. Compos Sci Technol 70:1742–1747

Kalia S, Kaith B, Kaur I (2009) Pretreatments of natural fibers and their application as reinforcing material in polymer composites—a review. Polym Eng Sci 49:1253–1272

Kaynak C, Kaygusuz I (2016) Consequences of accelerated weathering in polylactide nanocomposites reinforced with halloysite nanotubes. J Compos Mater 50:365–375

Kharas G, Sanchez-Riera F, Severson D (1994) Polymers of lactic acid. In: Mobley DP (ed) Plastics from microbes. Hanser Gardner Publications, Inc.

Klyosov AA (2007) Wood-plastic composites. Wiley

Lanzillotta C, Pipino A, Lips D (2002) New functional biopolymer natural fiber composites from agricultural resources. In: Antec-conference proceedings, unknown, pp 2185–2189

Li X, Tabil LG, Panigrahi S (2007) Chemical treatments of natural fiber for use in natural fiber-reinforced composites: a review. J Polym Environ 15:25–33

Lim L-T, Auras R, Rubino M (2008) Processing technologies for poly (lactic acid). Prog Polym Sci 33:820–852

Liu H, Liu D, Yao F, Wu Q (2010) Fabrication and properties of transparent polymethylmethacrylate/cellulose nanocrystals composites. Biores Technol 101:5685–5692

Lu J, Askeland P, Drzal LT (2008) Surface modification of microfibrillated cellulose for epoxy composite applications. Polymer 49:1285–1296

Mohammed L, Ansari MN, Pua G, Jawaid M, Islam MS (2015) A review on natural fiber reinforced polymer composite and its applications. Int J Polym Sci 2015

Mohanty AK, Misra M, Drzal LT (2005) Natural fibers, biopolymers, and biocomposites. CRC Press

Najafi N, Heuzey M, Carreau P (2012) Polylactide (PLA)-clay nanocomposites prepared by melt compounding in the presence of a chain extender. Compos Sci Technol 72:608–615

Nakagaito AN, Yano H (2008) Toughness enhancement of cellulose nanocomposites by alkali treatment of the reinforcing cellulose nanofibers. Cellulose 15:323–331

Nassiopoulos E, Njuguna J (2015) Thermo-mechanical performance of poly (lactic acid)/flax fiber-reinforced biocomposites. Mater Des 66:473–485

Ndazi BS, Karlsson S (2011) Characterization of hydrolytic degradation of polylactic acid/rice hulls composites in the water at different temperatures. Express Polym Lett 5

Nickel J, Riedel U (2001) Structural materials made of renewable resources (biocomposites), biorelated polymers. Springer, Place Published, pp 27–40

Ochi S (2008) Mechanical properties of kenaf fibers and kenaf/PLA composites. Mech Mater 40:446–452

Oksman K, Skrifvars M, Selin J-F (2003) Natural fibers as reinforcement in polylactic acid (PLA) composites. Compos Sci Technol 63:1317–1324

Oliveira M, Santos E, Araújo A, Fechine GJ, Machado AV, Botelho G (2016) The role of shear and stabilizer on PLA degradation. Polym Testing 51:109–116

Omrani E, Menezes PL, Rohatgi PK (2016) State of the art on tribological behavior of polymer matrix composites reinforced with natural fibers in the green materials world. Eng Sci Technol Int J 19:717–736

Pan P, Zhu B, Kai W, Serizawa S, Iji M, Inoue Y (2007) Crystallization behavior and mechanical properties of bio-based green composites based on poly (L-lactide) and kenaf fiber. J Appl Polym Sci 105:1511–1520

Qin L, Qiu J, Liu M, Ding S, Shao L, Lü S, Zhang G, Zhao Y, Fu X (2011) Mechanical and thermal properties of poly (lactic acid) composites with rice straw fiber modified by poly (butyl acrylate). Chem Eng J 166:772–778

Riedel U, Nickel J (1999) Natural fibre-reinforced biopolymers as construction materials–new discoveries. Die Angew Makromol Chem 272:34–40

Saeidlou S, Huneault MA, Li H, Park CB (2012) Poly (lactic acid) crystallization. Prog Polym Sci 37:1657–1677

Saheb DN, Jog JP (1999) Natural fiber polymer composites: a review. Adv Polym Technol J Polym Process Inst 18:351–363
Sakakibara A (1980) A structural model of softwood lignin. Wood Sci Technol 14:89–100
Salit MS, Jawaid M, Yusoff NB, Hoque ME (2015) Manufacturing of natural fiber reinforced polymer composites. Springer
Sanjay M, Madhu P, Jawaid M, Senthamaraikannan P, Senthil S, Pradeep S (2018) Characterization and properties of natural fiber polymer composites: a comprehensive review. J Clean Prod 172:566–581
Sarasua J-R, Prud'Homme RE, Wisniewski M, Le Borgne A, Spassky N (1998) Crystallization and melting behavior of polylactides. Macromolecules 31:3895–3905
Sepe R, Bollino F, Boccarusso L, Caputo F (2018) Influence of chemical treatments on mechanical properties of hemp fiber-reinforced composites. Compos B Eng 133:210–217
Shen L, Haufe J, Patel MK (2009) Product overview and market projection of emerging bio-based plastics PRO-BIP 2009. Report for European polysaccharide network of excellence (EPNOE) and European bioplastics 243
Shibata M, Oyamada S, Si K, Yaginuma D (2004) Mechanical properties and biodegradability of green composites based on biodegradable polyesters and lyocell fabric. J Appl Polym Sci 92:3857–3863
Shinzawa H, Nishida M, Tanaka T, Kanematsu W (2012) Accelerated weathering-induced degradation of poly (lactic acid) fiber studied by near-infrared (NIR) hyperspectral imaging. Appl Spectrosc 66:470–474
Siqueira G, Bras J, Dufresne A (2010) Cellulosic bionanocomposites: a review of preparation, properties, and applications. Polymers 2:728–765
Spinu M, Jackson C, Keating M, Gardner K (1996) Material design in poly (lactic acid) systems: block copolymers, star homo-and copolymers, and stereocomplexes. J Macromol Sci Part a Pure Appl Chem 33:1497–1530
Spiridon I, Paduraru OM, Zaltariov MF, Darie RN (2013) Influence of keratin on polylactic acid/chitosan composite properties. Behavior upon accelerated weathering. Ind Eng Chem Res 52:9822–9833
Tawakkal IS, Cran MJ, Bigger SW (2014) Effect of kenaf fiber loading and thymol concentration on the mechanical and thermal properties of PLA/kenaf/thymol composites. Ind Crops Prod 61:74–83
Thakur VK (2013) Green composites from natural resources. CRC Press
Thakur VK, Thakur MK (2016) Handbook of sustainable polymers: structure and chemistry. Jenny Stanford Publishing
Thakur VK, Thakur MK, Gupta RK (2014) Raw natural fiber-based polymer composites. Int J Polym Anal Charact 19:256–271
Thygesen A (2006) Properties of hemp fiber polymer composites-An optimization of fiber properties using novel defibration methods and fiber characterization.
Tokoro R, Vu DM, Okubo K, Tanaka T, Fujii T, Fujiura T (2008) How to improve mechanical properties of polylactic acid with bamboo fibers. J Mater Sci 43:775–787
Van Rijswijk K, Brouwer W, Beukers A (2001) Application of natural fibre composites in the development of rural societies. Delft University of Technology, Delft
Wambua P, Ivens J, Verpoest I (2003) Natural fibers: can they replace glass in fiber-reinforced plastics? Compos Sci Technol 63:1259–1264
Wollerdorfer M, Bader H (1998) Influence of natural fibers on the mechanical properties of biodegradable polymers. Ind Crops Prod 8:105–112
Yew G, Chow W, Mohd Z, Mohd A (2009) Natural weathering of poly (lactic acid): effects of rice starch and epoxidized natural rubber. J Elastomers Plast 41:369–382
Zaidi L, Kaci M, Bruzaud S, Bourmaud A, Grohens Y (2010) Effect of natural weather on the structure and properties of polylactide/cloisite 30B nanocomposites. Polym Degrad Stab 95:1751–1758

Nanocellulose Reinforced Polylactic Acid Bionanocomposites

Amina Hachaichi, Mohammad Jawaid, Mohammad Asim, and B. Kouini

Abstract Nanocellulose, is a bio-filler that can be extracted from several plant sources such as sisal, date palm, cotton, kenaf and bamboo, according to the method of extraction, size and experimental conditions. Nanocellulose is classified into three types: nanocrystals cellulose (CNC), nanofibers cellulose (CNF) and bacterial cellulose (BC), Nanocellulose can be used in wide applications such as packaging, medicine and pharmaceutical, due to its properties such as biodegradability, compatibility and high mechanical properties. This is the reason why most researchers in these recent years have focused on studying bionanocomposites based on nanocellulose and especially with polylactic acid (PLA) as a matrix, PLA is the most attractive biopolymer due to its properties that improve when enhanced with nanocellulose which qualify it to use in environmentally friend applications. In this chapter, a detailed study is made about the type of natural fibers that can be used to isolate nanocellulose, the isolation techniques of nanocellulose and their classifications, and some pre-studies on nanocellulose/PLA bionanocomposite.

Keywords Nanocellulose · Natural fiber · Treatment · Polylactic acid · Bionanocomposites

A. Hachaichi
Research Unit Materials, Processes and Environment (URMPE), Faculty of Technology, M'Hamed Bougara University, 35000 Boumerdes, Algeria
e-mail: a.hachaichi@univboumerdes.dz

M. Jawaid · M. Asim (✉)
Laboratory of Biocomposite Technology, Institute of Tropical Forestry and Forest Products (INTROP), Universiti Putra Malaysia, UPM, Serdang, 43400 Selangor, Malaysia
e-mail: khanfatehvi@gmail.com

B. Kouini
Laboratory of Coatings, Materials and Environment, M'Hamed Bougara University, 35000 Boumerdes, Algeria

M. Jawaid et al. (eds.), *Eco-Friendly Adhesives for Wood and Natural Fiber Composites*, Composites Science and Technology,
https://doi.org/10.1007/978-981-33-4749-6_10

1 Introduction

Bionanocomposite is one of the most advanced researches in nanotechnology field that has attracted the attention of industrialists and academics to produce it, as advanced environmental friendly materials that replace traditional non-biodegradable petroleum plastics (Bitinis et al. 2013). These compounds can be defined as a combination of polymer (Biopolymer) and organic or inorganic nanofillers (Size < 100 nm), it can be used for various applications, such as packaging, building construction, agriculture sector, aerospace, automotive parts, and sports materials (Arora 2018; Shchipunov 2012).

Nanocellulose particles can be obtained from macromolecule cellulose derived from natural fibers such as roselle fibers (Kianl et al. 2018), bamboo (Rasheed et al. 2020) and kenaf bast fibers (Kargarzadehl et al. 2012), by using different methods: Mechanical, Chemical and Biological treatment. Nanocellulose fibers characterized by low density, biodegradability, biocompatibility, low cost and ease of processing (Santosl et al. 2016), which qualify it to be used as reinforcing filler for biopolymer like polylactic acid (PLA), polyethylene glycol (PEG), polyglycolic acid (PGA). Among the biopolymer, PLA is widely used in different fields of application, because of its remarkable properties such as biodegradability, biocompatibility, ease of use, processability and renewability (Kianl et al. 2019; Scaffaro et al. 2017).

Nanocellulose particles can be classified to three types and that according to the extraction technique used: Cellulose Nano Crystals (CNC), Cellulose nanofibers (NFC) and Bacterial Cellulose (BC). The addition of these nanoparticles into PLA matrix can improve barrier, thermal and mechanical proprieties of bionanocomposites material. Thus, in order to use it in specific applications (Kianl et al. 2019; Nasir 2017). The aim of this chapter is to study the different techniques used for nanocellulose extraction from various natural fibers and their using as a reinforcing material for polylactic acid bionanocomposites.

2 Natural Fiber

Natural fibers can be extracted from different parts of plant, and it can be classified as stems fibers or trunks (bamboo, jute, flax, and date palm etc.), leaves (Sisal, pineapple leaf fiber, Abaca, ALF, sisal, banana etc.), fruit (OPEFB, Coconut etc.) and seeds (Cotton, kapok etc.) (Jawaid 2016). Figure 1, shows the commonly natural fibers used as reinforcing materials in polymers for preparation of composites and nanocomposites materials, and their origin plants.

Natural fibers consist three basic components, cellulose, hemicellulose, lignin and minor components such as pectic and waxes substances (Bitinis et al. 2013). The chemical composition of natural fibers depends on their origin, the type of extraction method used and the experimental conditions (Kabirl et al. 2012). The Table 1 shows some natural fibers and their chemical compositions. The selection of natural fibers

Fig. 1 Different types of plants (**a**) and their extracted fibers (**b**) (Shridhar 2019; Kıasıf 2018; Hakeeml et al. 2014; Ferreira et al. 2017; Suzuki and Arai 2015; Gondal et al. 2018; Goudenhooft et al. 2019; Gondall et al. 2018; Abdul Ajeez 2017)

depends mainly on its applications according to the requirements of different uses (Jawaid 2016). Natural fibers have mechanical properties that are depended on the amount of cellulose, which is defines strength and rigidity of fibers by its bonds as hydrogen bonds (Mishral et al. 2018), also, it depended to microfibril angle, their elongation and Young's modulus (Jawaid 2016; Goda and Cao 2007). Table 2 listed the physical and mechanical proprieties of several natural fibers.

Table 1 Chemical compositions of some natural fibers (Siakengl et al. 2019; Mwaikambo 2006)

Fibers	Cellulose	Hemicellulose	Lignin	Pectin	Wax
Bamboo	73.83	12.49	10.15	0.37	–
Cotton	82.7	5.7	–	–	1
Date palm	35–45.1	9.75–27.7	11.45-29.48	–	–
Oil palm	65	–	29	–	–
Jute	61–71	14–20	12–13	–	0.5
Kenaf	45–57	21.5	8–13	3–5	–
Flax	71	18.6-20.6	2.2	2.3	1.7
Pineapple	70–80	18.8	12.7	1.1–1.2	3.2–4.2
Ramie	68.6–76.2	13-16	0.6-0.7	1.9	0.3
Sisal	65	12	9.9	10	2

Table 2 Physical and mechanical proprieties of several natural fibers (Asim 2015)

Fibers	Density (g/cm^3)	Elongation (%)	Tensile strength (MPa)	Moisture absorption (%)	Young's modulus (GPa)
Cotton	1.5–1.6	3–10	287–597	8–25	5.5–12.6
Jute	1.3–1.46	1.5–1.8	393–800	12	10–30
Flax	1.4–1.5	1.2–3.2	345–1500	7	27.6–80
Ramie	1.5	2–3.8	220–938	12–17	44–128
Sisal	1.33–1.5	2–14	400–700	11	9–38

3 Chemistry of Cellulose

Cellulose is one of the most abundant and sustainable natural polymers on earth (Wang and Sain 2007), it can be extracted from plant biomass, such as wood, cotton, jute, ramie and sisal, cellulose is a semi-crystalline material formed from linear polysaccharide of –d glucopyranose units ($C_6H_{11}O_5$) n linked together by β (1,4) linkages, the repetitive unit is cellobiose (Abdul Khalil et al. 2014). Cellulose has several properties, such as low cost, low density, biodegradability, and renewability (Kianl et al. 2017), which especially qualifies it for use in several fields, as food and pharmaceutical industries (Kargarzadeh et al. 2017).

The physico-chemical properties of cellulose are depended on hydroxyl groups, which are capable to form two hydrogen bonds (Inter and Intramolecular bonds) as represents in Fig. 2 (Lin and Dufresne 2014), this relates to the location of glucose unit. Also, the properties of cellulose are affected by their degree of polymerization which is estimated mostly between 1500 and 3500 depending on the cellulose origin and the method of its extraction (Bhat 2017).

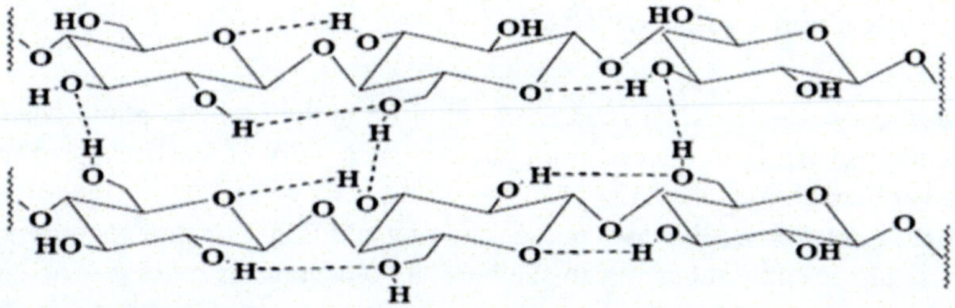

Fig. 2 Chemical structure of cellulose with inter and intramolecular hydrogen bonds (Dasanl et al. 2017)

4 Nanocellulose

Nanocellulose material, has received increasing attention in recent years due to its unique physical and mechanical properties which made it a good material for several applications, for instance, food, pharmaceutical and biomedical, Nanocellulose can be isolated from cellulose which is extracted from plant, it can be divided into three main types that related to the type of isolation technique, dimensions and shapes, cellulose nanocrystals (CNC), cellulose nanofibers (CNFs) and bacterial cellulose (BC) (Abdul Khalil et al. 2014; Kargarzadeh et al. 2018; Zinge and Kandasubramanian 2020).

4.1 Cellulose Nano Crystals (CNC)

CNC can be isolated from plants or from lignocellulose fibers by using acid hydrolysis process, generally, sulfuric acid (H_2SO_4) is the most acid used in this process, also there is other strong acids can be use it like phosphoric acid (H_3PO_4), and formic acid (CH2O2).The concentration of acid, time reaction and temperature, are considered as control factors on the degree of CNC crystallinity, Typical diameter of CNC is in the range of 4 to 55 nm and their range length is between 90 to 400 nm (Abdul Khalil et al. 2014; Zinge and Kandasubramanian 2020).

4.2 Cellulose Nanofibers (CNF)

CNF can be isolated by a mechanical process, also, it can be combined with an enzymatic and/or chemical treatment to obtain a high yield (Abdul Khalil et al. 2014). CNF fibers contain a crystalline and amorphous domain in contrast to CNC which contains 90% of crystalline domain. CNF fibers are characterized by a diameter ranging between 6 to 100 nm and length of (200–1800 nm) (Zinge and Kandasubramanian 2020).

4.3 *Bacterial Cellulose (BC)*

In addition to the abundant sources of cellulose in the biomass plant. Also, bacterial cellulose can be produced from bacteria such as Acetobacter, Agrobacterium, Pseudomonas (Lin et al. 2013).

The production of BC does not require any additional treatments to remove the non-cellulosic compounds, it can synthesis by biosynthesis process. The combination of the fibers with the gel membrane resulting from biosynthesis leads to the formation of bundles of fibers (Lin and Dufresne 2014; Jiang and Lo Hsieh 2013), with a three-dimensional structure ranging in length from 20 to 100 nm (Zinge and Kandasubramanian 2020). These bacteria exhibit very distinctive properties such as biocompatibility, high purity, high degree of crystallization (80–90%) and high mechanical properties, BC can be used in a variety of applications, medicine, cosmetics and food industry. high cost of BC production is the reason for its low production (Mishra 2018).

5 Isolation Techniques of Nanocellulose

5.1 *Mechanical Process*

5.1.1 High Pressure Homogenization (HPH)

HPH is an effective mechanical technique, used to isolate the cellulose nanofibrils (CNF) by treating them with a high-pressure homogenization method (150–200 MPa), without using to any organic solvents (Zinge and Kandasubramanian 2020), nanocellulose particles characterized by their insolubility in water and most solvents, this is what leads to clogging problems; therefore, Pretreatment process is necessary before HPH to reduce cellulose fiber size (Bhat 2017).

5.1.2 Microfluidizer

Micro-fluidization is a similar technique to HPH, the difference between them is that the micro fluidization works by the principle of a static shear while HPH works at constant pressure (Zinge and Kandasubramanian 2020). In addition, microfluidizer has unique proprieties, especially fast mixing, this what leads to a fast-chemical reaction, it also used to improve the reinforcement proprieties (de Moura et al. 2012).

5.1.3 Grinding

The simplicity of the grinding process, its low cost and low energy consumption, are among the main factors that made it a widely used process (Jossetl et al. 2014), grinding process works to break up cellulose to nanoscale, through breaking down of hydrogen bond. This process can be followed by chemical treatments to remove non-cellulosic compounds such as lignin, pectin and wax which improves the mechanical properties of nanocellulose (Iwamotol et al. 2007; Zhao et al. 2014).

5.1.4 Cryocrushing

Cryocrushing is another technique can be used to isolate nanocellulose by immersing them cellulose fiber in liquid nitrogen and then crushing by using a mortar and pestle (Chakraborty et al. 2005), the frozen cellulose fibers are affected by the pressure forces applied to them by the ice crystals, which leads to their rupture and thus transforming them into nanocellulose (Kalia 2011, 2011).

5.2 Chemical Hydrolysis

Hydrolysis acid is one of the most effective techniques used to isolate CNC by using strong concentrated acid such as sulfuric acid (H_2SO_4), hydrochloric acid (HCl), phosphoric acid (H_3PO_4) or nitric acid (HNO3) (Bhat 2017). H_2SO_4 is the most widely acid used in this process, it works to break down amorphous regions and segregates the crystalline regions (Luzi et al. 2019).

H_2SO_4 reacts with hydroxyl groups (OH) on the surface of cellulose to form sulfate ester groups, that help the dispersion of CNC in water (Börjesson and Westman 2015). Acid concentration, reaction time and temperature are important factors for governing the yield percentage of CNC (Zinge and Kandasubramanian 2020) (Fig. 3).

5.3 Biological Treatments

Biological treatment is ecotechnique that can be used to extract BC by using microorganisms such as bacteria and enzymes. in addition to the advantages of this process like high yield, low consummation of energy and higher selectivity, there are disadvantages like high cost and the long time it will take limits its use (Kargarzadeh et al. 2018).

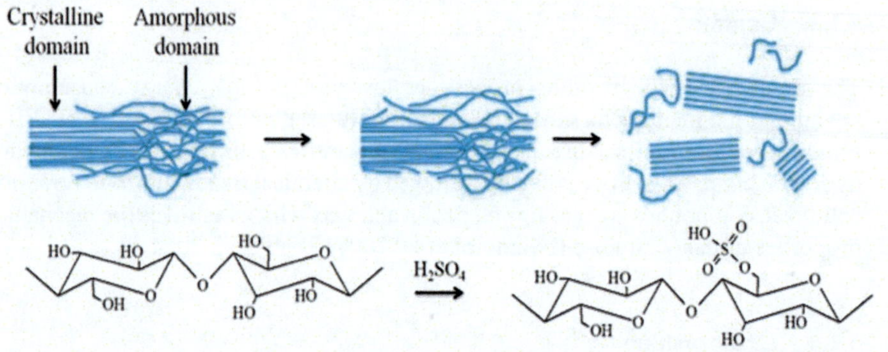

Fig. 3 Acid hydrolysis of cellulose into nanocrystals cellulose by using H_2SO_4 (M. Rajinipriya et all Importance of Agricultural and Industrial Waste in the Field of Nanocellulose and Recent Industrial Developments of Wood Based Nanocellulose: A Review ACS Sustain. Chem. Eng. 6 2807)

6 Poly Lactic Acid (PLA)

Biopolymers, are good alternatives to petroleum polymers due to their numerous characteristics of being environmentally friendly, renewable and biodegradable with exceptional processing capacity (Siakengl et al. 2019; Kalia 2011). Which qualifies it for use in a variety of fields such as packaging, pharmaceutical and medicine (Lasprillal et al. 2012).

PLA (2-Hydroxy propionic acid, CH_3–CHOHCOOH) is one of the most studied and researched biopolymers. It also occupies an important place in the biopolymers market in terms of their using (Saba 2017). PLA can be derived from renewable plant resources such as cornstarch and potato (Kianl et al. 2017), also it can be synthesized from Lactic acid monomers (L or D-lactic isomers) by using two different methods. Ring opening polymerization or direct polycondensation (Ahmed et al. 2011). The chirality of lactic acid monomer makes PLA an interesting polymer with physicochemical properties represented in Table 3.

7 Nanocellulose Reinforced PLA Bionanocomposite

Bionanocomposite based on PLA as a matrix and nanocellulose as biofiller is among the most sought-after compounds, due to its typical properties that combined the proprieties of nanocellulose and PLA, which will be higher than pure PLA or pure nanocellulose it used in three major areas Medicine, automobile industry and packaging applications (Saba 2017). Some research works on various forms of Nanocellulose reinforcing PLA bionanocomposites are tabulated in Table 4.

Table 3 Physicochemical properties of PLA (Jawaid 2016)

Proprieties	PDLLA Poly (D, L-lactic acid)	PDLA Poly (D-lactic acid)	PLLA Poly (L-lactic acid)
Structure	Amorphous	crystalline	Hemicrystalline
Melting temperature T_m/°C	120–170	120–150	173–178
Glass transition temperature (T_g)/°C	43–56	40–60	55–80
Density(g/cm^3)	1.25	1.248	1.290
Decomposition temperature/°C	185–200	200	200
Elongation at break/%	Variable	20–30	20–30
Breaking strength/(g/d)	Variable	4–5	5–6
Half-life in 37 °C normal saline	2–3 months	4–6 months	4–6 months
Solubility	PLLA solvents and acetone, ethyl lactate, tetrahydrofuran, ethyl Acetate, dimethyl sulfoxide, N, N xylene and dimethylformamide		Chloroform, Furan, Dioxaneanddioxolane

8 Conclusion and Future Perspectives

Natural fibers can be derived from deferent parts of plants such as trunks, leaves, fruit or seeds. Its selection depends on their proprieties and applications, it composed of three principal constituents: cellulose, hemicellulose and lignin, nanocellulose can be extracted from natural fiber by different methods like chemical treatment which mostly used to isolate CNC, mechanical which generally used for CNF extraction and biological for BC, it's possible to use both techniques to obtain a good propriety of nanocellulose. The type of nanocellulose depended on isolation technique, dimensions and shapes. Most of the previous studies focused on CNC/PLA Bionanocomposites, because the CNC extracting method by hydrolyzed acid is easy, and it has a good surface compatibility with PLA matrix, contrary to CNF that was extracted by mechanical method that consumes High energy, and also the high cost of BC is the reason behind its low production. Scientific studies are developing and research is continuing, so the future prospects for Nanocellulose/PLA compound may take new field based on the use of simple and effective techniques, that, can be open a new application for these bionanocomposites.

Table 4 Reported study on various forms of Nanocellulose reinforcing PLA bionanocomposites

Origin of nanocellulose	Type of nanocellulose	Extraction method	Bionanocomposite	Preparation method	References
Hemp	CNC	Acid hydrolysis (H_2SO_4)	CNC/PLA/Poly butylene succinate (PBS)	Solvent casting	Luzi et al. (2016)
Ramie	CNC	Acid hydrolysis (H_2SO_4)	CNC/PLA	Melt-extrusion	Raquez et al. (2012)
Bamboo	CNW	Acid hydrolysis (H_2SO_4)	CNC/PLA	Casting	Borkotokyl et al. (2018)
Oil palm	CNC	Acid hydrolysis (H_2SO_4)	CNC/PLA/Poly3-hydroxybutyrate-co-3-hydroxyvalerate (PHBV)	Casting	Dasanl et al. (2017)
Cotton pulp	CNC	Acid hydrolysis (H_2SO_4)	CNC/PLA	Solvent casting	Wang et al. (2020)
Soft wood kraft pulp	CNF	Mechanically refining	CNF/PLA/poly β-hydroxybutyrate (PHB)	Extrusion	Aydemir and Gardner (2020)
Kenaf	CNF	Chemical (NaOH-AQ) and mechanical (refining, cryo-crushing, and high-pressure homogenization)	CNF/PLA	Twin screw extrusion	Jonoobil et al. (2010, Jonoobi 2009)
Picea abies	Enzymatic-NF	–	CNF/PLA/Polyethylene oxide (PEO)	Electrospinning	Henrikssonl et al. (2007), Ghafaril et al. (2020)

(continued)

Table 4 (continued)

Origin of nanocellulose	Type of nanocellulose	Extraction method	Bionanocomposite	Preparation method	References
Gluconacetobacter xylinus	Bacterial cellulose nanofibers	–	CNF/PLA/Poly ethylene glycol (PEG)	Solvent casting and particulate leaching methods	Dahmanl et al. (2010), Abu Ghalia and Dahman (2017)

References

Abdul Ajeez AM (2017) Experimental investigation of mechanical properties of flax fibres reinforced composites with epoxy. Int J Res Appl Sci Eng Technol V:845–853

Abdul Khalil HPS et al (2014) Production and modification of nanofibrillated cellulose using various mechanical processes: a review. Carbohydr Polym 99:649–665

Abu Ghalia M, Dahman Y (2017) Fabrication and enhanced mechanical properties of porous PLA/PEG copolymer reinforced with bacterial cellulose nanofibers for soft tissue engineering applications. Polym Test 61:114–131

Arora B et al (2018) Bionanocomposites: green materials for a sustainable future. Elsevier Inc, pp 699–712

Asim M et al (2015) A review on pineapple leaves fibre and its composites. Int J Polym Sci

Aydemir D, Gardner DJ (2020) Biopolymer blends of polyhydroxybutyrate and polylactic acid reinforced with cellulose nanofibrils. Carbohydr Polym 250:116867

Bhat AH (2017) Cellulosic biocomposites: potential materials for future. In: Jawaid M, Salit M, Alothman O (eds) Green Biocomposites. Green Energy Technol, Springer, Cham

Bitinis N et al (2013) Poly (lactic acid)/ natural rubber/ cellulose nanocrystal bionanocomposites Part I. Processing and morphology 96:611–620

Börjesson M, Westman G (2015) Crystalline nanocellulose—preparation, modification, and properties. Cellul - Fundam Asp Curr Trends

Borkotokyl SS et al (2018) Biodegradable poly (lactic acid)/Cellulose nanocrystals (CNCs) composite microcellular foam: Effect of nanofillers on foam cellular morphology, thermal and wettability behavior. Int J Biol Macromol 106:433–446

Chakraborty A et al (2005) Cellulose microfibrils: a novel method of preparation using high shear refining and cryocrushing. Holzforschung 59:102–107

Dahmanl Y et al (2010) Potential of biocellulose nanofibers production from agricultural renewable resources: Preliminary study. Appl Biochem Biotechnol 162:1647–1659

Dasanl YK et al (2017) Development and material properties of poly (lactic acid)/poly(3-hydroxybutyrate-co-3-hydroxyvalerate)-based nanocrystalline cellulose nanocomposites. J Appl Polym, Sci, p 134

de Moura MR et al (2012) Microfluidizer technique for improving microfiber properties incorporated into edible and biodegradable films. Adv. Microfluid:219–240

Ferreira CR et al (2017) Comparative study about mechanical properties of structural standard concrete and concrete with addition of vegetable fibers. Mater Res 20:102–107

Ghafaril R et al (2020) Processing-structure-property relationships of electrospun PLA-PEO membranes reinforced with enzymatic cellulose nanofibers. Polym Test 81:106182

Goda K, Cao Y (2007) Research and development of fully green composites reinforced with natural fibers. J Solid Mech Mater Eng 1:1073–1084

Gondal P et al (2018) Experimental study of sisal and jute fiber based biocomposite 6:398–404

Goudenhooft C et al (2019) Flax (Linum usitatissimum L.) fibers for composite reinforcement: Exploring the link between plant growth, cell walls development and fiber properties. Front Plant Sci 10:1–23

Hakeeml KR et al (2014) Biomass and bioenergy: processing and properties. Biomass Bioenergy Process Prop 9783319076:1–367

Henrikssonl M et al (2007) An environmentally friendly method for enzyme-assisted preparation of microfibrillated cellulose (MFC) nanofibers. Eur Polym J 43:3434–3441

Iwamotol S et al (2007) Nano-fibrillation of pulp fibers for the processing of transparent nanocomposites. Appl Phys A Mater Sci Process 89:461–466

Jawaid M (2016) nanoclay reinforced polymer composites: natural fibre/nanoclay hybrid composites. Eng Mater 301

Jiang F, Lo Hsieh Y (2013) Chemically and mechanically isolated nanocellulose and their self-assembled structures. Carbohydr Polym 95:32–40

M. Jonoobi (2009) Chemical composition, crystallinity, and thermal degradation of bleached and unbleached kenaf bast (Hibiscus cannabinus) pulp and nanofibers. BioResources 4: 626–639
Jonoobil M et al (2010) Mechanical properties of cellulose nanofiber (CNF) reinforced polylactic acid (PLA) prepared by twin screw extrusion. Compos Sci Technol 70:1742–1747
Jossetl S et al (2014) Energy consumption of the nanofibrillation of bleached pulp, wheat straw and recycled newspaper through a grinding process 3D printing of cellulose reinforced composites view project porous nanofibrillated cellulose functional materials view project. Artic Nord Pulp Pap Res J 29:167–175
Kabirl MM et al (2012) Chemical treatments on plant-based natural fibre reinforced polymer composites: an overview. Compos Part B Eng 43:2883–2892
Kalia S et al (2011) Cellulose-based bio- and nanocomposites: a review. Int J Polym Sci
Kalia S (2011) (eds) Cellulose fibers: bio-and nano-polymer composites: green chemistry and technology. Springer Science & Business Media
Kargarzadeh H et al (2017) (eds) Handbook of nanocellulose and cellulose nanocomposites. Wiley-VCH Verlag GmbH & Company
Kargarzadeh H et al (2018) Recent developments in nanocellulose-based biodegradable polymers, thermoplastic polymers, and porous nanocomposites. Prog Polym Sci 87:197–227
Kargarzadehl H et al (2012) Effects of hydrolysis conditions on the morphology crystallinity and thermal stability of cellulose nanocrystals extracted from kenaf bast fibers. Cellulose 19:855–866
Kianl LK et al (2017) Isolation and characterization of microcrystalline cellulose from roselle fibers. Int J Biol Macromol 103:931–940
Kianl LK et al (2018) Isolation and characterization of nanocrystalline cellulose from roselle-derived microcrystalline cellulose. Int J Biol Macromol 114:54–63
Kianl LK et al (2019) A review on processing techniques of bast fibers nanocellulose and its polylactic acid (PLA) nanocomposites. Int J Biol Macromol 121:1314–1328
Kıasıf G (2018) The effects of natural environmental data in the traditional japanese house design. Int J Adv Res 6:361–371
Lasprillal AJR et al (2012) Poly-lactic acid synthesis for application in biomedical devices - A review Biotechnol. Adv 30:321–328
Lin SP et al (2013) Biosynthesis, production and applications of bacterial cellulose. Cellulose 20:2191–2219
Lin N, Dufresne A (2014) Nanocellulose in biomedicine: current status and future prospect. Eur Polym J 59:302–325
Luzi F et al (2016) Production and characterization of PLA_PBS biodegradable blends reinforced with cellulose nanocrystals extracted from hemp fibres Ind. Crops Prod 93:276–289
Luzi F et al (2019) Valorization and extraction of cellulose nanocrystals from North African grass: Ampelodesmos mauritanicus (Diss). Carbohydr Polym 209:328–337
Mishra R (2018) Materials chemistry and the futurist eco-friendly applications of nanocellulose: status and prospect. J Saudi Chem Soc 22:949–978
Mwaikambo L (2006) Review of the history, properties and application of plant fibres African. J Sci Technol 7:121
Nasir M et al (2017) Nanocellulose Preparation methods and applications In Cellulose-Reinforced Nanofibre Composites. J Woodhead Publishing:261–276
P Gondal et all (2018) Experimental study of sisal and jute fiber based biocomposite 6(4):398–404
Rajinipriya M et al (2018) Importance of agricultural and industrial waste in the field of nanocellulose and recent industrial developments of wood based nanocellulose: a review. ACS Sustain Chem Eng 6:2807–2828
Raquez JM et al (2012) Surface-modification of cellulose nanowhiskers and their use as nanoreinforcers into polylactide: a sustainably-integrated approach. Compos Sci Technol 72:544–549
M. Rasheed et al (2020) Morphological, chemical and thermal analysis of cellulose nanocrystals extracted from bamboo fibre 160. Elsevier B.V
Saba N et al (2017) Current organic synthesis current organic synthesis. Science bentham 14: 156–170

Santosl F et al (2016) The Use of cellulose nanofillers in obtaining polymer nanocomposites: properties, processing, and applications. MSA 7:257–294

Scaffaro R et al (2017) Polysaccharide nanocrystals as fillers for PLA based nanocomposites. Cellulose 24:447–478

Shchipunov Y (2012) Bionanocomposites: green sustainable materials for the near future. Pure Appl Chem 84:2579–2607

Shridhar M (2019) Preparation and analysis of mechanical properties of short sisal and glass fiber reinforced composite. Int J Adv Sci Eng 5(1009–1016):2019

Siakengl R et al (2019) Natural fiber reinforced polylactic acid composites: a review. Polym Compos 40:446–463

Suzuki K, Arai S (2015) On sandwich boards made with kenaf-fibers on sandwich boards made with kenaf-fibers

Wang B, Sain M (2007) The effect of chemically coated nanofiber reinforcement on biopolymer-based nanocomposites. BioResources 2:371–388

Wang K et al (2020) Properties of poly (L-lactic acid) reinforced by L-lactic acid grafted nanocellulose crystal Int. J. Biol. Macromol 156:314–320

Ahmed J, Varshney SK, Polylactides-chemistry, properties and green packaging technology: a review. Int J Food Prop 14:37–58

Zhao Y et al (2014) Fabrication and characteristics of cellulose nanofibril films from coconut palm petiole prepared by different mechanical processing. Ind Crops Prod 65:96–101

Zinge C, Kandasubramanian B (2020) Nanocellulose based biodegradable polymers. Eur Polym J 133:109758

The Challenge of Environment-Friendly Adhesives for Bio-Composites

Saman Ghahri, Ehsan Bari, and Antonio A. Pizzi

Abstract Different resins are used for composites manufacturing, with most of them being based on oil. Synthetic resins are the most commonly used oil-based products in the world. Synthetic resins are widely used to bond wood-based panels. The typical adhesives for this application are oil-based. Most of the constituents of these adhesives present a high risks due to their environmental and human impact. Some of the compounds used are even now classed as carcinogenic. For example, urea–formaldehyde (UF) resins, which are widely used as adhesive, provide emission of carcinogenic formaldehyde gas. But, today, eco-friendly materials are increasingly applied as awareness increases in recent years of the consequences of using oil-based products. Thus, this chapter tries to introduce the environment-friendly adhesives both suitable as well as accessible to use for bio-composites manufacturing by exposing their advantages and disadvantages.

Keywords Environment-friendly adhesives · Bio-composites manufacturing · Synthetic resins

1 Introduction

By using eco-friendly materials, it is possible to decrease the drastic exploitation of oil and petrochemical resources, reduce the amount of carbon dioxide emission and take a responsible approach to the environment and human health. This strategy

S. Ghahri
Department of Wood and Paper Sciences, Faculty of Natural Resources, Tarbiat Modares University, Noor, Iran

E. Bari (✉)
Department of Wood Sciences and Engineering, Technical Faculty of No. 1, Sari Branch, Technical and Vocational University (TVU), Mazandaran, Iran
e-mail: bari_lenzites@yahoo.com

A. A. Pizzi
Department of Industrial Chemistry, ENSTIB-LERMAB, University of Lorraine, BP 21042 Epinal Cedex 9, 88051 Nancy, France

M. Jawaid et al. (eds.), *Eco-Friendly Adhesives for Wood and Natural Fiber Composites*, Composites Science and Technology,
https://doi.org/10.1007/978-981-33-4749-6_11

implies the improvement of air quality as well as improving sustainable resources management. Nowadays, the wood industry makes efforts and is also willing to use natural-based and environmental friendly wood adhesives instead of their synthetics counterparts (Ghahri et al. 2016a, 2016b; Ghahri and Pizzi 2018).Urea–formaldehyde adhesives are the dominant ones by volume of all these (Whitfield et al. 2007). They are relatively inexpensive, they are colourless, present excellent cohesive strength once cured and they harden fast allowing fast hot press times for wood panels such as interior wood panel products. (Zeppenfeld and Grunwald 2005). When the final bonded products are used under higher humidity conditions, UF resins are usually modified with significantly more expensive compounds such as melamine, phenol or resorcinol (Pizzi 2014).

Formaldehyde emission from wood panel products has become a hot topic of concern since the early 1970s. For ever stricter emission regulations has caused formaldehyde emission to be continuously decreased and strictly regulated. Formaldehyde is now classified as an oncogenic material of category 1B (Solt et al. 2019). Today, world demand for safe products and its sustainability has led to an increase in research efforts to replace oil-based resins with eco-friendly adhesives. Well-known research centers and organizations in the world like the US Environmental Protection Agency (EPA) in 2017, the California Air Research Board (CARB), the International Agency for Research on Cancer (IARC) in 2006 and the European Union (EU) in 2015 reclassified Formaldehyde as a human carcinogen (Commission Regulation (EU) 2015/491, EPA 2017). In future decades, by increasing wood based composites consumption and production in the world, the challenges of the availability of fossil resources for producing oil-based resins and restrictions on formaldehyde emissions of wood based composites will be even more critical for the wood industry (Salthammer et al. 2010; Hemmilä et al. 2019). Recently, one of the most severe limits implemented against formaldehyde emission is the new emission test method proposed in Germany. The German Federal Ministry for the Environment, Natural Conservation and Nuclear Safety decided to decrease the legal limits for formaldehyde emissions (from the European emission level E1 of 0.10 to 0.05 ppm) from the 1st of January 2020 (BUM 2018).

2 Eco-Friendly Wood Adhesives: Adhesion Mechanism and Properties

The world adhesives market is dominated in volume by the adhesives used for wood and wood panels. They constitute about one third by volume of all the world adhesives production (Pizzi 2016). The volumes per year of synthetic wood adhesives used are simply staggering, i.e. urea–formaldehyde (UF) adhesives top eleven million tons of resin solids (Pizzi 2016). At such huge levels it becomes hard to find a way to substitute synthetic with renewable-sourced bioadhesives (Pizzi 2016). The low-cost and the adjustable properties of these adhesives have made it difficult for new bio-based alternatives (e.g. lignin, tannin, starch, protein) to enter this market. Sustainable adhesives should not only be available at low costs, be readily and abundantly available, but also need to be easily distributable, fast reacting, and have a long pot life (Dunky 2003). The mechanical strength of manufactured panels and especially their moisture tolerance are additional crucial parameters the new, bio-based adhesives have yet to fulfil completely satisfactorily (Zhao et al. 2011). Two main factors which cause the wood panels industry to move away from using formaldehyde-based resins are formaldehyde emissions problems, and restrictions and sustainability of raw materials and final products. In the panel industry, sustainability and petroleum independence cannot, as yet, justify the increase in cost due to new bio-based adhesives. Thus, concern about formaldehyde emissions from wood based panels, especially in indoor applications, is currently the most important driving factor (Navarrete et al. 2013). Any knowledge about environment friendly adhesives and their properties as well as adhesion mechanisms will help scientists and industry to find proper methods in adhesive preparation and to produce bonded products of acceptable specifications.

2.1 Tannin Based Adhesives

Tannin adhesives present per se a very low level of formaldehyde emission. Nonetheless, research is still, continuously going on to further decrease emission or even eliminate formaldehyde from them. This to ensure that the emission is at its lowest possible in the bonded wood panels (Pizzi 2016). The word tannin has been used loosely to represent two different classes of chemical compounds of mainly phenolic nature: hydrolysable tannins and condensed tannins. The former are mixtures of simple phenols such as gallic acid and ellagic acid and of esters of a sugar, mainly glucose, with gallic and digallic acids (Pizzi 2003).

Condensed tannins, on the other hand, constituting more than 90% of the total world production of commercial tannins (220,000 tons per year), are both chemically and economically more attractive for the preparation of adhesives and resins (Pizzi 2013). Condensed tannins are polyhydroxyphenols, polyflavonoids, which are soluble in water, alcohols and acetone and can coagulate proteins. They are mainly obtained commercially by water extraction from wood and bark. The other main components of the extracts, called non-tannins, are simple sugars and polymeric carbohydrates. The non-tannins can reduce the performance and water resistance of tannin-bonded joints. The polymeric carbohydrates also increase the viscosity of the extracts. (Pizzi 2013). Many companies, mainly in the southern hemisphere, have long-time used or they still use large amounts of these adhesives. Among them CSR Wood Panels, Australia, Bisonboard and Novobord, South Africa; Sekisui Chemical Co., Japan; Masisa, Chile; and Zimbord, Zimbabwe. Particularly in Chile pine bark tannins have been used for 10 years. This is of a main interest as procyanidin type tannins such as these can be obtained from many tree and other sources anywhere in the world. This is promising for further expansion of their production and use

(Pizzi 2016). One of the most acceptable hardeners now for tannin based adhesive is hexamine (hexamethylenetetramine). Hexamethylenetetramine has been used industrially already as a tannin adhesive hardener. Hexamine used as a hardener for a fast reacting species is not at all a formaldehyde-yielding compound, yielding only extremely low formaldehyde emissions in bonded wood joints (Pizzi 1994, 2013) because it does not decompose down to formaldehyde. ^{13}C NMR analysis has confirmed this mechanism (Pichelin et al. 1999; Kamoun and Pizzi 2000a, 2000b; Pizzi 2013). In presence of chemical species with very reactive nucleophilic sites, such as melamine, resorcinol and condensed flavonoid tannins, hexamine decomposition does not proceed up to formaldehyde and ammonia. It rather proceeds through reactive intermediates, hence mainly through the formation of reactive imines and iminoaminomethylene bases. Figure 1 shows the mechanism of Tannin based adhesive hardening by hexamine. The use of hexamine as a hardener of a tannin, hence a tannin-hexamine adhesive, is thus, a more environmentally friendly proposition than to use formaldehyde (Pizzi 2016, 2018).

Formaldehyde emission by UF resins is now the main concern for the production of industrial scale wood panels. That tannin based adhesives have already been

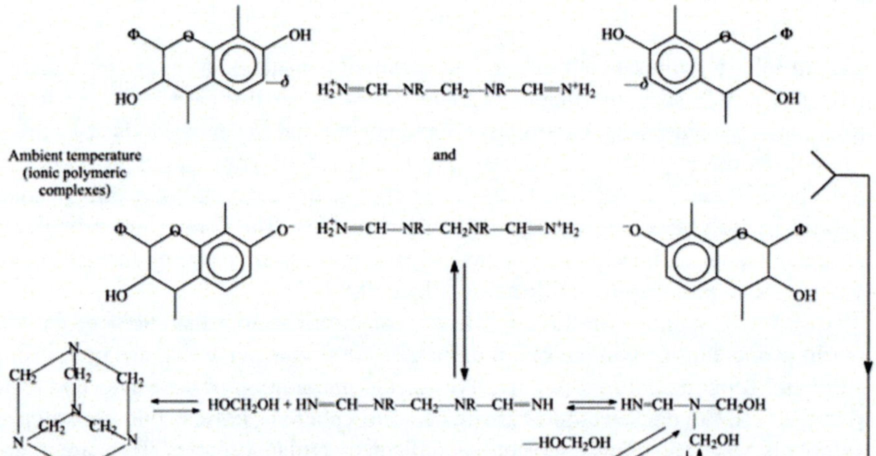

Fig. 1 Tannin based adhesive condensation by hexamine (Pizzi 2018)

successfully used in industrial scale is good news. Experimental results have shown their excellent performance as environment friendly materials and especially as non-formaldehyde-emitting adhesives. Several industrial applications of tannin adhesives are on record: for example, from the early 1970s until the middle 1990s two particleboard companies in South Africa used mimosa tannin adhesives. From the early 1970s to the present the same have been and are used by a company in Australia, mainly for housing suspended flooring (Pizzi, private communication); Since 1993 up to 2002 pine bark tannin adhesives (Pizzi et al. 1993) were continuously used for manufacturing industrially excellent quality exterior particleboard by a Chilean company, with MDF trials also giving good results (Valenzuela et al. 2012). The results of such a manufacturing production confirmed that pine bark tannin adhesives are usable industrially, at a reasonably low cost, for good wood panel production, satisfying both mechanical and emission requirements of world standards (Valenzuela et al. 2012).

Mimosa tannin adhesive were successfully used to produce eucalyptus particle board in industrial scale. The boards were produced on a Bison line Dieffenbacher single daylight press. The resultants boards were found to have excellent weather resistance properties and excellent durability compared to phenol-melamine-urea–formaldehyde bonded particle board (Van Niekerk and Pizzi 1994). Field tests of these tannin bonded particleboards for 15 years demonstrated the high performance of these board against unprotected exposure to the weather, even in the harsh, high altitude UV radiation of South Africa high plains (Van Niekerk and Pizzi 1994). Pine tannin based adhesives prepared by the reaction of polymeric MDI (4.4′-diphenylmethane diisocyanate) with sulphited pine tannin extract were used to produce industrial particleboard (Pizzi et al 1993). The reaction mechanism between tannin and MDI was demonstrated (Fig. 2). The results showed that internal bond (IB) strength and final properties as well as swelling parameters were improved compared to melamine formaldehyde control resins.

The use of tannins for industrial production has a long history. Thus, the use of tannin based fast-setting adhesive systems for exterior grade finger joints and glulam were developed and used industrially since 1973 to the present. Traditional adhesives for finger jointing are melamine-urea–formaldehyde (MUF) and phenol-resorcinol–formaldehyde (PRF). The results showed that tannin as an environment friendly material has a great potential for replacing synthetic resins in this case. The main advantages obtained from this research were considerable decrease in the adhesive cost and decreasing in time delay by improving the production flow (Pizzi et al. 1980; Mansouri et al. 2009). The industrial suitability of phenolic honeymoon fast-setting adhesives surface-grain gluing were proved by Pizzi et al. (1983) and the results showed that by using tannin in adhesive formulation acceptable results from delamination test and block shear test for glulam were achieved.

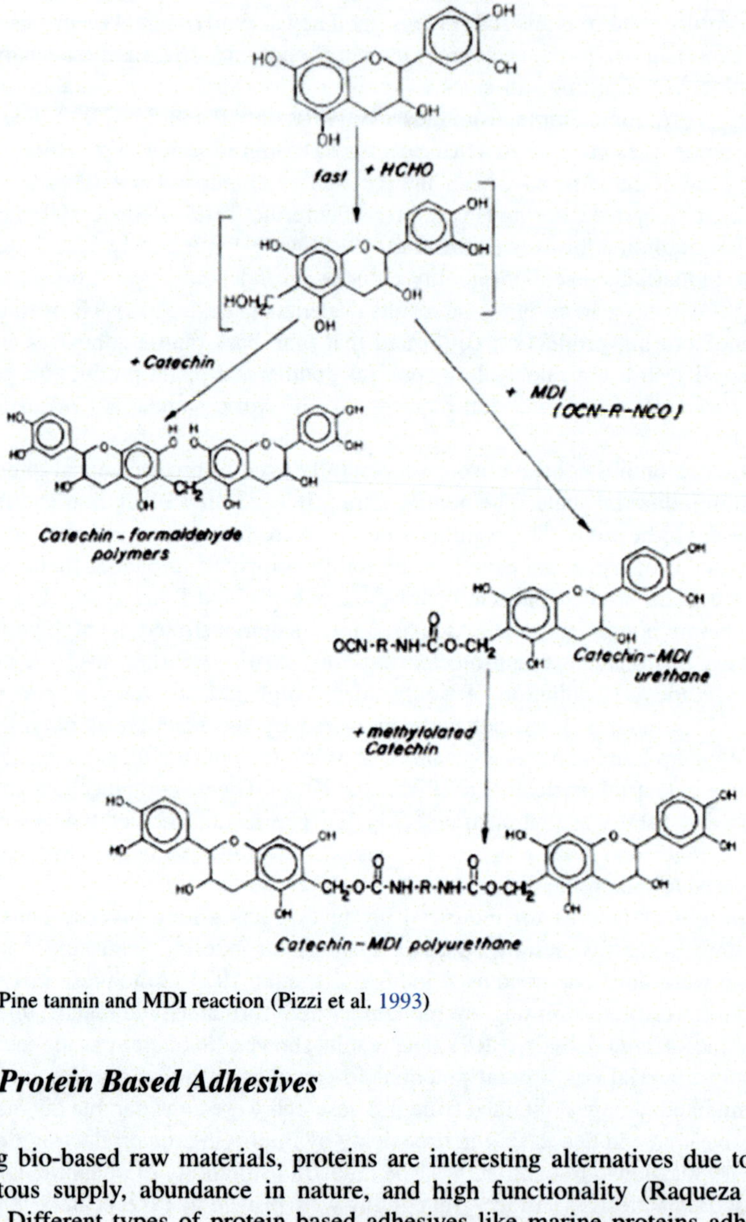

Fig. 2 Pine tannin and MDI reaction (Pizzi et al. 1993)

2.2 Protein Based Adhesives

Among bio-based raw materials, proteins are interesting alternatives due to their ubiquitous supply, abundance in nature, and high functionality (Raqueza et al. 2010). Different types of protein based adhesives like marine proteins adhesives containing high amounts of 3,4-dihydroxyphenylalanine (DOPA) (Rzepecki and Waite 1991), Casein based adhesives (Kehr and Sirch 1998; Krug and Tobisch 2010; van Herwijnen et al. 2016), marama protein adhesive (Amonsou et al. 2013), bone glue (Konnerth et al. 2009), Spent hen protein glue (Wang and Wu 2012) wheat protein (gluten) cold set adhesives (Amirou et al. 2019) and Dry blood glue (Lin and Gunasekaran 2010; Gunasekaran and Lin 2012; Garcia et al. 2013) have been used successfully to produce different bio-composites. For example, pure and cross-linked casein as binders for wood particleboard were investigated by some researchers. The results showed that achieving good internal bond (IB) strength, between 0.64 and 0.85 N/mm^2 (based on EN standard) is possible but, compared to traditional UF resin, press time is too long and unacceptable for industrial production (Kehr and Sirch 1998; Krug and Tobisch 2010; van Herwijnen et al. 2016) Also, bone glue as an animal protein was used as adhesive to produce wood particleboard. The results from this study indicated a 0.56 N/mm^2 IB strength in the particleboard prepared by using 10 to 20 percent of adhesive (Konnerth et al. 2009).

Industrial concerns for production speed and pressing factor seem to show that these adhesives appear not yet to be to the level of industrially acceptable hot press times and speed of production. However, to achieve faster press times appears to still be possible (Krug and Tobisch 2010). However, the high price of some bio proteins like pure casein and the need for replacing them with less expensive vegetable proteins has led to use soy protein as a priority for bio-adhesives. The by-product of the soybean oil extraction has a great potential to be used as a binder in the wood industry. Soy flour as well as soy protein based adhesives seem to have a great advantage over other bio-based ones because of their low price and availability at different quality levels, from cheap soy flour to the more expensive soy protein isolate. In 2018, the Food and Agriculture organization (FAO) of the United Nations reported the world production of soybeans to be approximately 312 million tons for the American continent, and 27 million tons for Asia with the clear soybean ability to be used as an eco-friendly adhesive in the world's wood industry.

Soy protein is composed of several amino acids, some of them being more reactive than others (Basso et al. 2015). Some technologies are able to cross-link proteins effectively. For example, thermostable networks have been obtained by reaction of protein skeletal amino groups, and also some amino and amido groups of casein reacted with formaldehyde or other aldehydes (Schwarzkopf et al. 2010). In such a process, methylene bridges are formed mainly by cross-linking with formaldehyde, whereas imine covalent bonds can be formed by reacting with glyoxal and glutaraldehyde (GA) (Schwarzkopf et al. 2010). According to previous research, soybean has a good potential for the preparation of bio-adhesives for industrial applications (Frihart and Satori 2013; Wescott et al. 2006). The technologies for soy adhesives use different crosslinking agents to provide potentials for adhesives to be used in plywood and particleboard and medium-density fiber board (MDFB) as well as other engineered wood products (Wolf 1970). Any knowledge about soybean flour composition and its protein structure as well as adhesion mechanisms will help scientists and industry to find proper methods in adhesive production and manufacture products with good specifications. Soybeans contain about 20% oil, 34% carbohydrates, 40% protein, and 4.9% ash (Wolf 1970). Soybean oil is mainly composed of saturated and unsaturated triglycerides (Kinsella 1979). Soybean carbohydrates consist of the complex polysaccharides including cellulose, hemicelluloses, and pectin. Based on protein content, the soybean products are classified as flour and grits (40–55 w/w% protein) and protein concentrate (70 w/w% protein) and soybean protein isolate (>90 w/w% protein) (Lundblad 2014). The composition of all three soybean products is given

Table 1 Composition of different defatted soybean products

Products (g/100 g)	SF	SPC	ISP
Protein	48	64	92
Carbohydrate	31–32	14–15	–
Fat	0.3	0.3	0.5
Fiber	3	4.5	<1
Ash	7	7	4
Moisture	10	10	<5

Adopted from Functional properties of soybean proteins [10]. SF: Soy flour; SPC: Soy protein concentrate; ISP: Soy protein isolate

in Table 1. Soybean-based adhesives, commonly called soy-based adhesives, have widely been used in the production of wood composites since the 1930s to 1960s (Frihart et al. 2010). Soy-based adhesives have many advantages such as low cost, easy handling, and low hot press temperature (Ken 1990).

Soybean protein is a storage protein. It has usually a very compact structure. Inside the compact structure, protein chains are held together by disulfide bonds, hydrogen bonds, electrostatic attraction between oppositely charged groups, and hydrophobic interactions (Ken 1990). Figure 3 shows a model for compact protein structure formation.

Several works have suggested that unfolding the soy protein structure significantly improves the strength and the water-resistance of wood composite panels bonded

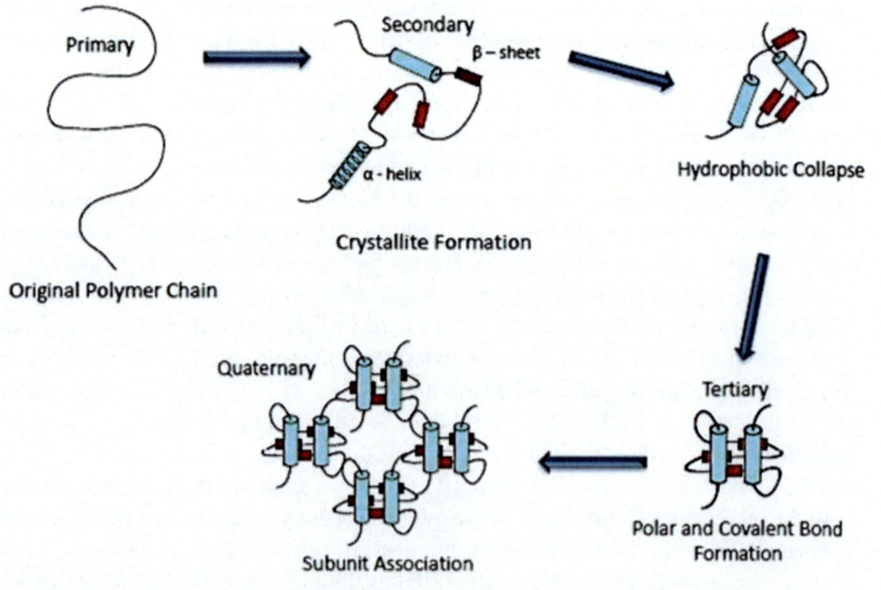

Fig. 3 Protein structure formation (Mo et al. 2003)

with the modified soy protein (Hettiarachchy et al. 1995a, 1995b). The mechanisms proposed for such an improvement were as follows: when the compact protein structure is unfolded, the protein chains can spread well on the wood surfaces and can easily penetrate into the wood (Huang and Sun 2000a). Figure 4 shows schematically the denaturation and unfolding of the protein structure.

All the functional groups; such as amino groups, carboxylic acid groups, as well as hydroxyl groups in the unfolded soybean protein structures which can react with the wood components and also other cross-linking agents will result in strong bonds. Soybean protein modification and unfolding is necessary for achieving good adhesion as well as good properties in the products bonded with soybean based adhesive. Different chemicals can be used to unfold soybean proteins. Alkalis, organic solvents, surfactants, urea, guanidine, and protease enzymes are usually used to modify/unfold soy proteins (Hettiarachchy et al. 1995a, 1995b; Hettiarachchy and Kalapathy 1998; Huang and Sun 2000b). Many researchers have already tried to prepare acceptable soy-based resin, but most of them have shown some drawbacks of same kind or other. For example, some of these adhesives, e.g., soybean flour and soy protein adhesive, are not sufficiently resistant against moisture absorption even to withstand indoor application. The higher viscosity of soy based adhesives is also another limitation for industrial application to produce fiber board and particleboard. The improvemet of the humidity resistance of these adhesives has been singled out for additional research efforts (Hettiarachchy et al. 1995a, 1995b; Sun and Bian 1999; Gu and Li 2011; Zhu and Damodaran 2014). From previous work, Polyamidoamine-epichlorohydrin

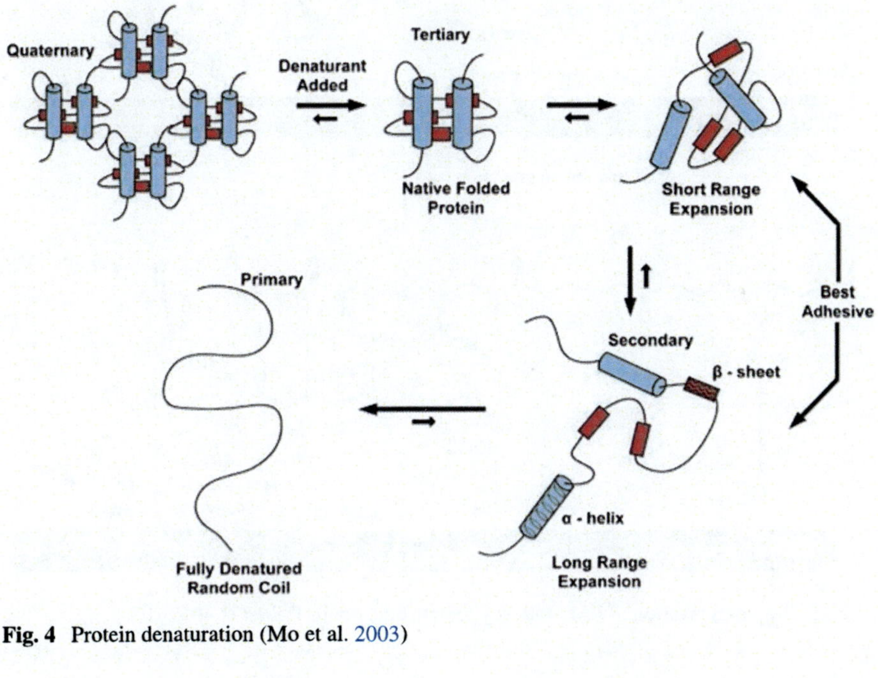

Fig. 4 Protein denaturation (Mo et al. 2003)

(PAE) is an excellent cross-linker for soy protein, while other protein cross-linkers include polyamides, polyethyleneimines (PEI), pMDI, and ketones (Li et al. 2004; Li 2007a, 2007b, 2007c; Hamarneh et al. 2010a, 2010b; Gu and Li 2011). Figure 5 shows the interaction and proposed adhesion mechanism of PAE and soy protein. In competition with the other bio-adhesives, there are several commercial PAE cross-linked soy proteins adhesive systems (Kymene®, SOYBABY®, OZERO®, Soyad™, Prolia™) in the world, although all these have the defect of relatively slow hot press times.

Thus, to decrease the humidity susceptibility of panels glued with these adhesives, several chemical treatments have been tried. Among these polyamidoamine-epichlorohydrin (PAE), sodium dodecyl sulfate (SDS) and even phenol formaldehyde (PF) (Huang et al. 2000a; Wescott et al. 2006; Allen et al. 2010a, 2010b). Because of the problems about environment pollution, there is great interest instead in substituting synthetic materials in adhesives to obtain a very high bio-based content. For this reason, using natural modifiers like vegetable tannins (chestnut and mimosa tannins) were considered (Ghahri et al. 2017; Ghahri and Pizzi 2018). The results showed

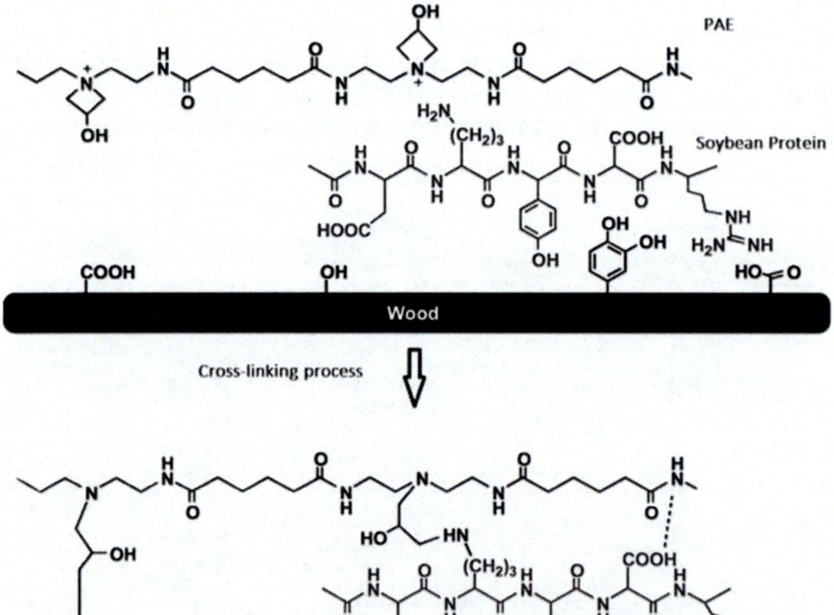

Fig. 5 Proposed reaction of PAE with soy protein and wood (Frihart et al. 2010)

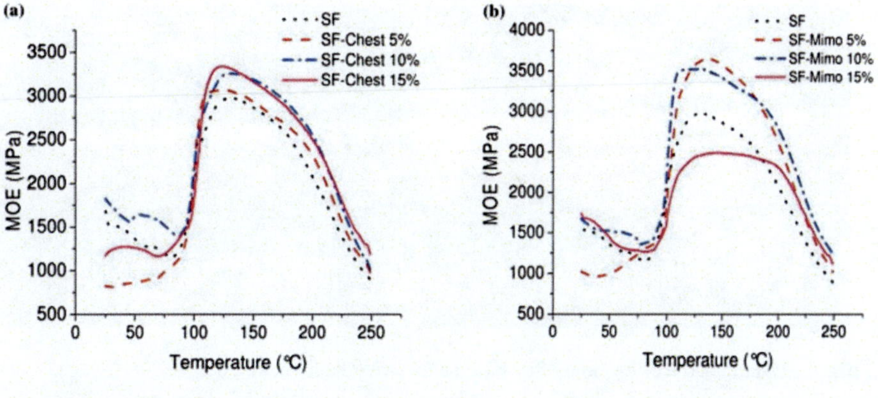

Fig. 6 TMA curves of the mean Young's modulus(MOE) variation as a function of the temperature for a beech joint bonded with soyflour/tannin adhesives during this latter hardening. **a** Soy flour. **b** Soy flour/tannin adhesives at 95:5; 90:10; 85:15 soy:tannin weight ratios, at 10°c/min rate of temperature rise. Soy flour = SF; Chestnut tannin = Chest; Mimosa tannin = Mimo (Ghahri and Pizzi 2018)

that natural tannins increased moisture resistance and mechanical strength of manufactured soy adhesives bonded bio composites, successfully. MALDI-TOF analysis has confirmed the reactions between soy amino acids and tannins components by ionic and covalent bonds. Thermo-mechanical analysis (TMA) also confirmed good reaction between soy flour and soy isolated protein and mimosa and chestnut tannins by the higher MOEs achieved.

A hydrolysable tannin, namely chestnut wood tannin extract, and a flavonoid tannin, namely wattle bark (*Acacia mearnsii* de Wildt) extract have been added to soy as modifiers to constitute the improved adhesives to bond particleboards. The TMA analysis (Fig. 6) proved that the improvement appeared to be due to a facile reaction occurring between the hydrolysable tannin and soy flour. Gel time and gel temperature were shortened by the addition of any of the two tannins (Ghahri and Pizzi 2018).

Soy based adhesive were successfully used to produce plywood with higher mechanical properties like dry and wet shear resistance and excellent water resistance (Fig. 7). Comparative delamination cyclic tests were performed on tannin modified soy based adhesive bonded poplar plywood versus urea formaldehyde bonded plywood. This showed that boards of excellent durability can be produced by modified soy adhesive (Ghahri et al. 2017).

Soy adhesive have excellent characteristics. Be it in the form of either soy flour or soy protein and its hydrolysate it bonds well, it is environment friendly, it is prepared with ease, it has no toxic emissions, it is easily available and abundant, and it is relatively inexpensive. It recommends soybean as suitable biomaterial for replacing synthetic and carcinogenic chemicals. However, besides these notable advantages, there are also some problems such as poor water-repellent properties, high viscosity, and low solids, and longer press times which are required to be improved. Several

Fig. 7 Manufactured plywood by Soy based adhesive (Ghahri et al. 2017)

methods and chemicals as cross-linking agents already exist for soybean-adhesive improvement. Recent studies have shown that modified soybean-based adhesives have the same properties as traditional applied synthetic resins in industry. The main future challenges for soybean-based adhesives will be to avoid their low solids and to try to decrease their modifiers dependency from petrochemicals.

Nowadays, commercial soy-based adhesives are used in the wood product industry. Soy protein-Kymene® adhesive system has become widely used for interior plywood and engineered wood flooring with some limited acceptance in particleboard (Li 2007a, 2007b, 2007c; Li et al. 2004; Allen et al. 2009; Wescott et al. 2006). The adhesive consist of soy flour (SF) and PAE resin. The bonded wood composites have shear strengths comparable to or higher than those bonded with commercial UF resins. The original studies on the soy-PAE adhesives used purified soy protein isolate (SPI), but the SF with much lower cost is now used commercially (Vnucec et al. 2016). In the USA it is since 2008 that the seven hardwood plywood factories of Columbia Forest Products, for the production of "PureBond®" interior-grade plywood (Malin 2005), have substituted UF resins with "SoyAd™" (Anonymous 2012), a soy-based adhesive developed and manufactured by Ashland in which 15% polyamide has been added to SPI (soy protein isolate) (Malin 2005). Other commercial soy based adhesive like SOYBABY® and OZERO® are successfully used industrially, and to a great extent, in China to bond MDF, HDF and laminated flooring (Gui et al. 2013; Yang et al. 2011; Vnucec et al. 2016). SOYBABY® has much better moisture resistance than UF resins. Cargill's Prolia™ industry which produce different type of wood panels like plywood and wood particleboard are using Soyad™ adhesives (Cargill 2015). Soyad™ adhesive is a kind of soy-PAE adhesive prepared with soybean flour and PAE resin (Solenis 2015). When SF and PAE blended together, the PAE reacts with the protein amino acids in the SF to form a durable and water resistant thermosetting adhesive that is comparable in strength and performance to that of UF-based resins. Soyad™ adhesives are used for hardwood plywood, particleboard, MDF and engineered wood flooring (Solenis 2015).

Soy protein isolate (SPI) and insoluble soy flour polymeric carbohydrates have been reacted with sodium periodate for specific oxidation of vicinal –OH groups to investigate the reactions involved in this approach to soy flour adhesives. The reactions have been shown to generate carbohydrate oligomer fractions presenting one, two or multiple aldehyde groups. With the exception of the small molecular weight heptanedial, the smaller molecular weight aldehydes generated from mono- and disaccharides by the same reaction do not appear to form from the insoluble soy flour carbohydrates, or have already reacted. Reaction of periodate with soy protein isolate has been shown to generate some aldehydes too (Fig. 8). When the mix of SPI and soy insoluble carbohydrates is treated with periodate the majority of the aldehyde carrying species observed appear to be higher molecular weight carbohydrate oligomer fractions (Frihart and Lorenz 2019). This approach has been shown to be well adaptable to soy flour to yield excellent plywood panels by the non-volatile non-toxic and higher molecular weight aldehydes produced by a small proportion of periodate specific oxidation of the carbohydrates in soy flour reacting with the soy protein in the soy flour itself (Frihart et al. 2019). Moreover, this approach has been successfully applied to tannin adhesives by generating the same aldehydes from the carbohydrate fractions in tannin extract or from added mono and disaccharides, as well as to lignin, with excellent results (Xi et al. 2020; Chen et al. 2020).

Wheat gluten protein hydrolysate has also been used as a biomass feedstock to prepare environmentally friendly protein-based adhesives. Glutaraldehyde was used to modify it to obtain a glutaradehyde-wheat protein (GP) adhesive (Xi et al. 2019). Polyethylenimine (PEI) was also used as a crosslinking agent. Plywood has been prepared and tested giving excellent performance. The results shows that modification by glutaraldehyde can effectively improve the bonding performance of wheat protein adhesives, the plywood bonded strength having been improved by its addition. The effect of PEI as a crosslinking agent became also evident.

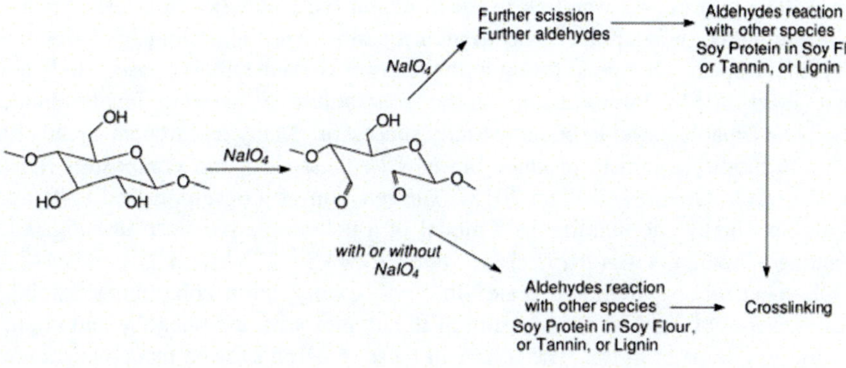

Fig. 8 Reaction of periodate with soy protein isolate (Frihart et al. 2019)

2.3 Lignin Based Adhesives

The term "lignin" for the first time was used by the Swiss botanist A. P. Candolle. Lignin, comes from the word "lignum", which in Latin means "wood". Lignin, after cellulose, is the second most abundant bio-polymer on earth and it is one of the main polymers of the secondary cell walls of woody plants (Kirk-Othmer 2006). Moreover, it is a material that has a main role in the carbon cycle because it is an excellent source of non-fossil carbon (Whetten and Sederoff 1995; Boerjan et al. 2003). Depending on its origin, it has a more or less complex structure. The most complete structural model of lignin was suggested by Adler in 1977 (Fig. 9) (Alder 1997).

There are three monolignols which enter in the constitution of all kinds of lignin in plants.: p-coumaryl alcohol (M1H), coniferyl alcohol (M1G), and sinapyl alcohol (M1S) are the three main monolignols which produce p-hydroxyphenyl, guaiacyl, and syringyl residues in the polymer (Vanholme et al. 2008). Figure 10 shows repeat units of lignin and its derivatives.

Lignin, due to its simil-phenolic nature, is reputed to be the most intensely researched biopolymer for wood adhesives. It is the second most abundant biopolymer on earth and is inexpensive, but it is also relatively unreactive towards aldehydes, formaldehyde included. Nimz (1983), Newman and Glasser (1985), Gardner and Sellers (1986), Shimatani et al. (1994), Calvé (1999), Viikari et al. (1999), Blanchet et al. (2000), Kim and Kim (2003), Lopez-Suevos and Riedl (2003), Pizzi (2016) and Chen et al. (2020) have written detailed reviews on lignin adhesives that merit to be consulted for further information.

Earlier on in the history of wood adhesives lignin or lignosulphonates have been added to synthetic phenolic adhesives, and even to aminoplastic thermoset adhesives. The aim of this was first of all to render the synthetic resin less expensive. Unfortunately, such a practice misfired as the more was the lignin added, the longer became the panels hot press times, with the lower cost advantage consequently being lost (Pizzi 2016). Moreover, none of the wood adhesives based on just lignin alone has been capable as yet to be industrially successful. Many researchers already have tried to modify lignin to produce bio-based adhesive for bio composites (Ghahri et al. 2016b; Ghahri and Pizzi 2018). The reaction of a desulphurized kraft lignin with hexamethylene diamine as a model of a polyamine has been investigated by Santiago-Medina et al. (2017). Solid state CP-MAS ^{13}C NMR, FTIR and MALDI-TOF spectroscopy studies revealed that polycondensation compounds leading to resins were obtained by the reaction of the amines with the phenolic and aliphatic hydroxyl groups of lignin. The results of these research showed that covalent bonds are formed by reaction between the amine and the aromatic rings of lignin by substitution of the lignin phenolic –OH groups and same bonds appears by reaction between the amine and the aliphatic side chain of lignin by substitution of the lignin alcohol –OH groups of the lignin side chain (Santiago-Medina et al. 2017).

Generally, two different types of lignins, namely Kraft lignin and Lignosulfonate, have been used to prepare lignin-based adhesives for bonding wood panels. However,

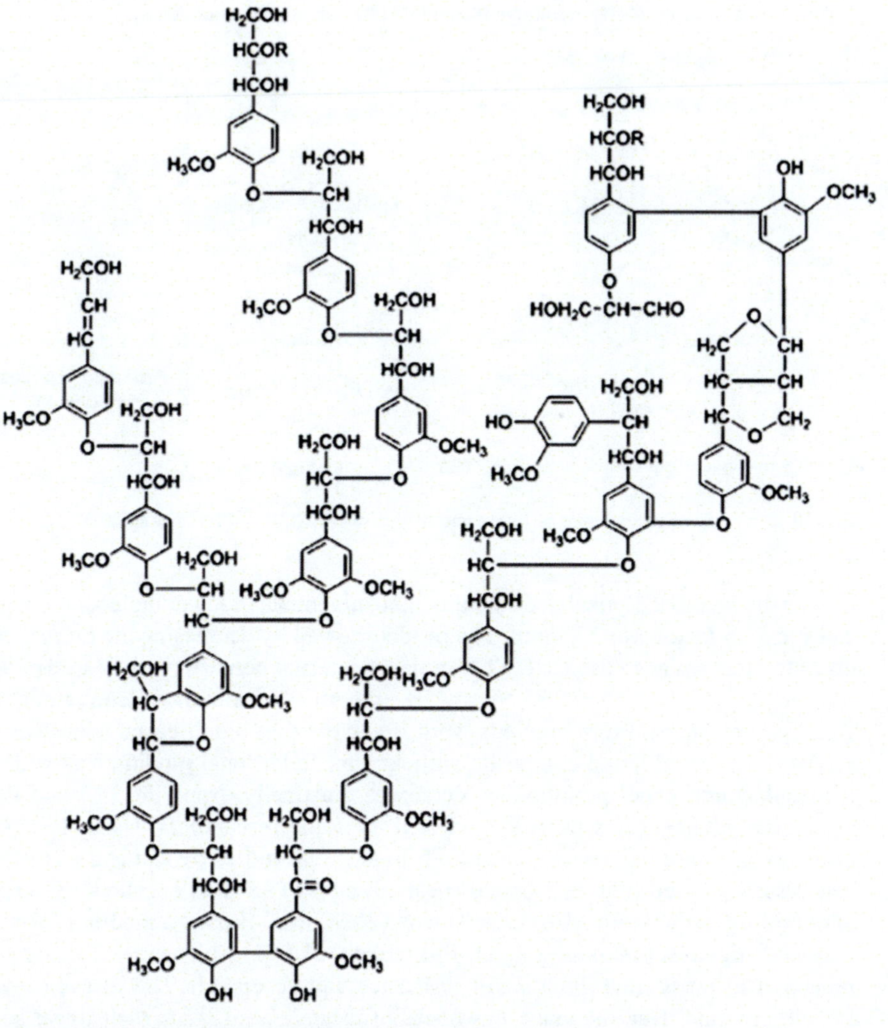

Fig. 9 Adler's structural model of lignin (Adler 1997)

their use has always led to adhesives with some serious problem or unacceptable limitations to meet industrial production requirements. Consequently, most research has been aimed to improving their performance. Thus, the addition of chemicals has been researched to improve the performance of lignin adhesives-bonded wood panels such as glyoxal, pMDI, furfuryl alcohol, polyethylenimine (PEI), polyamidoamine-epichlorohydrin (PAE), soy protein, tannin, synthetic phenolic resins, and even others (Philippou et al. 1982; Li and Geng 2004; Liu and Li 2006; El Mansouri et al. 2007a, b; Lei et al. 2008; Xiao et al. 2013; Hunt et al. 2017). El Mansouri et al. (2007a, b) within their research reached up to 0.81 N/mm^2 IB strength of wood particleboard prepared by cross-linking with polymeric diphenylmethane isocyanate a 10% total adhesive

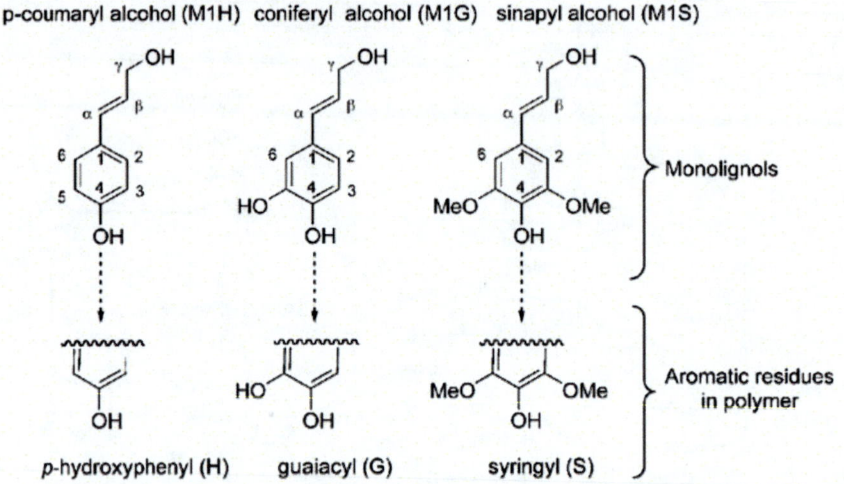

Fig. 10 Structure of repeat units of lignin and its derivatives (Calvo-Flores et al. 2010)

load on dry wood of glyoxalated calcium lignosulphonate of lower molecular weight. The pressing factor was 32 s/mm. The results showed by decreasing the press factor down to 15 s/mm decreased to 0.67 N/mm^2 the internal bond strength. Based on EN standards (EN 312-2 1996) 0.35 N/mm^2 IB strength is the minimum requirement for wood particleboard. From previous work it is clear that lignin-based adhesives do not work for wood based composite without some additional modifiers as without it manufactured panel performance decreases drastically. Applying PEI and PAE as a crosslinking agents were investigated by some researchers. Using PEI as a crosslinking agent for demethylated kraft lignin was studied by Liu and Li (2006). The results showed high dry tensile shear strength (TSS) and excellent resistance after boiling test as well. Also, Li and Geng (2004) used PAE as a modifier together with kraft lignin for preparing lignin-PAE adhesive. The results showed acceptable reduction of press time, high water resistance and acceptable TSS at even lower PAE/lignin ratio. But, the main drawbacks of these adhesives are their lower solid content (16–20% wt/wt) which cause a high adhesive consumption and long press time in wood particle composites produced in industrial scale.

One of the bio materials already used as a cross-linking agent for lignin is soy protein. Soy protein because of the several functional groups used with sorghum lignin by Xiao et al. (2013). The results of this study also showed high TSS (6.2–6.6 N/mm^2) for lap joint of 3 mm thick veneers. But the press time was again too long. For the first time, low molecular mass lignin was modified by natural phenolic compound to produce wood adhesive by Navarrete et al. (2010). They were used mimosa tannin extract (origin: Tanzania. Supplied by: Silva, S. Michele Mondovi', Italy) and hexamine as tannin hardener to cross-link glyoxalated lignin. This adhesive was used to produce particleboards (adhesive usage 10% based on dry wood) and two types of plywoods (adhesive usage 380 g/m^2 double glueline) and the results showed that all types of prepared composites met interior standards (Navarrete et al. 2010). The Schematic representation of the series of reactions occurring in the formation of the tannin-lignin hardened network were shown at Fig. 11.

Notwithstanding the major drawbacks of lignin adhesives, some industrial companies attempt to modify lignin and to use this adhesive to produce wood based composites. Because of its natural abundance, extremely low cost and phenolic structure, lignin does appear to be one of the best candidates for adhesives preparation. Lignin as a natural phenolic adhesive and as a byproduct of the wood pulping and paper industry is an attractive raw material. Forss (1976) described new lignin based adhesive, namely Karatex, for plywood, particleboard and fiberboard. According to Forss (1976) the amount of lignin in lignin-PF adhesives for particleboard, plywood and fiberboard can be increased to 40 to 70% if a high molecular weight fraction (molecular weight > 5000) of either lignosulfonate or black liquor, obtained from alkaline pulping of wood, is applied. Fractionation of spent sulfite liquor (SSL), black liquor, can be achieved by ultrafiltration (Pizzi 2003). According to the authors the higher effectiveness of high-molecular-weight lignin molecules is due to their higher level of cross-linking, which requires less pH for the formation of an insoluble copolymer than do low-molecular-weight lignin molecules. However, from previous research, the low-molecular-weight ammonium based SSL is more effective. Forss (1976) appear not to have checked the influence of inorganic salts in SSL or black liquor that are separated off during ultrafiltration. They assume that condensates between smaller lignin molecules are more reactive than are high molecular-weight molecules (Pizzi 2003). In the manufacture of particleboard, either high-frequency (HF) heating or combined contact/HF heating has been tried. In the latter case, the press platen temperature was 180 °C. German standard requirements for weather resistant particle boards to be met need pressing times between 10 and 12 s /mm and 8 to 12% lignin-PF adhesive, based on dry wood particles. One advantage inherent in the fractionation by ultrafiltration is that the lignin becomes more uniform and less dependent on variations in pulping conditions and of wood source, which sometimes cause serious problems in the application of technical lignins. Full-scale plywood mill tests, some of them running continuously for several weeks, appear to have been performed in

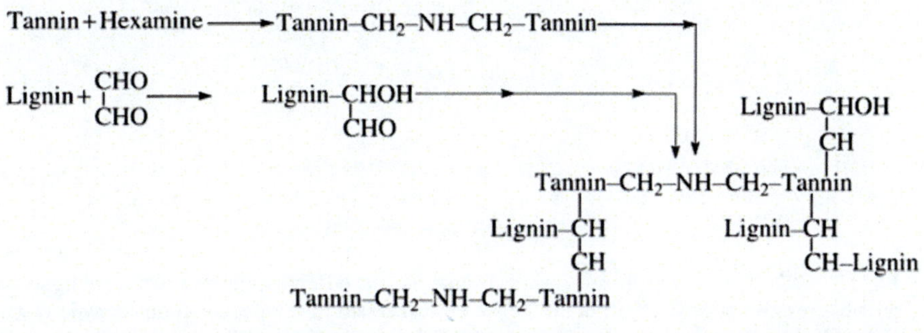

Fig. 11 Series of reactions occurring in the formation of the tannin/lignin adhesive (Navarrete et al. 2010)

two Finnish plywood mills. Again, this does not appear to be in operation anymore (Pizzi 2003).

Lignin, a completely bio-based, non-toxic, non-hazardous raw material is available, in industrial quantities, at constant and controlled quality. One of the most used synthetic resins to cross-link lignin are PF resins. Figure 12 shows the possible schematics of cross-linking reactions between lignin and PF resins. In recent decades, Commercial plywood adhesive based on PF-lignin have been developed by UPM biochemical (Helsinki, Finland). Also, UPM is the largest plywood producer company in Europe with approximately 1 million cubic meter production per year. UPM BioPiva™ claims that they have prepared and commercialized a plywood adhesive composed of 50% phenol and 50% lignin coreacted with formaldehyde. The recent achievement of UPM BioPiva™ is replacing 75% PF resin by lignin to produce lignin based phenolic resin. Also, the UPM Plywood company is producing

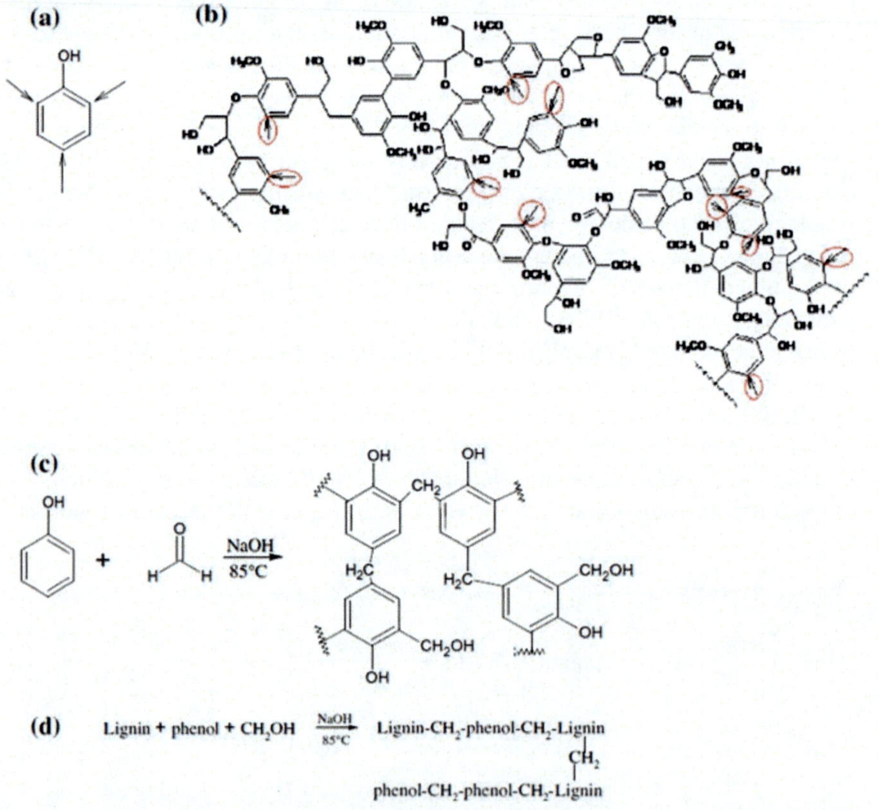

Fig. 12 **a** Reactive sites of phenol towards formaldehyde; **b** Representative structure of lignin with possible reactive sites (Parajuli et al. 2005); **c** Mechanism of the phenol–formaldehyde reaction under alkaline conditions (Pizzi and Mittal 2003; Wu et al. 2005); **d** Proposed mechanism of the phenol-lignin-formaldehyde reaction (Grishechko et al. 2013)

industrial birch plywood with panel thickness from 6.5 to 30 mm with 155 g/m^2 sgl glue factor and 0.75 min/mm press factor with their 50/50 lignin-PF adhesive. The results from shear test (based on EN 314–1 standard) after soaking in cold and boiling water showed higher quality of boards with acceptable formaldehyde release based on EN 717-2 and the ability to be used under outdoor conditions (Hunt et al. 2017).

2.4 Carbohydrate Based Adhesives

Carbohydrate-based adhesives have long been known and used with all sorts of raw materials, such as monosaccharides, gums and polysaccharides of all sorts. There are three known approaches in using these materials for wood adhesives: (i) directly as wood adhesives, (ii) as additives to synthetic thermosetting resins, (iii) by degrading them to elementary chemical building blocks to be used to build-up the adhesive, (Pizzi 2016). The approach (ii) above consists in preparing, under acid treatment conditions of carbohydrates, furfural and furfuryl alcohol as the building blocks of furan resins which classify them as bioresins too. They are already used for a long time indeed as foundry sand binders, and in wood preservation. For more detailed reviews on them cf. Belgacem and Gandini (2003). However, although extensively used for much more value added industrial processes such as in foundries, and the good performance as wood adhesives of furan resins, the two building blocks mentioned above are rather costly and almost black in color, both of these being real disqualifiers for their use as wood panel adhesives (Pizzi 2016). Carbohydrates-based wood adhesives can also be prepared by dissolving carbohydrates in strong alkali, this being to-day also an area of particular research interest. This is so because all sorts of lignocellulosic materials, even agriculture waste, can be dissolved and used as an adhesive with such an approach, and more detailed reviews on the subject do exist (Chen 1996; Pizzi 2016). Different types of Carbohydrate based adhesives like starch based adhesives (Eriksson et al. 2011; Amini et al. 2013; Selamat et al. 2014), corn flour based adhesives (Moubarik et al. 2009, 2013a), glucose based adhesives (Jackson et al. 2008, 2010; ECOSE 2017; Insulation 2017), sucrose based adhesives (Umemura et al. 2013; Widyorini et al. 2016a, 2016b) and chitosan (Umemura et al. 2003; 2010) are used successfully to produce different bio-composites.

Starch is the major carbohydrate reserve in plant tubers and seed endosperm where it is found in each as granules. Typically, it contains several million amylopectin molecules accompanied by a much larger number of smaller amylose molecules. By far the largest source of starch is corn (maize) with other commonly used sources being wheat, potato, tapioca and rice (Bodyfelt et al. 1988). Starch is a biopolymer and consists of two major components: amylose and amylopectin. Amylose that builds up to 15–35% of the granules in most plants, is a primarily linear polysaccharide with a-(1–4)-linked D-glucose units. Some amylose molecules, particularly those of large molecular weight may have up to ten or more branches (Hizukuri et al. 1981). Amylopectin, is a highly branched molecule, with a-(1–4)- linked D-glucose

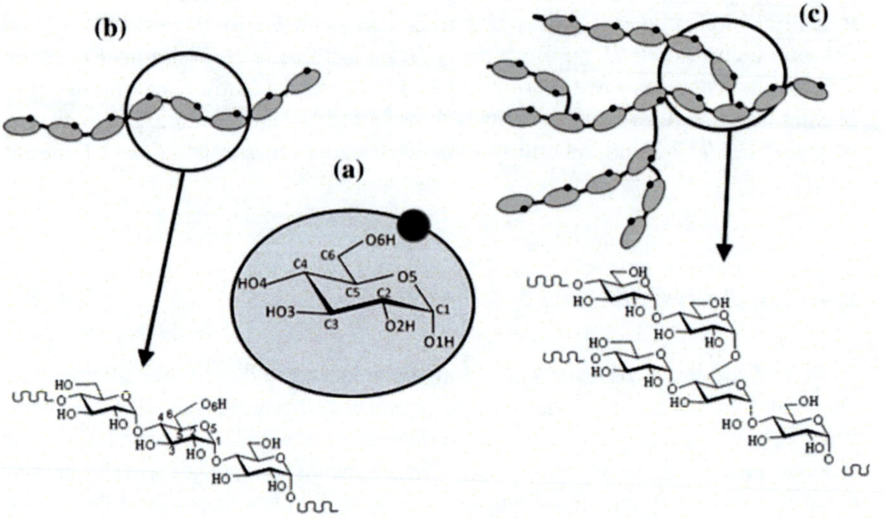

Fig. 13 Basic structural of **a** D-glucose, **b** amylose and **c** amylopectin (Pérez and Bertoft 2010)

backbones and exhibits about 5% of a-(1–6)-linked branches, which have a profound effect on its physical and biological properties (Pérez and Bertoft 2010) (Fig. 13).

Starch is rather inexpensive and easily available everywhere. Starch gelatinizes when heated in water. Thus, it forms a viscous but still fluid gel by cleaving amylopectins and amylose. The viscous fluid gel then "retrogrades" on cooling to an effective, not fluid gel due mainly to amylose internal rearrangements (Slot et al. 2019). According to research reports, it was indicated that the chemical treatment is very useful way to achieve good adhesion properties in the starch-based adhesives. Different methods and chemicals were used for starch adhesive modification. Several chemical modification strategies; such as using carboxymethylation (Selamat et al. 2014), glutaraldehyde (Amini et al. 2013), tannins (Moubarik et al. 2013a), urea formaldehyde (Moubarik et al. 2013b), phenol formaldehyde (Moubarik et al. 2009) polyvinyl amine (Eriksson et al. 2011) for starch were done to improve the adhesion properties of starch based adhesives. Both native and carboxymethylated starch in powder form, obtained from the trunk of an oil palm, were used by Selamat et al. (2014) to try to prepare adhesives. This adhesive was used to produce particleboard and the results showed excellent IB strength between 1 to 1.3 N/mm^2 but the major problems are the very long press time (240 s/mm) and high thickness swelling (Selamat et al. 2014). In other research, corn starch slurry was modified by glutaraldehyde and was used to produce particleboard. Notwithstanding the longer press times (240 s/mm), good IB strengths (0.6–0.9 N/mm^2) were obtained but, thickness swelling was relatively higher (30–35% for 2 h immersion) than EN standards requirements (Amini et al. 2013).

Moubarik et al. (2013a) used corn flour (73% starch) based adhesives to produce interior grade wood particleboard. They used mimosa tannin to modify corn flour. The laboratory results showed that the optimized IB strength of the particleboards were in the 0.48–0.50 N/mm^2 range. Moreover, the formaldehyde emission levels obtained from boards bonded with the corn flour-tannin modified adhesive were 0.4 mg/100 g, thus considerably lower than boards bonded with the control urea formaldehyde adhesive (3.7 mg/100 g). Based on the relevant EN standard, the IB strength must be 0.35 N/mm^2 or higher, and the formaldehyde emission must be 6.5 mg average with no value higher than 8 mg as perforator value for emission class E1 resins (Moubarik et al. 2013a). Moubarik et al. (2009) used cornstarch and quebracho tannin together with phenol–formaldehyde (PF) resins to manufacture plywood. The results showed that excellent shear strength and adhesion properties were obtained for plywood panels when bonded with resins with up to 20% substitution of PF by cornstarch and quebracho tannin. Moreover, small joint proportions of corn starch and quebracho tannin improved the PF resin boiling water resistance and decreased formaldehyde emission compared to the panels bonded with the control PF resin (Moubarik et al. 2009).

The use of starch based adhesive to produce industrial wood panels was investigated by Moubarik et al. (2013b). A corn flour–urea formaldehyde adhesive was used to produce wood particleboard. Industrial particleboard panels of dimension 8.2 m × 1.85 m × 19 mm were manufactured at the EGGER-ROL (France) Group. 9% (w/w based on dry particles) of adhesive resin solid was loaded, at a total pressing time of 2.6 min and pressing temperature of 220 °C. According to the results, physical and mechanical properties of industrial particleboard panels increased by using corn flour in the UF resin, and the particleboard panels formaldehyde emission decreased (Moubarik et al. 2013b). Other attempts to produce industrial corn starch based adhesive for wood particleboard were tried, again by Moubarik et al. (2013c). Laboratory cornstarch-mimosa tannin-based resins were introduced in a UF classic adhesive formulation to substitute part of the UF resin. Based on the results obtained, industrial PB panels bonded with optimal cornstarch-mimosa tannin-UF (10:4:86; mass ratio) resins showed comparable mechanical properties to those of boards bonded with commercial UF resins. Introducing corn starch-tannin into the UF resin causes a decrease in formaldehyde emission and improves water resistance of the panels. This research clearly showed the cornstarch and tannin potential to replace part of the UF resin (Moubarik et al. 2013c).

2.5 Sugar-Based Adhesive

In 2004, new carbohydrate based adhesive were prepared by Capps (2004) by mixing corn syrup, isocyanate, polyols and water. This resin was used to produce laboratory particleboard of IB strength of 0.8 N/mm^2. Two patents were lodged in 2008 and 2014 for new carbohydrate based adhesives by Jackson et al. In these patents they claimed new adhesive based on using citric acid, glucose and a nitrogen source. These adhesives are available as the ECOSE® commercial brand. These adhesives were successfully used to produce PB, MDF and oriented strand board (OSB) (Jackson et al. 2008, 2010; ECOSE 2017; Insulation 2017). Combination of 75% sucrose and 25% citric acid as an aqueous solution as an adhesive was used to produce particleboards. Umemura et al. (2013) produced high density (800 kg/m^3) wood particleboard with 66 s/mm press factor and 5 to 40% resin content. The results showed that 30% resin content yielded the highest IB strength and the lowest water swelling (12%). Kusumah et al. (2017) prepared sorghum bagasse particle boards in which the relative proportions of citric acid and sucrose used to prepare the adhesive differed. Namely, 90:10 and 85:15 ratios of citric acid to sucrose were used. These boards presented a higher IB strength (1.17 N/mm^2). But the main problem was the long press time (67 s/mm) and the higher moisture content of the wood particles after gluing. Based on the good crosslinking properties of condensed tannins, mimosa tannin was used as a modifier for sucrose-based adhesive. Zhao and Umemura (2014) used 10 to 40% mimosa tannin-sucrose adhesive to manufacturing PB. The results showed that 1 to 1.2 N/mm^2 IB strength were achieved for thin high density boards. However, the too long press time for the boards preparation are the main weakness of this adhesive for industrial application, this being due to the adhesive low solid content. Citric acid, was also found to be able to bond alone plywood and LVL veneers by the formation of cross-links directly with the wood veneers carbohydrates and even its lignin (del Menezzi et al. 2018) as well as improving sensibly by the same mechanism the water resistance of wood joints welded by mechanical friction (Amirou et al. 2017).

2.6 *Eco-Friendly Adhesives Industrial Challenges and Opportunities in Future*

Different types of adhesives are currently used in the woodworking industry. Synthetic resin based on the reaction with formaldehyde, especially UF resins, represent a huge volume of the adhesives for wood (Solt et al. 2019). These resins are prepared by the reaction of carcinogenic formaldehyde with various chemicals such as urea, melamine and phenol. The main impetus for today's renewed interest in bio-based adhesives is the acute sensitivity of the general public to anything that has to do with the environment and its protection. It is not even this concern per se that motivates such an interest. There are rather very strict, for some synthetic adhesives, almost crippling, government regulations which are just starting to be implemented to allay the environmental concerns of the public (Pizzi 2013).

Because of the world human health problems and legislation restriction laws about formaldehyde gas emission, environment friendly adhesives have a good opportunity for being used in industry. Generally, environment friendly adhesives do not have any toxic and carcinogenic gas emissions, their raw materials are abundant in nature and they can present good adhesion properties by little modification. These parameters cause such adhesives to become good candidates to replace common synthetic resins like UFs. However, all this notwithstanding, by scanning the literature mentioned above, these adhesives have great challenges to overcome to be used as industrial binders. Too long hot press times, high viscosity and low solid contents limit somewhat their application in some wood industry sectors such as fiberboard and particleboard.

2.6.1 Overview of Requirements Concerning Eco-Friendly Adhesives for the Woodworking Industry

For all industry everywhere the most important issue is the cost–benefit equation. Generally, this equation relates to the cost of energy, raw materials, production instruments, employees salaries etc. Industry, because of constant cost and benefit pressure to achieve high benefits needs to increase the products prices and improve production yields. Thus, price setting in the normal economy is related to the supply–demand chain. Consequently, industrial companies want to increase their benefits by reducing their production costs. The first and best solution to keep production costs low is by using a high rate of production hence increasing production yield. Besides management issues, different parameters are involved in the rate of production, like instruments version, operator proficiency and raw materials quality. However, adhesives as one of the main raw materials in the wood industry, play a significant role with industry decision makers for costs and performance. Apart other effective parameters in wood based composites production (especially fiberboard and particleboard production lines) such as hot press actual pressure, press temperature, press speed, raw board thickness, moisture content, raw material quality and other forming line parameters, a fast curing resin with a fast gel time significantly increases both production rate and production yield. Thus, it is paramount for industry to use fast curing and low cost resins. Resin curing depends on some resin parameters like resin molar ratio, free-formaldehyde content and its solid content. If environment-friendly adhesives want to take synthetic resin position in the wood industry the markets should prepare themselves.

2.7 Eco-Friendly Adhesives Operational Limitation for Industrial Scale Production

2.7.1 Glue and Gluing Parameters

To achieve good adhesion in wood products, especially wood-based composites, several parameters have an important role. Some of them are related to glue properties such as glue solids content, viscosity and gel time as well as curing in the hot press. One of the important issue for applying eco-friendly adhesives at industrial scale is their solids content. For example, UF resins curing during fiberboard formation in a continuous Conti-Roll hot press strongly depend on resin solids. Higher solids content leads to high adhesion and cohesion values as well as high IB strengths. But applying higher solid content resin depends on two main factors: adhesive viscosity and gluing equipment capability. Thus, adhesive solids content should be considered with its viscosity and the gluing equipment which will be used in the production line. Some of the equipment like blow line systems strongly depend on glue viscosity and water content in the glue-mix. In some cases, the glue-mix high viscosity results in high solid contents increasing the spraying pressure in the nozzles, risking to block them and so stopping the production. Thus, considering the issues about applying synthetic adhesives in the production especially when blowline systems are used, solids content and viscosity are the most important parameters for environment-friendly adhesives.

Most of environment-friendly adhesives are water based and they have low solids content and high viscosity as well. Indeed, such adhesives are a mix of different component with high hydrophilic nature like carbohydrates and which cause a high viscosity of their solutions. Thus, the major problem of applying an environment-friendly adhesive through a blow line system is a high nozzle blocking risk during production. To avoid such a blocking risk, the adhesive viscosity should be adjusted carefully.

Another challenges for the industrial use of environment-friendly adhesives applied by a blow line system at an industrial scale are spraying and storage ability because of its molecular particle size and the suspension form of the final adhesive. In other words, the major drawback of these adhesives is their high precipitation risk in the adhesive tank and in the blow line equipment. This can cause big problems in continuous production like fiberboard especially if suddenly the production stops. Compared to bio-adhesives, UF resins, because they are in more homogenous solution and equal molecular size have a much lower precipitation risk in the adhesive storage tanks and nozzles. Another issue that needs considering is the use of other chemicals like fire retardants, fungicides and paraffin in the dosing station. Good results, should be achieved by preventing any undesirable interaction between the adhesive and other chemicals. Any unwanted reactions between chemicals causes the reduction of physical and mechanical strength especially of the IB strength of boards. There are a lot of factors about using adhesive in the production such as adhesive degree of polymerization, adhesive tack, transparency and color, water compatibility, pH, buffering capacity and glue hardening that need to be controlled before using a bio adhesive at industrial level.

3 Operational Limitation in Industrial Scale Production

3.1 Adhesives Limitation

Marine proteins give adhesives of high strength and good water resistance. However, they are scarcely available and expensive. DOPA (dihydroxy phenylalanine) is the main constituent of interest in these proteins (Rzepecki and Waite 1991).

3.1.1 Soy Based Adhesives Limitation

However, SOYBABY® has a disadvantage of high hot pressing temperature (higher than 170 °C), while OZERO® has the disadvantage of high adhesive viscosity (more than 50,000 mPas) (Yang et al. 2011; Gui et al. 2013). Their acceptance for particleboards is limited, while the high amount of moisture introduced during adhesive application is considered a serious issue (Solt et al. 2019). The high viscosity of these adhesives is their most severe limiting factor for being sprayed for particleboard manufacturing. Vnučec et al. (2016) have reviewed the use of these adhesives for particleboards. Other reviews are also present in the literature. Thus two soy adhesive, namely OZERO® (Gui et al. 2016) and SOYBABY® (2011) are reputed to be used in China for particleboard and MDF, respectively; but results and technical data are not reported. Columbia Forest Products soy-bonded PureBond® plywood is the largest user of formaldehyde-free soy-based wood panel adhesives. However, this "SoyAd™" adhesive still contains determinant and essential compounds obtained from oil (Anonymous 2012). The most developed crosslinker for soy protein is polyamidoamine-epichlorohydrin (PAE), while other protein crosslinkers include polyamides, polyethyleneimines (PEI), pMDI, and ketones (Li et al. 2004; Li 2007a, 2007b, 2007c; Hamarneh et al. 2010a, 2010b; Gu and Li 2011). Unlike the other biopolymers, there are several commercial adhesive systems (Kymene®, SOYBABY®, OZERO®, Soyad™, Prolia™) based on soy proteins, and they are used mainly for interior plywood and engineered wood flooring (Vnučec et al. 2016). Their acceptance for particleboards is limited, while the high amount of moisture introduced during adhesive application is considered a serious issue (Solt et al. 2019).

3.1.2 Soybean Adhesives Challenges and Opportunities in Future

Nowadays, there is some interest for industry to replace with soy adhesives the more currently used synthetic thermosetting wood panel adhesives, due to their easy availability, no formaldehyde, large quantity, and low cost. Between all protein sources available in the nature, it seems that soybean flour is the most likely future protein resource for wood adhesives. Very large quantities of the soybean flour are available with a reasonable degree of consistency, at low cost and with little impact on human food supplies (Frihart 2015). It should also be taken into accounts that with the small proportion of cross-linking agent, water-resistant wood bonds can be achieved by soybean protein. Although, there are commercial products using soy flour and cross-linking agents, more research is needed to make this an economically acceptable approach at higher solids contents and for a greater range of products (Frihart 2015).

3.1.3 Lignin Based Adhesives Limitation

Some were tried industrially but for one reason or another, too long a pressing time, high corrosiveness for the equipment, etc., they did not meet with commercial success (Pizzi 2016). High molecular weight of lignin is the main drawback for adhesive preparation.

3.1.4 Carbohydrate Based Adhesives Limitation

Furfural, furfuryl alcohol and even the more recently interesting hydroxymethyl furfural, all derived from carbohydrates acid hydrolysis are too dark coloured and rather too costly to be used alone as furan resins for wood panel adhesives. Their niche is in adhesives for higher value added products, such as binders for foundry sands. Although they can be used very successfully for panel adhesives, for such an application they will remain limited to be used as reinforcing, reactive additives for more economical synthetic and bio adhesives.

3.1.5 Starch Limitations

Two approaches have been used. Eriksson et al. (2011) describe the addition of a synthetic polymer to starch deeming this necessary, but this was not able to upgrade its adhesive performance. This approach only succeeded in obtaining a reasonable internal bond strength of the panels by adding polyvinyl amine but at the expenses of a very long hot press time of 30 s per mm thickness. The reported IB strength of 0.6 MPa is sufficiently high to possibly allow for a shorter press time, but the work does not seem to have reported such shortening. Unfortunately, the internal bond strength of the board decreased of 80% if the small proportions of polyvinyl amine are not added to starch. Moreover, when the load on dry wood of starch alone was increased this did not alter the dismal results obtained.

The second approach, envisaging the addition of up to 30% sugar solution also yielded poor results when starch was used alone (Tondi et al. 2012). Addition of tannin and of NaOH improved the starch adhesive performance, but the performance appeared to be somewhat substandard anyway with only overall panel density higher than the average appearing to improve internal bond strengths somewhat. The contribution of reactive furan derivatives was hypothesized but without any scientific verification. However, such furan derivatives will mainly be generated at temperatures higher than 200 °C (Zhao and Umemura 2014) (cf. → 6.8). This type of conditions would exclusively occur for long press times of thin boards as for thermoplastic polymers (→6.12). Other approaches (Amini et al. 2013; Selamat et al. 2014) also suffer of the same drawback, namely too long hot press times, and thus irrelevant for industrial board manufacturing. Starch-PVAc latex-bonded (Nie et al. 2012) (→4.6.1) and maize starch-epoxies-modified resins-bonded veneer lap joints (→2.7) yielded good TSS values between 4.5 and 6 MPa, but at the very long press time of 25 min at 80–90 °C. However, under wet test conditions, the joints still presented TSSs of 2.5 MPa. Wang et al. (2011, 2012, 2013), Zhang et al. (2015) and Li et al. (2015) used maize starch to bod panels pressed instead at ambient temperature for a very long time of 24 h, but obtained sometime in this way a good performance 5–10 MPa). Equally, long press time were necessary for casein-based resins (Sauer and Hagenmüller 1938) and water-carried whey protein isolates/isocyanate wood adhesives (Gao et al. 2011).

4 Other New Bio-Adhesive Systems

Xylan dispersions cross-linked among others with trimethylolpropane triacetoacetate (AATMP) and as dispersing agents dissolved polyvinylamine (→2.6.3) or diluted polyvinylamine were used to bond veneer lap joints. TSS of 5.5–7.6 MPa were obtained by hot pressg at 120 °C and a fast press time of 2.5 min, this decreasing to 4.5 MPa under wet testing. Norström et al. (2014), under identical conditions, but at 10 min press time, used vegetable gum dispersions to obtain TSSs values up to 7.5 MPa. The gums used were locust bean gum, guar gum, xanthan gum and tamarind gum. Although the drawback of this system was the high viscosity of the adhesives, that limited their solids content to just 6% by weight, the results satisfied the interior grade, non-structural requirements of European Norm EN.

4.1 Polysaccharide Adhesives of Animal Origin (Chitosans)

The literature describes plywood bonded with chitosans as adhesives (Umemura et al. 2003. Three layers plywood was prepared by dissolving in 1% acetic acid either chitosan powder or the same added of glucose (Umemura et al. 2010). The Maillard-reacted chitosans presented a progressively increasing bond strengths as a function of the increase in glucose addition. There was, however, a marked dependence on the type of chitosan used. Miller and Shonfeld (2002), Pizzi (2006) prepared wood particleboards and lap joints by using an adhesive based on cashew nut shell liquid (CNSL) reduced by ozonolysis (→4.2.3), reporting internal bond strengths of 1.05 MPa. Preactivation of wood particles by hydrogen peroxide was used by Philippou et al. (1982) before bonding them at 7% resin load with just furfuryl alcohol (→2.4.4) obtaining an internal bond strength of 0.71 MPa, and a26% 24 h cold water swelling. Amini et al. (2017) used cellulose nanofibrils to produce particleboards in a wet process, at a binder load of 17.5% on total panel weight, pressing for 85 s/mm, and obtaining internal bond strengths of 0.45. A further addition of a 3% PAE crosslinker markedly improved the panels mechanical performance. Kowaluk and Fuczek (2009) used PVAc (→4.6.1), in water to reach a 26% solids content, to bond particleboards obtaining an internal bond strength of 0.27 MPa. Unfortunately, no detailed conditions under which this result was obtained were described although they reported they were similar to industrial ones. PVAc has two main drawbacks in being used as a binder for particleboard: a technical one, namely its high viscosity and its high cost. The former may cause process problems, for spraying the resin on the wood chips, and for the high amount of water added that can cause steam blisters.

Citric acid is a fairly recent and unusual newcomer to the field of binders for wood panel adhesives. Sorghum bagasse particleboards bonded with 20% of it, pressed at 67 s/mm yielded 0.9 MPa of internal bond strength (Kusumah et al. 2017, 2016). Equally, Widyorini et al. (2016a, 2016b, 2017) bonded with 15 and 30% citric acid some bamboo based particleboards. The press time was 85 s /mm obtaining IBs between 0.35 and 0.6 MPa, and cold water swelling between 2 and 9%, depending on particle type and resin content. No-adhesives particleboard might well be another interesting approach. In this respect, Suhasman et al. (2017a, 2017b) premodified wood particles by oxidizing them with hydrogen peroxide and iron (II) sulfate. The drawback was again the too long press time at 100 s/mm, notwithstanding the 0.49 MPa IB obtained. Polyketone adhesives with proteins (*Jatropha curcas* L.) and soy as reactive components were also used (Hamarneh et al. 2010a, 2010b). 4 mm thick veneers were coated with a chemically modified thermosetting aliphatic polyketone. water emulsions. They yielded at 200 °C, 5 min press time, a rather low TSS between 2.7 and 3.4 MPa. According to the protein used, with the panels hot pressed for 5 min at 200 °C they yielded a rather low TSS of 2.7–3.4 MPa. Polylactic acid prepared wood fiber composites. Hu and Guo (2015) prepared bonded wood fiber boards using polylactic acid (PLA) fibers. They also added ammonium lignosulfonate, with the optimized conditions in relation to the long pressing time of 84 s/mm at 170 °C resulting in IB values of up to 1.7 N/mm^2 when 70% wood fiber, 30% PLA and addition of a further 10% ammonium lignosulfonate was used.

5 Conclusion and Future Perspective

The field of wood composite binders is a huge field of research of vast economic importance, fast moving in the conception and development of newer and sometimes even revolutionary types of binders. While the wood composite industry, mainly for reason of supply, is still dominated by traditional oil-derived adhesives, both in these fields as well as in the strongly upcoming field of bio-based adhesives, there has been almost incredible progress as well as developments dictated by the intellectual ferment induced by a number of outside constraints. These are the stricter government regulations to reduce and even eliminate formaldehyde and other materials that are to some extent toxic, consumer awareness and the consequent drive of industry to favor more environment-friendly materials. Finally the drive of industry to decrease or even eliminate their dependence on petrochemicals, due to the real or imagined future decrease of oil reserves with its consequent increase in the price of raw materials for purely traditionally manufactured wood binders, as equally contributed to the trend. As it stands, what presented in this review is only an overview of what has happened and is happening in this field, with possibly lesser or more difficult to implement approaches not being mentioned or not mentioned enough, or not yet appeared. Progress in this fascinating field of primary economic importance has been accelerating, and the number of new ideas, approaches, and new proposed binder systems is continuously increasing, providing a glimpse of an exciting and interesting research future.

References

Alder E (1977) Lignin chemistry: past, present and future. Wood Sci Technol 11:169–218

Amini MHM, Hashim R, Hiziroglu S, Sulaiman NS, Sulaiman O (2013) Properties of particleboard made from rubberwood using modified starch as binder. Compos B Eng 50:259–264

Amini E, Tajvidi M, Gardner DJ, Bousfield DW (2017) Utilization of cellulose nanofibrils as a binder for particleboard manufacture. BioResources 12:4093–4110

Amirou S, Pizzi A, Xi X, Delmotte L (2017) Citric acid as waterproofing additive in butt joints linear wood welding. Eur J Wood Wood Prod 75(4):651–654

Amirou S, Pizzi A, Xi X (2019) Wheat protein hydrolysates-resorcinol-aldehydes as potential cold setting adhesives. Eur J Wood Wood Prod 77:453–463

Amonsou EO, Taylor JRN, Minnaar A (2013) Adhesive potential of marama bean protein. Int J Adhes Adhes 41:171–176

Anonymous (2012) Soy-based adhesives provide formaldehyde-free alternative for interior wood products. Biobased Solutions 14. Available from https://soynewuses.org/wp-content/uploads/Adhesives.pdf

Allen AJ, Marcink JJ, Wagler TA, Sosnowick AJ (2010a) Investigations of the molecular interactions of soy-based adhesives. For Prod J 60:534–540

Allen AJ, Spraul BK, Wescott JM (2010b) Improved CARB II-compliant soy adhesives for laminates. In: Frihart CR, Hunt CG, Moon RJ (Eds) International conference on wood adhesives

Allen AJ, Spraul BK, Wescott JM (2010) Improved CARB II-compliant soy adhesives for laminates. In: Frihart CR, Hunt CG, Moon RJ (Eds) Proceedings of the international conference on wood adhesives 2009, Sep 28–30. Forest Product society, Lake Tahoe, Nevada, Madison, WI, USA, pp 176–184

Basso C, Pizzi A, Delmotte L (2015) A new approach to environmentally friendly protein plastics and foams. Bioresources 10:8014–8024

Belgacem MN, Gandini A (2003) Furan-based adhesives. In: Pizzi A, Mittal KL (eds) Handbook of adhesive technology, 2nd edn. Marcel Dekker, New York, pp 615–634

Blanchet P, Cloutier A, Riedl B (2000) Particleboard made from hammer milled black spruce bark residues. Wood Sci Technol 34(1):11–19

Bodyfelt FW, Tobias J, Trout GM (1988) The sensory evaluation of dairy products. Van Nostrand Reinhold

Boerjan W, Ralph J, Baucher M (2003) Lignin biosynthetic. Annu Rev Plant Biol 54:519–546

Bundesministerium für Umwelt Naturschutz und nukleare Sicherheit (BUM) (2018) Bekanntmachung analytischer Verfahren für Probenahmen und Untersuching für die in Anlage 1 der

Chemikalien-Verbotsverordnung genannten Stoffe und Stoffgruppen, Bundesanzeiger, BAnz, AT 26.11.2018 B2

Cargill (2015) Soy flour [Internet]. Cargill Incorporated, Minneapolis, MN, USA. May 15. Available from https://www.cargill.com/products/industrial/adhesives/soyflour/Viewpublication

Calvé LR (1999) Canadian Pat. 2042476, Canadian Patent Office, Ottawa, Canada

Calvo-Flores FG, Dobado JA (2010) Lignin as renewable raw material. Chem Sus Chem 3(11):1227–1235

Capps CL (2004) Saccharide-based resin for the preparation of composite products. Available from https://patents.google.com/patent/WO2003035740A1/en

Chen CM (1996) State of the art report: adhesives from renewable resources. Holzforschung Holzverwertung 48:58–60

Chen X, Xi X, Pizzi A, Fredon E, Du G, Gerardin C (2020) Oxidized demethylated lignin as a bio-based adhesive for wood bonding. J Adhesion. https://doi.org/10.1080/00218464.2019.1710830

Commission Regulation (EU) 2015/491 of 23 March 2015 amending Regulation (EU) No 485 605/2014 amending, for the purposes of introducing hazard and precautionary 486 statements in the Croatian language and its adaptation to technical and scientific 487 progress, Regulation (EC) No 1272/2008 of the European Parliament and of the 488 Council on classification, labelling and packaging of substances and mixtures, 489. https://eur-lex.europa.eu/legal-490content/EN/TXT/PDF/?uri=CELEX:32015R0491&from=PL. Accessed Dec 491 10, 2019

Del Menezzi C, Amirou S, Pizzi A, Xi X, Delmotte L (2018) Reactions with wood carbohydrates and lignin of citric acid as a bond promoter of wood veneer panels. Polymers 10(33):1–13

Dunky M (2003) Adhesives in the wood industry. In: Pizzi A, Mittal KL (eds) Handbook of adhesive technology, 2nd edn. CRC Press, Marcel Dekker, New York, pp 887–956

ECOSE (2017) www.ecosetechnology.com

El Mansouri NE, Pizzi A, Salvadó J (2007a) Lignin-based wood panel adhesives without formaldehyde. Holz Roh Werkst 65:65–70

El Mansouri NE, Pizzi A, Salvadó J (2007b) Lignin-based polycondensation resins for wood adhesives. J Appl Polym Sci 103:1690–1699

EN 312-3 (1996) Particleboards—specifications—Part 3: requirements for board for interior fitments (inculuding furniture) for use dry conditions. Europan Committee for standardization, Brussels, Belgium

EPA (2017) Code of Federal Regulations (CFR) 40 CFR § 770 on Formaldehyde Standards for Composite Wood Products Act, Title VI to the Toxic Substances Control Cct, TSCA, US Environmental Protection Agency

Eriksson AP, Fare J, Furberg AK, Khabbaz F (2011) Adhesive system and method of producing a wood based product. Available from https://patents.google.com/patent/US20070298274A1/en

Forss KJ (1976) Karatex-the lignin based adhesive for plywood, particleboard and fiberboard. Pap Puu 11:817–824

Frihart CR, Birkeland MJ, Allen AJ, Wescott JM (2010) Soy adhesives that can form durable bonds for plywood laminated wood flooring, and particleboard. Proceedings of the International Convention of Society of Wood Science and Technology and United Nations Economic Commission for Europe—Timber Committee, 11–14 October, Geneva, Switzerland, pp 1–12

Frihart CR, Satori H (2013) Soy flour dispersibility and performance as wood adhesive. J Adhes Sci Technol 27:2043–2052

Frihart CR (2015) Wood adhesives: past, present, and future. Forest Prod J 65:4–8

Frihart CR, Lorenz L (2019) Specific oxidants improve the wood bonding strength of soy and other plant flours. J Polym Sci a Polym Chem 57:1017–1023

Frihart CR, Pizzi A, Xi X, Lorenz L (2019) Reactions of soy flour and soy protein by non-volatile aldehydes generation by specific oxidation. Polymers 11(1478):1–18

Gao Z, Yu G, Bao Y, Guo M (2011) Whey-protein based environmentally friendly wood adhesives. Pig Resin Technol 40:42–48

Garcia D, Bacigalupe A, Ferre O (2013) Dry blood powder based adhesive as a renewable and free formaldehyde alternative for wood industry, European Coatings conference "Biobased Adhesives. Dusseldorf, Germany

Gardner D, Sellers T Jr (1986) Formulation of a lignin-based plywood adhesive from steam-exploded mixed-hardwood lignin. Forest Prod J 36(5):61–67

Ghahri S, Mohebby B, Mirshokraie A, Mansouri HR (2016) The effect of soy-flour mesh size and its adhesive acidity changes on shear strength of plywood. Iran J Wood Paper Ind 7:377–386

Ghahri S, Pizzi A, Mohebby B, Mirshokraie A, Mansouri HR (2016) Soy-based, tannin-modified plywood adhesives. J Adhes 94(3):218–237

Ghahri S, Mohebby B, Pizzi A, Mirshokraie A, Mansouri HR (2017) Improving water resistance of soy-based adhesive by vegetable tannin. J Polym Environ 26(5):1881–1890

Ghahri S, Pizzi A (2018) Improving soy-based adhesives for wood particleboard by tannins addition. Wood Sci Technol 52(1):261–279

Grishechko LI, Amaral-Labat G, Szczurek A, Fierro V, Kuznetsov BN, Celzard A (2013) Lignin–phenol–formaldehyde aerogels and cryogels. Micropor Mesopor Mat 168:19–29

Gu K, Li K (2011) Preparation and evaluation of particleboard with a soy flour-polyethylenimine-maleic anhydride adhesive. J Am Oil Chem Soc 88(5):673–679

Gunasekaran S, Lin H (2012) Glue from slaughterhouse animal blood

Gui C, Liu X, Wu D, Zhou T, Wang G, Zhu J (2013) Preparation of a new type of polyamidoamine and its application for soy flour-based adhesives. J Am Oil Chem Soc 90:265–272

Gui C, Zhu J, Zhang Z, Liu X (2016) Research progress on formaldehyde-free wood adhesive derived from soy flour. In: Rudawska A (Ed) Adhesives—applications and properties. InTech, Janeza Trdine9, 51000 Rijeka, Croatia

Hamarneh AI, Heeres HJ, Broekhuis AA, Picchioni F (2010a) Extraction of Jatropha curcas proteins and application in polyketone-based wood adhesives. Int J Adhesion Adhes 30:615–625

Hamarneh AI, Heeres HJ, Broekhuis AA, Sjollema KA, Zhang Y, Picchioni F (2010b) Use of soy proteins in polyketone-based wood adhesives. Int J Adhesion Adhes 30(7):626–635

Hemmilä V, Meyer B, Larsen A, Schwab H, Adamopoulos S (2019) Influencing factors, repeatability and correlation of chamber methods in measuring formaldehyde emissions from fiber- and particleboards. Int J Adhes Adhes 95:102420

Hettiarachchy NS, Kalapathy U, Myers DJ (1995a) Alkali-modified soy protein with improved adhesive and hydrophobic properties. J Am Oil Chem Soc 72:1461–1464

Hettiarachchy NS, Kalapathy U, Myers DJ (1995b) Modified soy proteins and their adhesive properties on woods. J Am Oil Chem Soc 72:507–510

Hettiarachchy NS, Kalapathy U (1998) Functional properties of soy proteins. ACS Symp Ser 708:80–95

Hizukuri S, Takeda Y, Yasuda M, Suzuki A (1981) Multibranched nature of amylose and the action of de-branching enzymes. Carbohydr Res 94:205–213

Hu JP, Guo MH. (2015) Influence of ammonium lignosulfonate on the mechanical and dimensional properties of wood fiber biocomposites reinforced with polylactic acid. Ind Crops Prod 78:48–57

Huang W, Sun X (2000a) Adhesive properties of soy proteins modified by sodium dodecyl sulfate and sodium dodecylbenzene sulfonate. J Am Oil Chem Soc 77:705–708

Huang W, Sun X (2000b) Adhesive properties of soy protein modified by urea and guanidine hydrochloride. J Am Oil Chem Soc 77:101–104

Hunt CG, Smith GD, Yan N (2017) Proceedings of the 2017 international conference on wood adhesives, Oct 25–27, Atlanta, GA. Published by Forest Products Society, Peachtree Corners, GA USA

Insulation K (2017) www.knaufinsulation.at

Jackson R, Aindow T, Baybutt G (2008) Inventors; Knauf Insulations Limited, assignee. Composite wood board. World Intellectual Property Organization Patent WO 2,008,089,847 A1, Jul 31

Jackson R, Aindow T, Baybutt G (2010) Composite wood board. United States Patent US 20,100,087,571 A1, Apr 8

Kamoun C, Pizzi A (2000a) Mechanism of hexamine as a non-aldehyde polycondensation hardener, part 1: mechanism. Holzforschung Holzverwertung 52(1):16–19

Kamoun C, Pizzi A (2000b) Mechanism of hexamine as a non-aldehyde polycondensation hardener, part 2: recomposition of intermediate reactive compound. Holzforschung Holzverwertung 52(3):66–67

Kehr E, Sirch H-J (1998) Entwicklung, Herstellung und Prüfung von Spanplatten unter Einsatz von natürlichen Bindemitteln. ihd-Dresden, Dresden, Germany

Ken AD (1990) Dominate forces in protein folding. Biochemistry 29:7133–7155

Kim S, Kim HJ (2003) Curing behavior and viscoelastic properties of pine and wattle tannin-based adhesives studied by dynamic mechanical thermal analysis and FT-IR-ATR spectroscopy. J Adhe Sci Technol 17(10):1369–1384

Kinsella JE (1979) Functional properties of soy proteins. J Am Oil Chem Soc 56:242–258

Kirk-Othmer (2006) Encyclopedia of chemical technology, 5th edn, vol 15. Wiley, Hoboken, pp 1–25

Krug D, Tobisch S (2010) Einsatz von Proteinen als Bindemittel für Holzwerkstoffe. Eur J Wood Prod 68:289–301

Konnerth J, Hahn G, Gindl W (2009) Feasibility of particle board production using bone glue. Eur J Wood Prod 67:243–245

Kowaluk G, Fuczek D (2009) PVAc glue as a binding agent in particleboards. Drewno 182:17–24

Kusumah SS, Umemura K, Yoshioka K, Miyafuji H, Kanayama K (2016) Utilization of sweet sorghum bagasse and citric acid for manufacturing of particleboard I: effects of pre-drying treatment and citric acid content on the board properties. Ind Crops Prod 84:34–42

Kusumah SS, Arinana A, Hadi YS, Guswenrivo I, Yoshimura T, Umemura K, Tanaka S, Kanayama K (2017) Utilization of sweet sorghum bagasse and citric acid in the manufacturing of particleboard. III: influence of adding sucrose on the properties of particleboard. BioResources 12:7498–7514

Lei H, Pizzi A, Du G (2008) Environmentally friendly mixed tannin/lignin wood resins. J Appl Polym Sci 107:203–209

Li K, Geng X (2004) Investigation of formaldehyde-free wood adhesives from kraft lignin and a polyaminoamide–epichlorohydrin resin. J Adhes Sci Technol 18:427–439

Li K, Peshkova S, Geng X (2004) Investigation of soy protein-kymene® adhesive systems for wood composites. J Am Oil Chem Soc 81(5):487–491

Li K (2007a) Formaldehyde-free lignocellulosic adhesives and composites made from the adhesives. U S Patent No. 7252735B2

Li K (2007b) Inventor, State of Oregon, assignee. Formaldehyde-free lignocellulosic adhesives, their manufacture and composites made from the adhesives. United States patent US 7,252,735 B2. Aug 7

Li K (2007c) Formaldehyde-free lignocellulosic adhesives and composites made from the adhesives. U S Patent No. 7252735B2

Li Z, Wang J, Li C, Gu Z, Cheng L, Hong Y (2015) Effects of montmorillonite addition on the performance of starch-based wood adhesive. Carbohydr Polym 115:394–400

Lin H, Gunasekaran S (2010) Cow blood adhesive: characterization of physicochemical and adhesion properties. Int J Adhesion Adhes 30:139–144

Liu Y, Li K (2006) Preparation and characterization of demethylated lignin-polyethylenimine adhesives. J Adhes 82:593–605

Lopez-Suevos F, Riedl B (2003) Effects of pinus pinaster bark extracts content on the cure properties of tannin-modified adhesives and on bonding of exterior grade MDF. J Adhe Sci Technol 17(11):1507–1522

Lundblad RL (2014) Chemical reagents for protein modifications, 4th edn. CRC Press, Chapel Hill, North Carolina, USA

Malin N (2005) Columbia forest products launches a revolution in plywood adhesives. Environ Build News 14:9

Mansouri HR, Pizzi A, Fredon E (2009) Honeymoon fast-set adhesives for glulam/fingerjoints of higher natural materials content. Holz Roh Werkstoff Eur J Wood Prod 67:207–210

Miller R, Shonfeld U (2002) Company literature, preform Raumgliederungssysteme GmbH. Feuchtwangen, Germany

Mo X, Cheng E, Wang D, Susan SX (2003) Physical properties of medium-density wheat straw particleboard using different adhesives. Ind Crops Prod 18:47–53

Moubarik A, Pizzi A, Allal A, Charrier F, Charrier B (2009) Cornstarch and tannin in phenol-formaldehyde resins for plywood production. Ind Crop Prod 30(2):188–193

Moubarik A, Mansouri HR, Pizzi A, Charrier F, Allal A, Charrier B (2013) Corn flour-mimosa tannin-based adhesives without formaldehyde for interior particleboard production. Wood Sci Technol 47:675–683

Moubarik A, Mansouri HR, Pizzi A, Allal A, Charrier F, Badia MA, Charrier B (2013) Evaluation of mechanical and physical properties of industrial particleboard bonded with a corn flour–urea formaldehyde adhesive. Compos Part B 44:48–51

Moubarik A, Pizzi A, Charrier F, Allal A, Badia MA, Mansouri HR, Charrier B (2013) Mechanical characterization of industrial particleboard panels glued with cornstarch-mimosa tannin-urea formaldehyde resins. J Adhes Sci Technol 27(4):423–429

Navarrete P, Mansouri HR, Pizzi A, Tapin-Lingua S, Benjelloun-Mlayah B, Pasch H, Rigolet S (2010) Wood panel adhesives from low molecular mass lignin and tannin without synthetic resins. J Adhes Sci Technol 24:1597–1610

Navarrete P, Kebbi Z, Michenot F, Lemonon J, Rogaume C, Masson E, Rogaume Y, Pizzi A (2013) Formaldehyde and VOCs emissions from bio-particleboards. J Adhes Sci Technol 27(7):748–762

Newman WH, Glasser WG (1985) Engineering plastics from lignin—XII: Synthesis and performance of lignin adhesives with isocyanate and melamine. Holzforschung 39(6):345–353

Nie Y, Tian X, Liu Y, Wu K, Wang J (2012) Research on starch-g-polyvinyl acetate and epoxy resin-modified corn starch adhesive. Poly Compos 34(1):77–87. https://doi.org/10.1002/pc.22379

Nimz HH (1983) Lignin-based wood adhesives. In: Pizzi A (ed) Wood adhesives: chemistry and technology. Marcel Dekker Inc., New York, pp 247–288

Norström E, Fogelström L, Nordqvist P, Khabbaz F, Malmström E (2014) Gum dispersions as environmentally friendly wood adhesives. Ind Crops Prod 52:736–744

Parajuli D, Inoue K, Ohto K, Oshima T, Murota A, Funaoka M, Makino K (2005) Adsorption of heavy metals on crosslinked lignocatechol: a modified lignin gel. React Funct Polym 62(2):129–130

Pérez S, Bertoft E (2010) The molecular structures of starch components and their contribution to the architecture of starch granules: a comprehensive review. Starch/Staerke 62(8):389–420

Philippou JL, Zavarin E, Johns WE, Nguyen T (1982) Bonding of particleboard using hydrogen peroxide, lignosulfonates, and furfuryl alcohol: effects of chemical composition of bonding materials. For Prod J 32:55–61

Pichelin F, Kamoun C, Pizzi A (1999) Hexamine hardener behaviour—Effects on wood glueing, tannin and other wood adhesives. Holz Roh Werkstoff 57(5):305–317

Pizzi A (1994) Hardening mechanism of tannin adhesives with hexamine. Holz Roh Werkstoff 52:229

Pizzi A (2003) Natural phenolic adhesives II: Lignin. In: Pizzi A, Mittal KL (Eds). Handbook of adhesive technology, 2nd ed. CRC Press, Marcel Dekker, New York, Chap 28

Pizzi A (2006) Recent developments in eco-efficient bio-based adhesives for wood bonding: opportunities and issues. J Adhes Sci Technol 20:829–846

Pizzi A (2013) Bioadhesives for wood and fibres. Rev Adhes Adhes 1(1):88–113

Pizzi A (2014) Synthetic adhesives for wood panels: chemistry and technology—a critical review. Rev Adhes Adhes 2:85–126

Pizzi A (2016) Wood products and green chemistry. Ann for Sci 73:185–203

Pizzi A (2018) Biosourced thermosets for lignocellulosic composites. In: Anish Khan A, Showkat Ahmad Bhawani AB, Abdullah M. Asiri AM, Imran Khan I (Eds). Thermoset composites: preparation, properties and applications. Materials Research Forum LLC, Millersville, PA, USA. pp 81–112

Pizzi A, Mittal KL (2003) Handbook of adhesive technology, 2nd edn. CRC Press, Marcel Dekker, New York

Pizzi A, Rossouw Ddu T, Knuffel W, Singmin M (1980) "Honeymoon" phenolic and tannin-based fast setting adhesive systems for exterior grade finger joints. Holzforschung Holznerwirt 32:140–151

Pizzi A, Cameron FA, Goulding TM, Kes E, v.d. Westhuizen PK (1983) "Honeymoon" fast-setting adhesives for timber laminating. Holz Als Roh-Und Werkstoff 41(2):61–63

Pizzi A, Von Leyser EP, Valenzuela J, Clark JG (1993) The chemistry and development of pine tannin adhesives for exterior particleboard. Holzforschung 47:164–172

Raqueza JM, Deleglise M, Lacrampea MF, Krawczak P (2010) Thermosetting (bio)materials derived from renewable resources: a critical review. Prog Polym Sci 35:487–509

Rzepecki LM, Waite JH (1991) DOPA proteins: versatile varnishes and adhesives from marine fauna. Bioorg Mar Chem 4:119

Salthammer T, Mentese S, Marutzky R (2010) Formaldehyde in the indoor environment. Chem Rev 110(4):2536–2572

Santiago-Medina FJ, Pizzi A, Basso MC, Delmotte L, Abdalla S (2017) Polycondensation resins by lignin reaction with (poly) amines. J Renew Mater 5(5):388–399

Sauer E, Hagenmüller K (1938) Über den Einfluß des Fettgehalts im Kasein bei dessen Verwendung zur Herstellung von Kaltleimen. Colloid Polym Sci 83:210–217

Schwarzkopf M, Huang J, Li K (2010) A formaldehyde-free soy-based adhesive for making oriented strandboard. J Adhes 86:352–364

Selamat ME, Sulaiman O, Hashim R, Hiziroglu S, Nadhari WNAW, Sulaiman NS, Razalia MZ (2014) Measurement of some particleboard properties bonded with modified carboxymethyl starch of oil palm trunk. Measurement 53:251–259

Shimatani K, Sano Y, Sasaya T (1994) Preparation of moderate-temperature setting adhesives from softwood kraft lignin: part 2: effect of some factors on strength properties and chracteristics of lignin-based adhesives. Holzforschung 48(4):337–342

Solenis (2015) soyad™ adhesive technology [Internet]. Wilmington: Solenis, Dec 12. Available from https://solenis.com/en/industries/specialties-wood-adhesives/innovations/soyad-adhesive-technology/

Solt P, Konnerth J, Gindl-Altmutter W, Kantner W, Moser J, Mitter R, van Herwijnen HWG (2019) Technological performance of formaldehyde-free adhesive alternatives for particleboard industry. Int J Adhes Adhes 94:99–131

Suhasman S, Yunianti D, Saad S, Baharuddin B (2017a) Characteristics of binderless particleboard made of three species of sulawesi bamboos. Wood Res J 4:68–71

Suhasman S, Massijaya MY, Hadi YS, Santoso A (2017b) Particle oxidation time for the manufacture of binderless particleboard. Wood Res J 2(1):27–33

Sun X, Bian K (1999) Shear strength and water resistance of modified soy protein adhesives. J Am Oil Chem Soc 76:977–980

Tondi G, Wieland S, Wimmer T, Schnabel T, Petutschnigg A (2012) Starch-sugar synergy in wood adhesion science: basic studies and particleboard production. Eur J Wood Prod 70:271–278

Umemura K, Inoue A, Kawai S (2003) Development of new natural polymer-based wood adhesives I: dry bond strength and water resistance of konjac glucomannan, chitosan, and their composites. J Wood Sci 49:221–226

Umemura K, Mihara A, Kawai S (2010) Development of new natural polymer-based wood adhesives III: effects of glucose addition on properties of chitosan. J Wood Sci 56:387–394

Umemura K, Sugihara O, Kawai S (2013) Investigation of a new natural adhesive composed of citric acid and sucrose for particleboard. J Wood Sci 59:203–208

van Herwijnen HWG, Krug D, Mäbert M, Büttner K, Jacob M, Pietzsch M (2016) Renewable wood adhesive based on crosslinkable protein. Holztechnologie 57:12–16

Van Niekerk IA, Pizzi A (1994) Characteristic industrial technology for exterior Eucalyptus particle board. Holz Als Roh- Und Werkstoff 52(2):109–112

Vanholme R, Morreel K, Boerjan JRW (2008) Lignin engineering. Curr Opin Plant Biol 11:278–285

Valenzuela J, von Leyser E, Pizzi A, Westermeyer C, Gorrini B (2012) Industrial production of pine tannin-bonded particleboard and MDF. Eur J Wood Prod 70(5):735–740

Viikari L, Hase A, Quintus-Leina P, Kataja K, Tuominen S, Gadda L (1999) European patent EP 95030 A1. European Patent Office, Brussels

Vnučec D, Kutnar A, Gorsek A (2016) Soy-based adhesives for wood-bonding—a review. J Adhes Sci Technol 31(8):910–931

Wang Z, Gu Z, Hong Y, Cheng L, Li Z (2011) Bonding strength and water resistance of starch-based wood adhesive improved by silica nanoparticles. Carbohydr Polym 86:72–76

Wang C, Wu J (2012) Preparation and characterization of adhesive from spent hen proteins. Int J Adhes Adhes 36:8–14

Wang Z, Li Z, Gu Z, Hong Y, Cheng L (2012) Preparation, characterization and properties of starch-based wood adhesive. Carbohydr Polym 88:699–706

Wang Z, Gu Z, Li Z, Hong Y, Cheng L (2013) Effects of urea on freeze–thaw stability of starch-based wood adhesive. Carbohydr Polym 95:397–403

Wescott JM, Frihart CR, Traska AE (2006) High-soy-containing water-durable adhesives. J Adhesion Sci Technol 20:859–873

Wescott JM, Birkeland MJ, Yarvoski J et al (2010) Recent advances in soy containing PB and MDF. In: Frihart CR, Hunt CG, Moon RJ (Eds). Proceedings of the International conference on wood adhesives 2009, Sep 28–30. Forest Product society, Lake Tahoe, Nevada, Madison, WI, USA, pp 136–141

Whitfield RM, Brown FC, Low R (2007) Socio-economic benefits of formaldehyde to the European Union (EU 25) and Norway. Global Insight, Lexington, MA

Whetten R, Sederoff R (1995) Lignin biosynthesis. Plant Cell 7:1001–1013

Wolf WJ (1970) Soybean proteins: their functional, chemical, and physical properties. J Agri Food Chem 18:969–976

Wu D, Fu R, Sun Z, Yu Z (2005) Low-density organic and carbon aerogels from the sol–gel polymerization of phenol with formaldehyde. J Non-Cryst Solids 351:915–921

Widyorini R, Nugraha PA, Rahman MZA, Prayitno TA (2016a) Bonding ability of a new adhesive composed of citric acid-sucrose for particleboard. Bioresources 11:4526–35

Widyorini R, Umemura K, Isnan R, Putra DR, Awaludin A, Prayitno TA (2016b) Manufacture and properties of citric acid-bonded particleboard made from bamboo materials. Eur J Wood Prod 74:57–65

Widyorini R, Umemura K, Kusumaningtyas AR, Prayitno TA (2017) Effect of starch addition on properties of citric acid-bonded particleboard made from bamboo. BioResources 12:8068–8077

Xi X, Pizzi A, Gerardin C, Liao J, Amirou S, Abdalla S (2019) Glutaradehyde-wheat protein-based adhesives for plywood. J Adhesion. https://doi.org/10.1080/00218464.2019.1630821

Xi X, Pizzi A, Frihart CR, Lorenz L, Gerardin C (2020) Tannin plywood adhesives by non-volatile aldehydes generation from specific oxidation of mono- and disaccharides. Int J Adhesi Adhes 98(102499):1–9

Xiao Z, Li Y, Wu X, Qi G, Li N, Zhang K et al (2013) Utilization of sorghum lignin to improve adhesion strength of soy protein adhesives on wood veneer. Ind Crops Prod 50:501–509

Yang G, Yang B, Yuan C et al (2011) Effects of preparation parameters on properties of soy protein based fiberboard. J Polym Environ 19:146–151

Zeppenfeld G, Grunwald D (2005) Klebstoffe in der Holz- und Möbelindustrie. DRWVerlag. About formaldehyde problems and restriction

Zhang Y, Ding L, Gu J, Tan H, Zhu L (2015) Preparation and properties of a starch-based wood adhesive with high bonding strength and water resistance. Carbohydr Polym 115:32–37

Zhao LF, Liu Y, Xu ZD, Zhang YZ, Zhao F, Zhang SB (2011) State of research and trends in development of wood adhesives. For Stud China 13:321–326

Zhao Z, Umemura K (2014) Investigation of a new natural particleboard adhesive composed of tannin and sucrose. J Wood Sci 60:269–277

Zhu D, Damodaran S (2014) Chemical phosphorylation improves the moisture resistance of soy flour-based wood adhesive. J Appl Polym Sci. https://doi.org/10.1002/APP.40451

Investigating the Recent Development of Bio-based Polyurethane Foam Produced from Renewable Resources

Atika Alhanish and Mustafa Abu Ghalia

Abstract The development of biobased polyurethane foam (PUF) from renewable resources of polyol against traditional PUF has attracted many researchers and investors. The reduction or elimination of petrochemical material has had an increased contribution to producing more sustainable PUF. This chapter aims to present an overview of the work accomplished to synthesize green chemistries from plant-based biomass by using the green polyol to produce a renewable foam. The effect of different bio-polyol grades, catalysts, chain extenders, surfactants, and blowing agents were discussed. However, despite their advantages, degradable polyesters have lack of optimum properties and fail to meet the desired mechanical and physical properties. Therefore, bio-additives require to modify and compensate in mechanical properties with increase environmentally friendly foam. Moreover, reinforced bio-additives have several key advantages over other synthesis methods, such as reduced fossil-based feedstock reliance, decreased CO_2 emissions, renewable foam, and high productivity. Recent advances in chemistry and technology of sustainable routes for the synthesis of NIPUs were also described.

Keywords Polyurethane foam · Renewable resources · Bio-based polyol · Bio-based plasticizer · natural materials · Non-isocyanate

1 Introduction

Nowadays, polyurethane foams (PUFs) have become an essential material in modern life with a favorable cost-performance balance. They have been used in almost every sector of manufacturing, including automotive, infrastructure, aerospace, medical device, consumer products, etc. The demand on PUFs has been superbly growing

A. Alhanish
Natural Gas Department of Chemical Engineering, Faculty of Petroleum and Engineering, University of Zawia, Zawia, Libya

M. Abu Ghalia (✉)
Department of Chemical Engineering, Ryerson University, Toronto M5B2K3, Canada
e-mail: mabughal@ryerson.ca

M. Jawaid et al. (eds.), *Eco-Friendly Adhesives for Wood and Natural Fiber Composites*, Composites Science and Technology,
https://doi.org/10.1007/978-981-33-4749-6_12

over the past years (Gama et al. 2018). In 2017, the latest report of plastic market revealed that the demand for polyurethane were 7.5% of the total demand of plastics (3.78 MTn), were 70% is in the form of foam (Gómez-Rojo et al. 2019). Nevertheless, they are still produced from petrochemicals, raising concerns about their negative influence on the ecosystem and human health (Gama et al. 2018).

Isocyanate and polyol, which are the main constituents in formulated PUFs have been subjected to intensive research due to their potential threat to environment and human health as they are basically derived from petroleum. Thus, continuous efforts on a global scale to replace petroleum-based PUFs counterparts by eco-friendly alternatives i.e. renewable materials is an attractive solution (Vollmer 2010).

Several attempts to produce high percentage of PU based on renewable materials were reported since the last decade (Ionescu 2005). It was found that using renewable materials results in less greenhouse gas emission as well as increased economic feasibility. In addition, the disposal of those materials is less toxic compared to products based on petrochemicals.

Currently, many countries focus their researches on utilizing the bio-based additives and special emphasis on bio-based polyols, from renewable materials. Several reports carried out using different renewable materials in formulating PUFs such as palm oil (Ng et al. 2017), rapeseed oil (Prociak et al. 2018), soybean oil (Herrán et al. 2019), liquefied wheat straw (Li et al. 2018), polymeric wastes (Paruzel et al. 2017), cottonseed oil (Pawar et al. 2016b), waste cooking oil (Salleh et al. 2018); and different types of hybrid renewable materials as the mixture of castor oil, kraft lignin, and residual glycerol (Carriço et al. 2016) and virgin castor oil with crude glycerol (Carriço et al. 2017). Some of these studies confirmed that the mechanical and thermal properties were comparable to those obtained by materials from fossil resources.

1.1 *Fundamental Concepts in the Synthesis of Polyurethane Foam*

In general, PUFs are obtained by the reaction of polyols with poly-isocyanates (usually two functional groups interacted to each other) producing repeating units of urethane group (gelling reaction) with an exothermic reaction (Fig. 1).

This process is followed by the expansion reaction (blowing reaction) of isocyanate and water to produce carbamic acid, which decomposes to form amine and

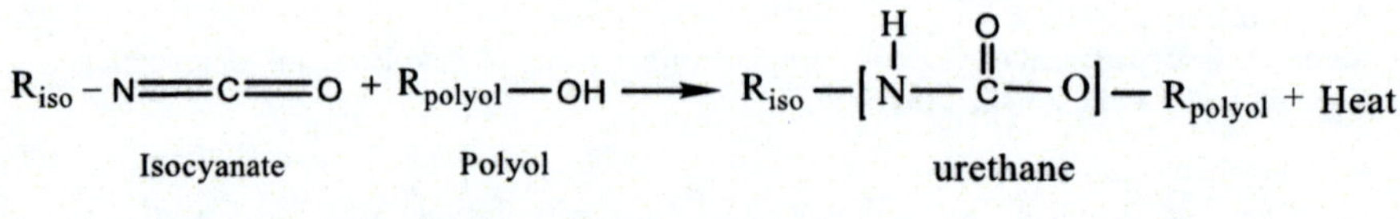

Fig. 1 Production of urethane linkage (Ionescu 2005)

(a) $R{-}NCO + H_2O \longrightarrow R{-}NHCOOH \longrightarrow R{-}NH_2 + CO_2$

Isocyanate Water Carbamic acid Amine Carbon dioxide

(b) $R{-}NCO + R'{-}NH_2 \longrightarrow R'{-}NHCONH{-}R$

Isocyanate Amine Urea

Fig. 2 Reactions involved in PUFs production (Li et al. 2015)

CO_2 bubbles (Fig. 2a). Furthermore, amine reacts with another isocyanate to produce urea (Fig. 2b). In addition, other side reactions may occur; due to the presence of other additives in the synthesis process in order to manipulate the physio-mechanical properties.

To control the foam properties, other chemicals are required such as blowing agents, surfactants beside other additives (Gama et al. 2018). By adjusting the chemicals contents and the type of those ingredients, different bio-based materials with various properties of PUFs can be achieved. For example, the preparation of the main types of PUFs (flexible and rigid foams), chemically follow the same manner and the difference of their properties comes from altering the content of the basic components (polyols and the isocyanates with additives). The basic reaction form two phases: soft and hard segment, connected through urethane group. A higher amount of isocyanate produces a rigid and hard segment, whereas flexible foam can be obtained via increasing the soft segments of polyols or by using plasticizers with manipulating iso-index.

Generally, additives play a crucial role in controlling the desired properties of the final PUF. The most used additives are blowing agents that are used to cause foaming, while surfactants can be added to improve miscibility between nonhomogeneous reactants. Catalysts are always needed to accelerate the speed of the reaction. Chain extenders (crosslinkers) are mainly used to modify the polymer structure by enhancing hardness. In addition to other additives such as bio-based fillers, plasticizers and pigments may be required to tailor the mechanical properties.

1.2 Structure Properties Relations in Polyurethane Foam

To exhibit a specific property, a specific molecular weight of polyol is needed. The nature of the chemical structures defines the properties of the final PUFs. For instance, using two functional groups (diols/diisocyanates) produces a linear polyurethane, while using higher functionality of polyol produce crosslinked or branched structure (Szycher 2013). Moreover, several properties can be adjusted by using certain type of isocyanate reactivities. Using aliphatic isocyanate, such as Hexane Diisocyanate (HDI) and Methylene Dicyclohexyl Diisocyanate (MDI) in the production of PUF slightly decreases the mechanical properties including modulus, tensile strength, and hardness, but present lower impact resistance and elongation at break in the PUF. On the other hand, PUFs produced by using aromatic isocyanates such as Toluene Diisocyanate (TDI) boost mechanical properties; tensile strength and higher elongation at break, besides being a rubbery (Gama et al. 2018).

In addition, the effect of varying polyols functionality such as molecular weight and hydroxyl groups have significant impact as well. Raising the molecular weight up to 3000 (Mwt) with equivalent the OH/ISO (1:1) functionality increases the elongation and tensile strength with improved resiliency. In turn, reducing the molecular weight while increasing polyols functionality (hydroxyl group) led to an increase in tear and tensile strength with slight decline in elongation and hardness in the foam (Gama et al. 2018).

2 Polyol Based Renewable Resources

Polyols compounds comprise of polyfunctional alcohols (two or more −OH group), usually along with other functionalities. There are different types of polyols used in the synthesis of PUFs; mainly derived from petroleum-based materials such as polyether and polyesters. Recently, the demand for polyols are globally heightened, especially in China, North America and the Europe, as well as growing concerns over the environment, price and scarcity of petroleum has motivated the researches in the field to develop sustainable polyols from renewable resources as represented in Fig. 3.

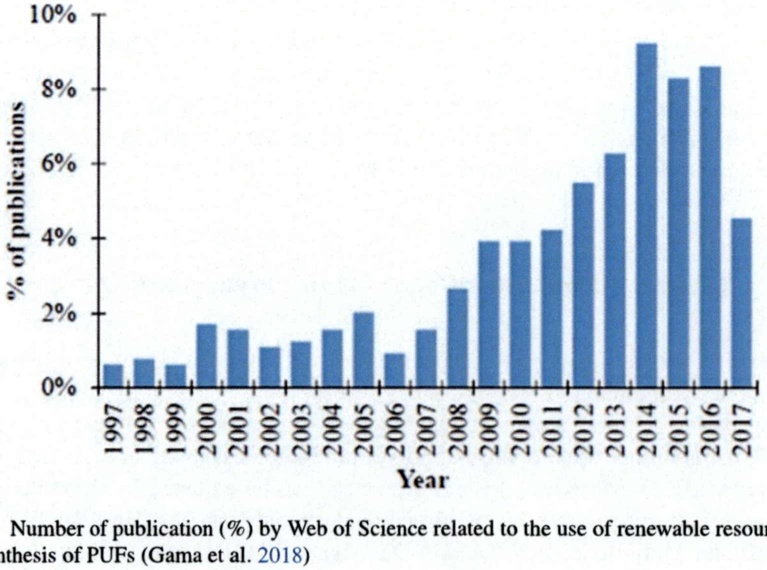

Fig. 3 Number of publication (%) by Web of Science related to the use of renewable resources in the synthesis of PUFs (Gama et al. 2018)

Today, bio-based polyols are commercialized by companies like BASF and Dow Chemical, Panolam and the global market is expected to reach 4.7 billion USD in the near future as reported by Markets and Markets™ (Gama et al. 2018).

2.1 Synthesis Polyol Based Renewable Resources

Synthesizing polyols from renewable materials has been more successful than that of isocyanates; and several production pathway are currently used at industrial scale (Herrán et al. 2019).

Many types of natural resources have been employed in synthesizing bio-polyols by different manufacture routes. Depending on the nature of the feedstock, the synthetic technique used; different derived polyols with various structures and properties could be achieved (Li et al. 2015). Vegetable oils, biomass, wastes and protein-based feedstock have been extensively studied for developing bio-polyols.

Vegetable oils are heterogeneous compounds comprising of tri-esters of glycerol and different fatty acids; the diversity of these materials depends on the type and the origin of the fatty acids. In general, the absence of hydroxyl group in all vegetable oils structure (except of lesquerella and castor oils) requires a chemical manipulation to introduce hydroxyl group before being used as a feedstock for synthesis of bio-polyols (Petrovic 2008).

There are many approaches utilized to derive polyols from vegetable oils. These methods work in functionalizing either the reactive sites as an example; double bonds between carbon atoms or ester linkage. Thus, the preparation methods could be classified into two categories: (i) reaction through double bonds and (ii) reaction through ester linkage (Fig. 4).

The properties of bio-polyols derived from vegetable oils depending on several factors including the type of vegetable oil used and the production process. For instance, the polyols produced from epoxidation followed by ring opening route contain a mixture of primary and secondary hydroxyl group, whereas hydroformylated polyols contain only primary hydroxyl group; furthermore, ozonolysis based polyols possess a terminal primary hydroxyl functions (Li et al. 2015). Meanwhile, polycondensation with combination of hydroformylation produces a hyperbranched polyols with high molecular weight (Petrović et al. 2012).

These reactions usually require at least two-step reactions. For example, the functionality of bio-polyols synthesized by hydroformylation is similar to the virgin vegetable oil where one hydroxyl group exists per double bond. Thus, a structural modification is applied. In contrast, castor oil can be employed directly in formulating PUFs with semi-rigid and semi-flexible properties due to the presence of secondary hydroxyl group in the structure. Despite, this secondary hydroxyl group the reactivity and functionality of castor oil for the production of rigid PUFs (Ionescu et al. 2016). To functionalize castor oil as a precursor in polyols production; two methods were applied widely through the ester moieties using transesterification; and the other is through the hydroxyl group via alkoxylation (Li et al. 2015). Table 1 summarize the

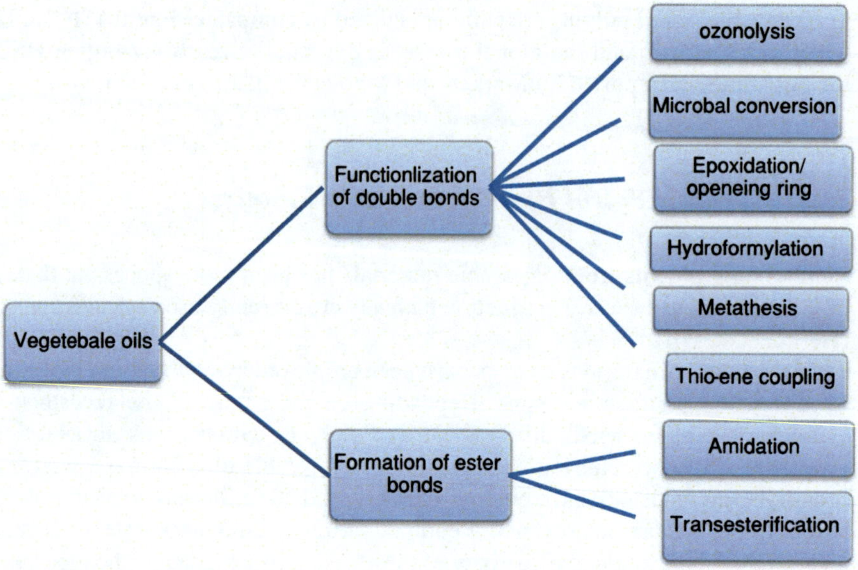

Fig. 4 Common methods for converting vegetable oils to bio-polyols

Table 1 Some basic properties of bio-polyols based vegetable oils (Furtwengler and Avérous 2018)

Feedstock	Chemical modification	OH value range (mg KOH/g)	Viscosity range (mPa.s, 25 °C)
Rapeseed Oil	Epoxidation, transamidation, transesterfication	77–390	150–650
Soybean oil	Epoxidation	150–333	550–8900
	Epoxidation, glycolysis, ethoxylated	71–260	6000–16,000
Castor Oil	Transesterfication, Michael thiol-ene	220–465	480–19,000
Palm oil	Cross-metathesis, transesterfication	136–460	260–35,000
	Epoxidation, polycondensation	98–116	320–5000
	Epoxidation	110	1470
Tung oil	Tung oil blend with short diol	223	n/a
	Alkaline transesterfication	163	n/a
Cottonseed oil	Epoxidation	365–370	200–210
Linseed oil	Hydroxylation	216	n/a
Coconut oil	Transesterfication	270–333	3500–9500

properties of various bio-polyols derived from different vegetable oils by different approaches.

For developing polyol with high functionality, several authors are dedicated to combine different methods based on the aforementioned methods; in order to utilize the merits offered by double bond modification and the merits offered by modification of ester linkage (Zhang et al. 2014). For instance, by using transesterification of triglycerides and ring opening of epoxidized vegetable oils, Kong et al. (Kong et al. 2013) obtained polyols with hydroxyl group higher than 200 mg KOH/g; while in other work high functional polyols with hydroxyl group in the range of (355–434 mg KOH/g) were obtained from vegetable oils by using epoxidation followed by reduction (Zhang et al. 2014).

At industrial scale, bio-based polyol is widely synthesized by ring opening polymerization as the most economical route to produce polyols from corn, soybean, olive, rapeseed, linseed, safflower, melon seed, karanja, and cotton seed (Karak 2012a). This method is simple and economically favorable; and the hydroxyl value of the polyol produced usually above 250 mg KOH/g which could be used to produce rigid PUFs (Prociak et al. 2018). All the aforementioned synthetic approaches are conventionally used in batch reactor; where some drawbacks could be considered such as oligomerization, unavoidable side reaction, long reaction time, and polyols with high viscosity could be achieved due to the consumption of newly formed hydroxyl groups, which will eventually effect on the performance of PUFs produced (Ji et al. 2015).

Using micro-flow system is another way gaining attention via microreactors. In comparative study, He et al. (He et al. 2013) investigated epoxidation of soybean in microreactor and in batch reactor. The results revealed promising findings in a short time by using a microreactor compared to the batch. A polyol produced with hydroxyl number of 300 mg KOH/g, acid value of 1.5 mg KOH/g and viscosity of 7550 mPa.s. Later on, a similar study has been conducted with a treatment step of soybean by allylic oxidation processed by epoxidation (Fang et al. 2016). The properties of polyol were hydroxyl group of 378 mg KOH/g and acid value of 0.6 mg KOH/g and viscosity of 8825. The alteration of hydroxyl group and viscosity had been interpreted as a result from oligmers due to the increase of hydroxyl groups as the authors confirmed (Fang et al. 2016).

On the other hand, in order to synthesis polyols with different architectures, thiol-ene click reaction seems to be a feasible choice. This reaction is well known for high yield, short time reaction, selectivity and simple conditions (Desroches et al. 2012). Recently, the development of new synthetic paths for bio-polyols production using thiol-ene click reaction is gaining attention (Yang et al. 2017). Ionescu and his co-workers (Ionescu et al. 2016) prepared high functional polyols from three samples of castor oils using thiol-ene click reaction with different initiators. In other study, Alagi et al. (2016) investigated using photoinduced thiol-ene click reaction to functionalize castor oil and soybean oil. Interestingly, about 99% of the double bonds converted to hydroxyl group. However, the research in this area is still limited (Furtwengler and Avérous 2018).

Many attempts for producing 100% PUFs were reported. In instance, an eco-friendly approach using catalytic/solvent free method via epoxidation technique successfully produced bio-polyols from both castor oil and soybean oil (Zhang et al. 2013). Another sustainable approach is using thiol–ene photo-click reaction with no solvent to produce polyol from soybean oil (Yang et al. 2017). The performance results of PUFs for both studies were superior.

Another interesting study used unmodified corn oil to synthesis bio-polyol via thiol-ene reaction in one step with no solvent (Ramanujam et al. 2019). Polyol with hydroxyl number of 176 mg KOH/g, acid value of 1.77 mg KOH/g and 99% yield was obtained.

Vegetable derivatives i.e. fatty acids and fatty acid esters are another renewable source of bio-polyols. Beside, using the same methods of converting vegetable oils; bio-polyol could be synthesized by dimerization of fatty acids and hydrosiylation of fatty acid esters (Li et al. 2015). Although, most of polyols derived from vegetable oils were synthesis using a massive amount of solvents, which raises concerns regarding the economic and environmental aspect (Zhang et al. 2013).

In addition, crude glycerol (CG) is economically a promising renewable source of bio-polyols. A short polyol with high OH value could be obtained as by product from a lot of reactions such as hydrolysis of triglycerides, and saponification (Tan et al. 2013). It was reported that CG had been used in PUFs production either as a sole polyols (Gama et al. 2018) neither in combination with polyols derived from petroleum (Hu et al. 2012). Also, it could be added into blends to increase the OH value of bio-polyols with low OH values such as castor oil (Carriço et al. 2017).

Generally, CG has impurities like water, fatty acids, inorganic salts, among others (Tan et al. 2013). However, it was utilized directly in PUFs production (Muller et al. 2018) and indirectly by going through treatment process prior to the synthesis process of PUFs (Gama et al. 2018).

Also, a self-reaction of CG has been reported through catalyzed reaction and uncatalyzed reaction producing polyols with high OH values (350 and 550 mg KOH/g) (Furtwengler and Avérous 2018). In other work, polyol with OH (230.30 mg KOH/g) has been obtained through converting CG to microbial oil by fermentation followed by chemical transformation to produce polyol (Uprety et al. 2017).

From an economical perspective, a simple, one-pot thermochemical process is favorable for producing polyol from CG; where the results obtained are efficient. One such study were conducted by Cui et al. (2017), a bio-polyols were produced by conversion CG through one-pot thermochemical approach. The properties of the product under the optimized conditions (temperature at 220 °C, molar ratio RCG/FA of 1.5, and reaction time of 5 h) were a hydroxyl number of 322 mg KOH/g, acid number of 1.7 mg KOH/g, and viscosity of 25 Pas. In earlier work, CG based polyols produced also by one-pot thermochemical process with hydroxyl number of 481 mg KOH/g and an acid number of about 5 mg KOH/g under specified conditions (Guo et al. 2000).

There has also been a focus on using biomass such as forestry and agricultural residues for deriving bio-polyols. These materials are rich with different valuable ingredients such as starch, cellulose, hemicellulose, rosin, and terpenes, among others. The most ubiquitous biomass source is lignocellulosic, such as crop residues and forest waste, which is rich in hydroxyl groups (Li et al. 2015). Using these materials as a partial or totally replacement of petroleum derived polyols in PUFs production demonstrated comparable properties to petroleum based analogues (Niu et al. 2011; Aniceto et al. 2012).

Oxypropylation (Aniceto et al. 2012) and acid liquefaction method (Niu et al. 2011) are the most widely processes applied to produce bio-polyols from biomass materials for PU production (Gama et al. 2018).

Oxypropylation has been employed to synthesis many biomass based polyols such as chitosan (Fernandes et al. 2008), corn starch (de Menezes et al. 2007), rapeseed cake (Serrano et al. 2010), data seeds (Briones et al. 2011), olive stones (Matos et al. 2010), sugar beet pulp (Pavier and Gandini 2000), cork residues (Evtiouguina et al. 2000), cashew nut shell (Ionescu et al. 2012) and lignin (Nadji et al. 2005). This method conveys the privilege of extracting the hydroxyl group wrapped inside the substrate (Gandini and Belgacem 2002). Also, it was demonstrated that oxypropylation process increases hydroxyl functionality, however it does not increase the number of hydroxyl groups (Belgacem and Gandini 2008).

This technique could be carried out through two reaction patterns: in one step and two-step processes (Gama et al. 2018). The product from both approaches comprise of oxyproplyted biomass, polypropylene oxide and unreacted biomass; and their concentration in the product depend on biomass/catalyst ratio, biomass source, biomass/propylene oxide, temperature, pressure and functionalization condition (Gandini and Belgacem 2002). It was also reported that a few highly acidic polyols could be derived from oxypropylation process; and through a total oxypropylation liquid polyols produced, which mainly used to prepare rigid PUF (Li et al. 2015).

Depending on the nature of biomass and certain oxypropylation factors; the produced bio-polyols could contain hydroxyl numbers ranging from about 82 to 610 mg KOH/g, residue contents ranging from 0 to 41%, viscosities in the range of 2.3–2860 Pa.s, and poly(propylene oxide) homopolymer contents (2–83%) (Li et al. 2015). Table 2 presents the properties of some lignocellulosic biomass based bi-polyols.

Table 2 General properties of polyols derived from lignocellulosic biomass (Li et al. 2015)

	Polyol properties			
Lignocellulosic biomass	Homopolymer content (wt.%)	Solid residue content (wt.%)	Hydroxyl number (mg KOH/g)	Viscosity (Pa.s)
Kraft Lignin	75–83	33–41	148–180	2.3–4.2
Organosolv Lignin	2–42	0	198–305	5.2–2860
Soda lignin	18–30	0	82–120	15.2–165
Sugar beet pulp	Not available	0–10	260–418	49–710
Rapeseed cake	Not available	Not available	610	84.2
Cork	68.7–75.8	0.2–4.9	~131	2.9–13.0

Liquefaction is another main technique to produce biomass based polyols; by using a specific liquefaction solvent and under specific conditions of time, temperature and catalyst (Hu et al. 2012). However, this method depends highly on petroleum solvents and excess amount of solvent is required to overcome the viscosity of biomass material which raises the cost of production (Furtwengler and Avérous 2018). From economical aspect, CG has been used as un-expensive substitution. In the work of (Wang et al. 1), CG was used as liquefier in the production of polyol derived from corn stover. A polyol contains hydroxyl number of 270–310 mg KOH/g was obtained. Also, an increase of viscosity and decrease in the acid number of the product was observed.

Generally, biomass liquefaction is considered as a sophisticated process in which numerous reactions occur (Hu et al. 2014). Commonly, the bio-polyols derived via liquefaction have some residues depending on the efficiency of the process. Although, the polyol derived from this process is rich in hydroxyl group and could be used directly in PUFs production (Li et al. 2015). General properties of polyols derived from different biomass via liquefaction are represented in Table 3.

The properties reported in the above mentioned table are for polyols produced from acid-catalyzed liquefaction using petroleum based polyhydric alcohols as solvents (Li et al. 2015). Acid liquefaction is an effective procedure to derive bio-polyols from biomass because the reaction could be carried out in moderate temperature and low pressure. Moreover, the product from this route possess higher hydroxyl group than the one produced from oxypropylation. However, many reactions occur in the acid liquefaction of biomass such as dehydration, degradation, hydrolysis, decarboxylation, deoxygenation, dehydrogenation; which leads to formation of new compounds (Li and Sun 2010).

It is worth to mention that polyols derived from acid catalyzed liquefaction contain high acid numbers up to 40 mg KOH/g. Thus, it is not a preferred route to derive bio-poly for PUFs production (Li et al. 2015). On the other hand, feedstocks with high protein content have been reported in synthesizing bio-polyols for PUFs production, where many functional groups such as carboxyl, amino and hydroxyl exist in these materials, which can react with other functional groups to produce different types of PU (Li et al. 2015). Typically, there are two methods to derive polyols from protein: (1) liquefaction of protein feedstock; and (2) modification of protein. Many studies investigated the synthesis of various protein based bio-polyols; one example is the use of soy proteins by functionalizing the protein with glycerol propylene oxide polyether (Park and Hettiarachchy 1999).

Other sources to derive bio-polyols are biomass derivatives. One such study is carried out by Pawar et al. (2016a). An algae oil extracted from microalgae was used to prepare bio-polyols for rigid PUF production via epoxidation /opening ring reactions. Exploring more sustainable pathways, a promising study investigated the use of new non- edible vegetable oils. In instance, an A C. salicifolia (CS) and C. integerrima (CI) extracted from flowering plants were used as bio-feedstock through one step approach using modified method of the one proposed by Monteavaro et al. (2005). A bio-polyols were obtained with properties from CI and CS as following: the hydroxyl value were 225.0 and 240.8 mg KOH/g, molecular weight of 1122.4

Table 3 General properties of biomass based bio-polyols via liquefaction (Li et al. 2015)

	Polyol properties			
Lignocellulosic biomass	Liquefaction efficiency (%)	Acid value (mg KOH/g)	Hydroxyl number (mg KOH/g)	Viscosity (Pa.s)
Enzymatic hydrolysis lignin	98	Not available	249	Not available
Cellulose or waste paper	55–99	19–30	360–396	2.6–3.9
Agriculture crop waste such as wheat straw, rice straw, corn stover, cotton stalks, corn cobs, etc	60–95	15–30	109–430	1.0–1.7
Biorefinery residue such as including bagasse, dried distillers grains (DDG), corn bran, date seeds, rapeseed cake residue, apple pomace, olive stone, etc	84–98	28–34	137–586	0.4–3.0
Wood such as including both softwood and hardwood species	80–98	12–38	200–435	0.3–31.6
Birch wood such as base-catalyzed liquefaction using petroleum-derived polyhydric alcohols as liquefaction solvent	>99	24–41	112–204	Not available
Soybean strew such as base-catalyzed liquefaction using biodiesel-derived crude glycerol as liquefaction solvent	65–75	<5	440–540	16–45

and 1166.1 g/mol and functionality of 4.50 and 5.00, respectively (Abril-Milán et al. 2018).

In recent years, waste cooking oil was investigated as a source to derive polyols through different approaches such as epoxidation and hydroxylation reaction. In one such study, the properties of produced polyol were hydroxyl value of 184 mg KOH/g and acid value of 0.68 mg KOH/g which is comparable with commercial polyols (Salleh et al. 2018). The earlier study investigated converting waste cooking oil via transesterification reaction to bio-polyol, and the product properties were 46.3 hydroxyl value and 2.985 mg KOH/g acid value (Enderus and Tahir 2017).

Polymeric waste is also considered as a good source for deriving bio-polyols for PUFs production. A lot of work had been done to recycle polymer wastes using different chemical recycling methods to obtain polyols; several of these works uses these polyols for the synthesis of PUFs with a promising result. As an example, Paruzel and his coworkers prepared a rigid PUF by converting a scraps of polyurethane and polycarbonate from automotive waste using glycerides extracted from coconut oil to liquid polyols with hydroxyl number of 300 mg KOH/g (Paruzel et al. 2017). Obviously, considerable efforts have been devoted for synthesizing of bio-polyols from renewable materials and the work still sustained in this area. Many attractive solutions were proposed and some of them had been covered in this chapter and some were not.

2.2 *The Effect of Different Polyol on the Polyurethane Foam Properties*

The properties of PUF are greatly tied to the properties of the polyols used (Sawpan 2018). Table 4 highlights the relationship between polyols properties and the behavior of PUFs.

In addition to the aforementioned table, PUFs derived from polyols with hydroxyl number greater than 150 mg of KOH/g were glassy, whereas those derived from polyols with hydroxyl number less than 150 were rubbery (Wang et al. 2009a). High functional polyols could be produced from vegetable oils with high degree of unsaturation by using different chemical reactions to convert double bond to hydroxyl group (Ionescu 2005), subsequently PUFs with good properties such as high glass transition and tensile strength are produced (Li et al. 2015).

Table 4 Relationship between polyols properties and PUFs behaviour (Furtwengler and Avérous 2018)

	Polyol properties		
PUF behavior	OH value (mg KOH/g)	Functionality	M_n (g mol^{-1})
Rigid	350–800	3–8	300–1000
Semi-rigid	100–200	3–3.5	500–2000
Flexible	5–10	2–3	3000–6000

On the other hand, polyol based on the synthesized method have different effects on PUF properties. For instance, it was reported that vegetable oil-based polyols obtained via hydroformylation or epoxidation have molecular weight of 40% higher than the one obtained by ozonolysis processes. However, the PUF synthesized by polyols derived from ozonolysis show excellent mechanical properties (Petrovic 2008). Also, a rigid PUF with very high specific compressive strength were produced from polyol with functionality of 12.4 that was synthesized by ring-opening reaction of epoxidized soybean oil using lactic acid as non-toxic agent (Herrán et al. 2019).

Rojek and Prociak (Rojek and Prociak 2012) also investigated the effect of two bi-polyols (I and II) that differed in hydroxyl values (114 and 196 mg KOH/g) and functionality (2.5 and 5.2), respectively, derived from rapeseed oil via epoxidation/ring opening reaction on the mechanical properties of flexible PUF. They found that foams prepared by polyol with low hydroxyl and functionality had higher elongation at break and resilience, while foams derived from polyol with higher hydroxyl value and functionality possessed higher strength, tensile, compression and excellent cell structure.

Also, rigid PUF with good physic-mechanical properties have been obtained from lignin based polyols, but with limited applications due to its tendency to sinterisation and being very dark in color (Ionescu 2005). In turn, it was used in hybrid polyols (lignin, castor oil and crude glycerol) to synthesize semi-rigid PUF and it was observed that foams with 17.5% lignin showed the best thermal and dimensional properties (Carriço et al. 2016). Furthermore, it was demonstrated that increasing the lignin content increases the Young's moduli, compression strength, and tear strength of lignin based PUF (Xu and Ferdosian 2017). However, beyond 3% lignin, the tensile and elongation at break dramatically decreased.

In recent study, microbial oil produced from CG were converted to polyol and used to synthesize rigid and semi-rigid PUF successfully, but the study did not represent any properties of the products (Uprety et al. 2017). CG based polyol were also used to produce PUF for wood adhesives applications with properties comparable to petroleum based analogies (Cui et al. 2017). Additionally, CG were used as liquefier solvent to synthesis PUFs and compared to pure glycerol; the performance of CG were remarkably better and the produced PUF possessed a density of 0.034 g/cm^3 and a compressive strength of 150 kPa (Hu et al. 2012).

Pawar and his team investigated the thermal properties of rigid PUF produced from bio-polyols derived from algae oil, and the results prevailed that the thermal properties obtained were similar to the properties of PUFs obtained from petroleum analogue (Pawar et al. 2016a). In addition, PUFs with good properties were prepared successfully from polyols derived from wastes. In the work of Paruzel and his team (Paruzel et al. 2017) low density rigid PUFs were prepared using recycled polyols derived from automotive waste which replaced 50% of the petroleum analogue used in the production process. The results showed foam with high compressive strength (>350 kPa), water absorption (~0.35 vol %), high content of closed cells (>91 vol %) and thermally stable up to 220 °C. Also, rigid PUF were successfully produced from waste cooking oil with maximum stress and load of 0.03 MPa and 0.09 kN (Enderus and Tahir 2017).

Sustainability proves to be a good choice not only from an economical perspective but also from an environmental and industrial prospective. Many studies reveal the excellent properties provided by high purity, high functional bio- polyols obtained by using more sustainable routes. Polyols produced via thiol-ene click used to synthesize PUF with excellent performance. Unmodified corn oil were converted to bio-polyols with 99% yield, without using solvent and in one step; a higher closed cell content (>95%), uniform cellular structure, appropriate density (~35 kg/m^3), moderate compression strength (120–81 kPa) and with high flame retardant PUF were obtained (Ramanujam et al. 2019). Additionally, a hyperbranched thermoplastic PUF with high toughness, excellent transparency, and good elastomeric properties were obtained by converting vegetable oils via thiol-ene click reaction (Alagi et al. 2016). Furthermore, synthetic route with no solvent nor catalyst of a hybrid castor- soybean based polyols via epoxidation /ring opening reaction, produced PUF exhibiting high glass transition temperatures, thermal stability, Young's modulus, and tensile strength (Zhang et al. 2013).

These attempts will pave the way toward a promising green PUF with great potentials. However, many areas such as using microreactors require further investigation. Attempts for exploring renewable alternatives for other constitutes that participating in synthesizing PUFs are reported. This chapter briefly covers the efforts of using renewable materials for deriving the main constitutes used in fabrication of PUFs.

3 Chain Extender and Plasticizer Based Renewable Resources

For improving PUFs properties, chain extenders are one of the essential materials that are used mainly for this purpose. They are a low molecular weight (up to 300 g/gmole), aliphatic or aromatic compound with terminated hydroxyls or terminated amines and short chain lengths compared to polyol. The vital roles of chain extender are increasing the molecular weight and extending the chain of the hard segment, where aromatic chain extenders offer harder material than aliphatic chain extender (Karak 2012b).

Bio based chain extenders such as 1,4-butanediol, 1,9-nonanediol,1,3-propanediol and 2,3-propanediol were derived from different biomasses successfully by fermentation (Furtwengler and Avérous 2018). Based on the available literature, petroleum-based chain extenders could be replaced by bio-based chain extenders without sacrificing PUFs properties. For instance, Rashmi et al. (2013) used 1,3-propanediol derived from corn as bio chain extender in order to synthesis thermoplastic PUFs. The properties of glass transition temperature, compression strength, modulus, and cell size of the product were unaltered in the case of using bio chain extender compared to petroleum-based chain extender.

In more sustainable work, Pawar et al. (2016b) explored the usage of ethylene glycol and lactic acid as bio based chain extenders and bio-polyol for synthesizing

PUFs and different blowing agents (water and cyclopentane) were also employed. Comparable results of decomposition behavior and thermal stability under the same condition were obtained, whereas glass transition temperature was higher in lactic acid and ethylene glycol samples with water as blowing agent compared to the samples used cyclopentane blowing agent and petroleum analogous.

Plasticizer is another functional additive widely used in the synthesis of PUFs in order to improve flexibility, processability, plasticity, and elongation. There are different bio based plasticizers derived successfully from various renewable resources such as vegetable oils and agriculture wastes (Bocqué et al. 2016). Based on the literature, ideal bio based plasticizer should satisfy certain points: (1) being nontoxic, (2) good miscibility with polymer, (3) the performance is comparable to the conventional plasticizers, (4) resistance to leaching from polymer and (5) low cost (Bocqué et al. 2016).

Table 5 summarize some important industrial bio-based plasticizers derived from different vegetable oils.

Table 5 Industrial bio based plasticizers (Bocqué et al. 2016)

Plasticizer trade name	Manufacturer	Feedstock	Bio based content (%)	Chemical name
Grinsted® SOFT-N-SAFE	Danisco	Castor oil	80	Acetylated castor oil
Kalflex® 14A	Varterco	Soybean oil	98	ESBO
Nexo® EO1	Nexoleum	Soybean oil	95	Methyl epoxy soyate
Resiflex® K50	Resypar	Soybean oil	100	Amyl epoxy soyate
PLS GreeV® 9	Petrom	Soybean oil	66	Nonyl epoxy soyate
DCS®	Jayant-Agro-Organics	Castor oil	100	Di-caprylsebacate
Radia®7295	Novance/Oleon	Rapeseed oil	78	Acetylated ester
Lankroflex® ED6	Akcros Chemicals	Soybean oil	68	Octyl epoxy stearate
—	Gerflor	Tall oil Sunflower oil	—	Tall-oil fatty esters Epoxidized
Virkoflex® 7010	Arkema	Soybean oil	—	Epoxidized soybean oil fatty esters
Hallgreen®	Hallstar	Linseed, soybean, castor oil	38–100	Fatty ester

Based on the aforementioned facts, PUFs can be synthesized in a more sustainable manner without scarifying the conventional properties of PUF based on petroleum analogues. Thus, further investigations for bio-based additives is a promising research area.

4 The Effect of Catalysis and Blowing Agent on the Polyurethane Foam Properties

Catalysts play fundamental role in precisely controlling the basic reactions i.e. gelling and blowing reactions in PUFs production, where the imbalance between those reactions produces foam with inappropriate cells (Dworakowska et al. 2014).

Generally, each catalyst has different catalytic activity toward gelling and blowing reactions. Two kinds of catalysts are widely used in synthesizing PU i.e. organometallics and amine catalysts. It was reported that amines catalyze blowing reactions better than gelation reaction, whereas organometallics are mainly used as gel catalysts although they influence blowing reactions (Dworakowska et al. 2014).

In PU production, tertiary amines are used widely to catalyze both basic reactions, where different types of tertiary amines are employed in synthesizing flexible and rigid PUFs such as 1, 4-diazobicyclo [2, 2, 2] octane (DABCO), dibutyltindilaurate (DBTDL), 1,2-dimethylimidazole 70% in EG (DM70), triethylenediamine (TEDA), tetra-metylethylenediamine (TE), bis(2-dimethylaminoethyl) ether (ETS), among others. The catalytic activity balances between both reactions for typical catalysts employed in PUFs applications are illustrated in Fig. 5.

Although, it was reported that amine catalysts are highly volatile, create toxic vapor and possess offensive odor (Dworakowska et al. 2014). On the other hand, the content of the catalyst has major effect on the properties of the final foam. It was found that using higher amount of organometallic catalyst dibutyl tin dilaurate (DBTDL) causes a slight increase of compressive strength, Young modules, and thermal conductivity in bio based rigid PUFs, whereas a small decrease in density was obtained in different samples with various types of blowing agents (Carriço et al. 2017).

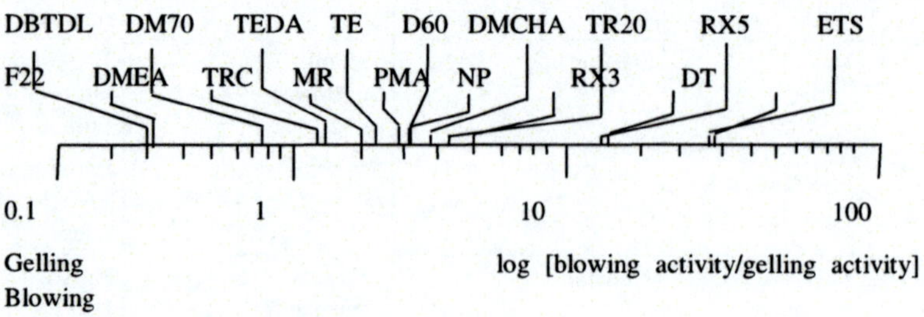

Fig. 5 Blowing/gelling catalytic activity ratio

In essence, the cellular structure of PUFs is produced by employing blowing agent through the foaming process. They are either externally added into reaction or produced in situ (Dutta 2018). Chlorofluorocarbons (CFCs) were used for a long time as typical blowing agent in production of PUFs, especially trichlorofluoromethane (CFC-11). Employing of CFC-11 produced PUF with incomparable properties that other plastic foam could not possess, such as low-density, closed cells, and good mechanical properties. However, it was reported that CFC-11 have negative impact on ozone depletion, which restricted their usage according to Montreal protocol. Subsequently, various types of chemical and physical blowing agents were developed as replacement to CFCs compounds. The most common blowing agents used are water and hydrocarbons, such as liquid carbon dioxide (CO_2), n-pentane, cyclopentane, and isopentane (Creazzo et al. 1995). Each blowing agent is used for synthesizing a specific type of PUFs, although it was reported that CO_2 and pentane could be used in synthesizing all types of foams (International 2012). Also, using different types of blowing agents may alter the final foam properties. For instance, a study investigated the impact of different physical and chemical agents (n-pentane, cyclopentane and water) in bio based rigid PUFs. It was observed that employing physical blowing agents (n-pentane and cyclopentane) produced foams with smaller cells and higher densities in comparison with the foams obtained with water (Carriço et al. 2017). The study also revealed that the type of blowing agent had significant effect on decreasing the density. In addition, the study found that increasing water level causes a decrease in compressive strength, Young's modulus, and tear strength.

5 Bio-additives

Despite the notable properties of PUFs, it is common to add bio-additives to enhance the properties for expanding their applications with increase PUF bio-based percentage. Numerous types of additives were investigated but will not be covered in this chapter, while using bio-based additives will be briefly demonstrated.

Different types of bio based additives such as egg shell wastes, walnut, cellulose, lignocellulosic fibers, textile wastes, hazelnut shells, and glass wool, just to mention a few were used in formulating PUFs to improve particular mechanical properties (Gama et al. 2018). Besides being eco-friendly, most of those materials are naturally available with low cost and unique properties. For instance, cellulose and walnut shells have a high surface area, low weight and low thermal conductivity as well as bearing –OH group (De Luca Bossa et al. 2020). In a recent study, cellulose and walnut shells as well as diatomite were used as natural additives for the synthesis of sustainable PUFs. The presence of cellulose and walnut shell enhanced the thermal stability and mechanical properties of the PUFs produced, where the stiffness in case of walnut shell was similar to flexible PUFs behavior. Meanwhile, diatomite showed reduction in thermal conductivity compared to other additives used, while an increase in stiffness was obtained (De Luca Bossa et al. 2020).

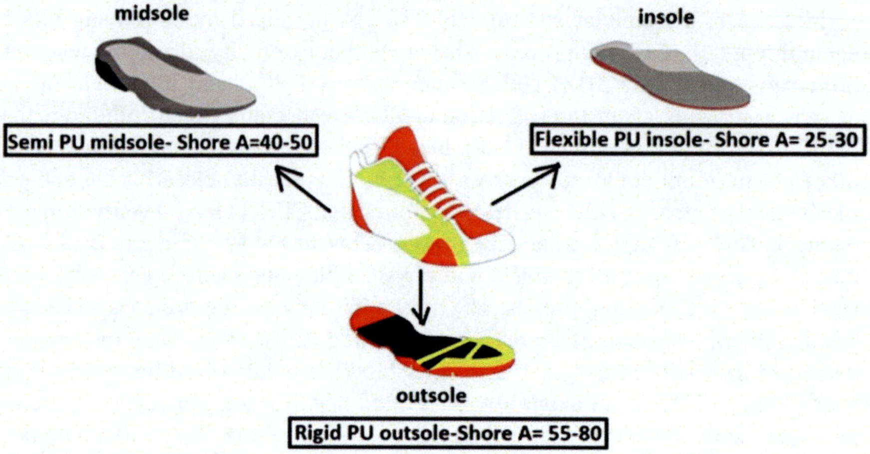

Fig. 6 Applications of TPU in footwear

Bio-additives are used in outsole footwear applications, where the material requires particular hardness and abrasion properties. Meanwhile, the PU foam is mainly dedicated to employ in flexible and semi-rigid materials. Adjusting parameters such as water, catalyst, and iso index are vital key to achieve specific properties for insole, semi-rigid. However, for outsole, the formulation is different as the chain extender, and polyol, isocyanate, catalyst, and additives are the main ingredients to tailor the mechanical properties. Figure 6 represented different hardness values for insole, midsole, and outsole.

6 Non-isocyanate

Safety and health concerns are associated with the use of isocyanate and so, there is a growing demand for synthesizing non-isocyanate PUs (NIPUs). Generally, there are three main approaches that have been investigated for this purpose: transurethanization, A–B type azide condensation, and aminolysis. (Furtwengler and Avérous 2018). NIPUs can be classified according to the original composition of materials to: renewable NIPUs, a hybrid NIPUs network, and chemically modified NIPUs (Gama et al. 2018).

To achieve 100% bio-based PU, a lot of attention towards the development of bio-based isocyanates with renewable resources will need to be investigated. For this purpose, two main routes that are explored include: (1) a self-condensation route of A-B type monomers, possess –OH with in situ formed isocyanate by Curtis rearrangement, and (2) ring opening of bi-functional cyclic carbonates, frequently via reaction of CO_2 with oxiranges such as limonene or epoxidized soybean with diamines (Sawpan 2018). It was reported that these materials possess a low permeability and an increased chemical resistance as well as improved thermal stability and waster absorption compared to conventional PUs (Gama et al. 2018).

Based on the literature, studies describing the synthesis of NIPU foams (NIPUFs) is still limited were three foaming processes described based on aminolysis reaction, and the main drawback of foaming NIPU is the absence of gas generation during the chemical process compared to conventional PUFs (Furtwengler and Avérous 2018). Also, it was reported that foam kinetic in NIPUFs requires long curing time compared to conventional PUFs. However, a study revealed that renewable self-blowing NIPUFs from sorbitol and lysine are used to produce rigid PUFs. The product proved to have low toxicity, was 100% bio-based and inexpensive with potential applications in packing and insulation as shown on Fig. 6 (Clark et al. 2018). The properties of NIPUFs is presented briefly in Table 6. In addition, PUFs cover different types of applications based on hardness and mechanical properties (Fig. 7). Adjusting demold time is an important part to use in mould technique.

There are several ways to replace chemical isocyanate by using NIPU (Fig. 8), and four synthetic pathways are designed, including (1) step growth polymerization (SGP) of bis(cyclic carbonates) and bis(cyclic amines), (2) (SGP) of linear activated decarbonates and diamines, (3) (SGP) of dicarbamates and diols, (4) ring opening polymerization of cyclic carbamates (Sardon et al. 2015).

Table 6 NIPUFs properties (Furtwengler and Avérous 2018)

Blowing agent nature	Foaming temperature (°C)	Foams' density (kg/m^3)	Foams' cell size (μm)	Foams' thermal conductivity (mW/m. K)
Chemical	80–120	197–295	Not available	Not available
Chemical	Room temperature	271–303	140–1300	100.4–115.5
Physical	50–80	142–219	135–171	Not available
Physical	80–100	110–176	2.6–10.8	50.1–64.7

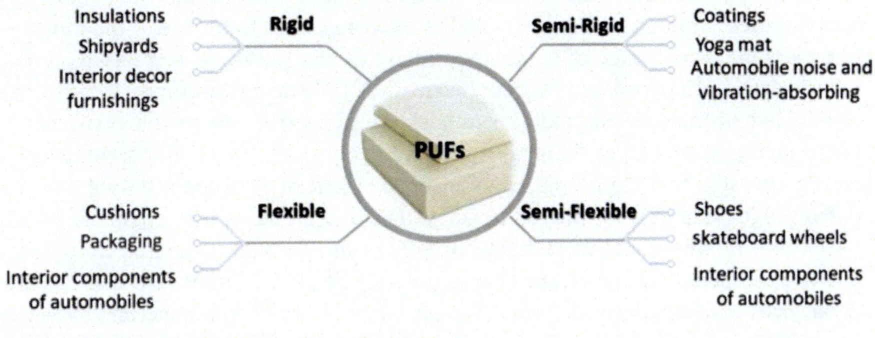

Fig. 7 Applications of different types of PUFs

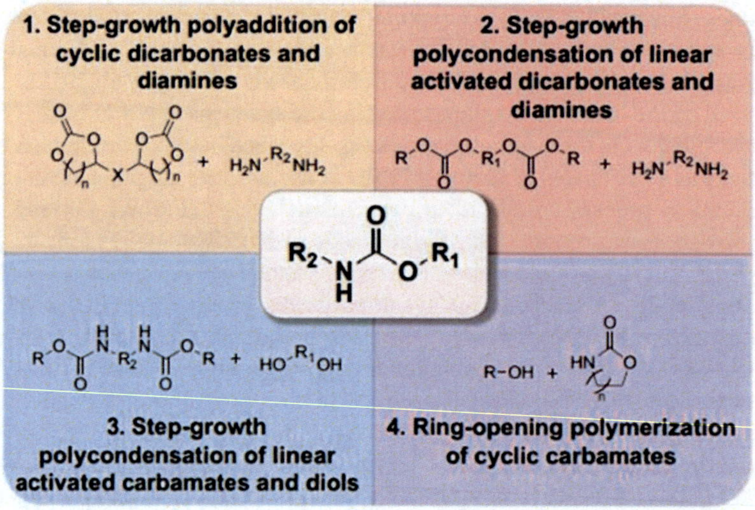

Fig. 8 Main synthetic routes toward non-isocyanate polyurethanes (NIPUs) adapted from (Sardon et al. 2015)

However, the synthesis of hybrid coating containing NIPU based on aliphatic amines and multicyclic carbonates with trade name Green Polyurethane® was reported, which is isocyanate and phosgene free (Rokicki et al. 2015). For further readings in this subject, the authors would like to suggest a review that comprehensively covers this topic (Rokicki et al. 2015).

7 Conclusion and Future Perspectives

Polymer engineering society is focusing these days in finding sustainable approaches for designing renewable, eco-friendly and biodegradable polymers with similar properties of their analogues. Currently, PUFs are considered to be one of the most used polymers in a wide range of applications and as other polymers it is based on materials derived from petroleum for their production. Following the current trend, several routes have been demonstrated for synthesizing PUFs from renewable resources and based on the results all of them are strongly encouraged. Two major approaches are widely investigated: (1) partial or total replacement of petroleum polyol with bio-based polyol, and (2) elimination of isocyanate and replaced with chemical systems (NIPUFs). However, the work with NIPUFs is still limited compared to bio-based polyol approaches. Currently, bio-based polyols are used at industrial scale as partial replacement, which derived from different types of renewable materials by various approaches. Also, in order to achieve greener PUFs other constituents participating in the production process such as chain extender, plasticizers, and additives were subjected for exploring bio replacements with encouraged results.

For the near future and for the benefit for mankind and the eco-system, all toxic and harmful chemicals will be banned similar to CFCs blowing agents. Thus, finding renewable alternatives with comparable properties is considered a crucial priority in PUFs industry and the efforts in this area is sustained.

Acknowledgements The authors would like to acknowledge the Department of Chemical Engineering at Ryerson University in Toronto and the Department of Chemical Engineering, Faculty of Petrolum and Natural Gas Engineering, University of Zawia, Libya.

References

Abril-Milán D, Valdés O, Mirabal-Gallardo Y et al (2018) Preparation of renewable bio-polyols from two species of Colliguaja for Rigid polyurethane foams. Materials (Basel) 11. https://doi.org/10.3390/ma11112244

Alagi P, Choi YJ, Seog J, Hong SC (2016) Efficient and quantitative chemical transformation of vegetable oils to polyols through a thiol-ene reaction for thermoplastic polyurethanes. Ind Crops Prod 87:78–88. https://doi.org/10.1016/j.indcrop.2016.04.027

Aniceto JPS, Portugal I, Silva CM (2012) Biomass-based polyols through oxypropylation reaction. Chemsuschem 5:1358–1368. https://doi.org/10.1002/cssc.201200032

Belgacem MN, Gandini A (2008) Monomers, polymers and composites, first. Elsevier Ltd., Oxford

Bocqué M, Voirin C, Lapinte V et al (2016) Petro-based and bio-based plasticizers: chemical structures to plasticizing properties. J Polym Sci Part a Polym Chem 54:11–33. https://doi.org/10.1002/pola.27917

Briones R, Serrano L, Ben YR et al (2011) Polyol production by chemical modification of date seeds. Ind Crops Prod 34:1035–1040. https://doi.org/10.1016/j.indcrop.2011.03.012

Carriço CS, Fraga T, Carvalho VE, Pasa VMD (2017) Polyurethane foams for thermal insulation uses produced from castor oil and crude glycerol biopolyols. Molecules 22. https://doi.org/10.3390/molecules22071091

Carriço CS, Fraga T, Pasa VMD (2016) Production and characterization of polyurethane foams from a simple mixture of castor oil, crude glycerol and untreated lignin as bio-based polyols. Eur Polym J 85:53–61. https://doi.org/10.1016/j.eurpolymj.2016.10.012

Clark JH, Farmer TJ, Ingram IDV et al (2018) Renewable self-blowing non-isocyanate polyurethane foams from lysine and sorbitol. Eur J Org Chem 2018:4265–4271. https://doi.org/10.1002/ejoc.201800665

Creazzo JA, Hammel HS, Cicalo KJ, Schindler P (1995) Zero-ODP blowing agents for polyurethane foams

Cui S, Liu Z, Li Y (2017) Bio-polyols synthesized from crude glycerol and applications on polyurethane wood adhesives. Ind Crops Prod 108:798–805. https://doi.org/10.1016/j.indcrop.2017.07.043

De Luca Bossa F, Santillo C, Verdolotti L et al (2020) Greener nanocomposite polyurethane foam based on sustainable polyol and natural fillers: investigation of chemico-physical and mechanical properties. Materials (Basel) 13. https://doi.org/10.3390/ma13010211

de Menezes AJ, Pasquini D, Curvelo AAS, Gandini A (2007) Novel thermoplastic materials based on the outer-shell oxypropylation of corn starch granules. Biomacromol 8:2047–2050. https://doi.org/10.1021/bm070389j

Desroches M, Escouvois M, Auvergne R et al (2012) From vegetable oils to polyurethanes: synthetic routes to polyols and main industrial products. Polym Rev 52:38–79. https://doi.org/10.1080/15583724.2011.640443

Dutta AS (2018) Polyurethane Foam Chemistry. Elsevier Inc.

Dworakowska S, Bogdał D, Zaccheria F, Ravasio N (2014) The role of catalysis in the synthesis of polyurethane foams based on renewable raw materials. Catal Today 223:148–156. https://doi.org/10.1016/j.cattod.2013.11.054

Enderus NF, Tahir SM (2017) Green waste cooking oil-based rigid polyurethane foam. IOP Conf Ser Mater Sci Eng 271. https://doi.org/10.1088/1757-899X/271/1/012062

Evtiouguina M, Margarida Barros A, Cruz-Pinto JJ et al (2000) The oxypropylation of cork residues: preliminary results. Bioresour Technol 73:187–189. https://doi.org/10.1016/S0960-8524(99)00158-3

Fang Z, Yang Z, Ji D et al (2016) Novel synthesis of a soy-based polyol for a polyurethane rigid foam. RSC Adv 6:90771–90776. https://doi.org/10.1039/C6RA20855J

Fernandes S, Freire CSR, Neto CP, Gandini A (2008) The bulk oxypropylation of chitin and chitosan and the characterization of the ensuing polyols. Green Chem 10:93–97. https://doi.org/10.1039/b711648a

Furtwengler P, Avérous L (2018) Renewable polyols for advanced polyurethane foams from diverse biomass resources. Polym Chem 9:4258–4287. https://doi.org/10.1039/c8py00827b

Gama NV, Ferreira A, Barros-timmons A (2018) Polyurethane foams: past, present, and future https://doi.org/10.3390/ma11101841

Gandini A, Belgacem MN (2002) Recent contributions to the preparation of polymers derived from renewable resources. J Polym Environ 10:105–114. https://doi.org/10.1023/A:1021172130748

Gómez-Rojo R, Alameda L, Rodríguez Á et al (2019) Characterization of polyurethane foam waste for reuse in eco-efficient building materials. Polymers (Basel) 11. https://doi.org/10.3390/POLYM11020359

Guo A, Javni I, Petrovic Z (2000) Rigid polyurethane foams based on soybean oil. J Appl Polym Sci 77:467–473. https://doi.org/10.1002/(SICI)1097-4628(20000711)77:2%3c467::AID-APP25%3e3.0.CO;2-F

He W, Fang Z, Ji D et al (2013) Epoxidation of soybean oil by continuous micro-flow system with continuous separation

Herrán R, Amalvy JI, Chiacchiarelli LM (2019) Highly functional lactic acid ring-opened soybean polyols applied to rigid polyurethane foams. J Appl Polym Sci 136:1–13. https://doi.org/10.1002/app.47959

Hu S, Luo X, Li Y (2014) Polyols and polyurethanes from the liquefaction of lignocellulosic biomass. Chemsuschem 7:66–72. https://doi.org/10.1002/cssc.201300760

Hu S, Wan C, Li Y (2012) Production and characterization of biopolyols and polyurethane foams from crude glycerol based liquefaction of soybean straw. Bioresour Technol 103:227–233. https://doi.org/10.1016/j.biortech.2011.09.125

International P (2012) Natural foam blowing agents sustainable ozone-and climate-friendly alternatives to HCFCs proklima international. Deutsche Gesellschaft für Internationale Zusammenarbeit (GIZ) GmbH

Ionescu M (2005) Chemistry and technology of polyols for polyurethanes. Rapra Technology Limited, Shawbury, UK

Ionescu M, Radojčić D, Wan X et al (2016) Highly functional polyols from castor oil for rigid polyurethanes. Eur Polym J 84:736–749. https://doi.org/10.1016/j.eurpolymj.2016.06.006

Ionescu M, Wan X, Bilić N, Petrović ZS (2012) Polyols and rigid polyurethane foams from cashew nut shell liquid. J Polym Environ 20:647–658. https://doi.org/10.1007/s10924-012-0467-9

Ji D, Fang Z, He W et al (2015) Synthesis of soy-polyols using a continuous microflow system and preparation of soy-based polyurethane rigid foams. ACS Sustain Chem Eng 3:1197–1204. https://doi.org/10.1021/acssuschemeng.5b00170

Karak N (2012a) Vegetable oil-based polyurethanes. Veg Oil-Based Polym 146–179. https://doi.org/10.1533/9780857097149.146

Karak N (2012b) Vegetable oil-based polyesters. Veg Oil-Based Polym 96–125. https://doi.org/10.1533/9780857097149.96

Kong X, Liu G, Qi H, Curtis JM (2013) Preparation and characterization of high-solid polyurethane coating systems based on vegetable oil derived polyols. Prog Org Coatings 76:1151–1160. https://doi.org/10.1016/j.porgcoat.2013.03.019

Li H, (Charles) Xu C Yuan Z Wei Q, (2018) Synthesis of bio-based polyurethane foams with liquefied wheat straw: Process optimization. Biomass Bioenerg 111:134–140. https://doi.org/10.1016/J.BIOMBIOE.2018.02.011

Li M-F, Sun R-C (2010) Liquefaction and gasification of cereal straws. In: Sun R-CBT-CS as a R for SB and B (ed), Chapter 7.4. Elsevier, Amsterdam, pp 253–265

Li Y, Luo X, Hu S (2015) Springer briefs in molecular science bio-based polyols and polyurethanes

Matos M, Barreiro MF, Gandini A (2010) Olive stone as a renewable source of biopolyols. Ind Crops Prod 32:7–12. https://doi.org/10.1016/j.indcrop.2010.02.010

Monteavaro LL, Da Silva EO, Costa APO et al (2005) Polyurethane networks from formiated soy polyols: synthesis and mechanical characterization. JAOCS, J Am Oil Chem Soc 82:365–371. https://doi.org/10.1007/s11746-005-1079-0

Muller LC, Marx S, Vosloo HCM et al (2018) Rigid polyurethane foams from unrefined crude glycerol and technical lignins. Polym from Renew Resour 9:111–132. https://doi.org/10.1177/2041247918803187

Nadji H, Bruzzèse C, Belgacem MN et al (2005) Oxypropylation of lignins and preparation of rigid polyurethane foams from the ensuing polyols. Macromol Mater Eng 290:1009–1016. https://doi.org/10.1002/mame.200500200

Ng W, Lee C, Chuah C, Cheng S (2017) Preparation and modification of water-blown porous biodegradable polyurethane foams with palm oil-based polyester polyolNo Title. Ind Crops Prod 97:65–78. https://doi.org/10.1016/J.INDCROP.2016.11.066

Niu M, Zhao G, jie, Alma MH, (2011) Polycondensation reaction and its mechanism during lignocellulosic liquefaction by an acid catalyst: a review. for Stud China 13:71–79. https://doi.org/10.1007/s11632-011-0109-7

Park SK, Hettiarachchy NS (1999) Physical and mechanical properties of soy protein-based plastic foams. J Am Oil Chem Soc 76:1201–1205. https://doi.org/10.1007/s11746-999-0094-3

Paruzel A, Michałowski S, Hodan J et al (2017) Rigid polyurethane foam fabrication using medium chain glycerides of coconut oil and plastics from end-of-life vehicles. ACS Sustain Chem Eng 5:6237–6246. https://doi.org/10.1021/acssuschemeng.7b01197

Pavier C, Gandini A (2000) Urethanes and polyurethanes from oxypropylated sugar beet pulp. I. Kinetic study in solution. Eur Polym J 36:1653–1658. https://doi.org/10.1016/S0014-3057(99)00245-1

Pawar MS, Kadam AS, Dawane BS, Yemul OS (2016a) Synthesis and characterization of rigid polyurethane foams from algae oil using biobased chain extenders. Polym Bull 73:727–741. https://doi.org/10.1007/s00289-015-1514-1

Pawar MS, Kadam AS, Singh PC et al (2016b) Rigid polyurethane foams from cottonseed oil using bio-based chain extenders: a renewable approach. Iran Polym J (English Ed) 25:59–68. https://doi.org/10.1007/s13726-015-0401-9

Petrovic ZS (2008) Polyurethanes from vegetable oils. Polym Rev 48:109–155. https://doi.org/10.1080/15583720701834224

Petrović ZS, Cvetković I, Milić J et al (2012) Hyperbranched polyols from hydroformylated methyl soyate. J Appl Polym Sci 125:2920–2928. https://doi.org/10.1002/app.36232

Prociak A, Kurańska M, Cabulis U, Ryszkowska J, Leszczyńska M et al (2018) Effect of bio-polyols with different chemical structures on foaming of polyurethane systems and foam properties. Ind Crops Prod 120:262–270. https://doi.org/10.1016/J.INDCROP.2018.04.046

Ramanujam S, Zequine C, Bhoyate S et al (2019) Novel biobased polyol using corn oil for highly flame-retardant polyurethane foams. C 5:13. https://doi.org/10.3390/c5010013

Rashmi BJ, Rusu D, Prashantha K et al (2013) Development of water-blown bio-based thermoplastic polyurethane foams using bio-derived chain extender. J Appl Polym Sci 128:292–303. https://doi.org/10.1002/app.38183

Rojek P, Prociak A (2012) Effect of different rapeseed-oil-based polyols on mechanical properties of flexible polyurethane foams. J Appl Polym Sci 125:2936–2945. https://doi.org/10.1002/app.36500

Rokicki G, Parzuchowski PG, Mazurek M (2015) Non-isocyanate polyurethanes: synthesis, properties, and applications. Polym Adv Technol 26:707–761. https://doi.org/10.1002/pat.3522

Salleh WNFW, Tahir SM, Mohamed NS (2018) Synthesis of waste cooking oil-based polyurethane for solid polymer electrolyte. Polym Bull 75:109–120. https://doi.org/10.1007/s00289-017-2019-x

Sawpan MA (2018) Polyurethanes from vegetable oils and applications: a review. J Polym Res 25. https://doi.org/10.1007/s10965-018-1578-3

Serrano L, Alriols MG, Briones R et al (2010) Oxypropylation of rapeseed cake residue generated in the biodiesel production process. Ind Eng Chem Res 49:1526–1529. https://doi.org/10.1021/ie9016732

Szycher M (2013) Syzycher's handbook of polyurethanes, 2nd edn. Taylor & Francis Group, LLC

Tan HW, Abdul Aziz AR, Aroua MK (2013) Glycerol production and its applications as a raw material: a review. Renew Sustain Energy Rev 27:118–127. https://doi.org/10.1016/j.rser.2013.06.035

Uprety BK, Reddy JV, Dalli SS, Rakshit SK (2017) Utilization of microbial oil obtained from crude glycerol for the production of polyol and its subsequent conversion to polyurethane foams. Bioresour Technol 235:309–315. https://doi.org/10.1016/j.biortech.2017.03.126

Vollmer G (2010) Use of renewable raw materials with special emphasis on chemical industry, pp 1–58

Wang CS, Yang LT, Ni BL, Shi G (2009a) Polyurethane networks from different soy-based polyols by the ring opening of epoxidized soybean oil with methanol, glycol, and 1,2-propanediol. J Appl Polym Sci 114:125–131. https://doi.org/10.1002/app.30493

Wang Y, Wu J, Wan Y et al (2009b) Liquefaction of corn stover using industrial biodiesel glycerol. Int J Agric Biol Eng 2:32–40. https://doi.org/10.3965/j.issn.1934-6344.2009.02.032-040

Xu C, Ferdosian F (2017) Conversion of lignin into bio-based chemicals and materials

Yang Z, Feng Y, Liang H et al (2017) A solvent-free and scalable method to prepare soybean-oil-based polyols by thiol-ene photo-click reaction and biobased polyurethanes therefrom. ACS Sustain Chem Eng 5:7365–7373. https://doi.org/10.1021/acssuschemeng.7b01672

Zhang C, Ding R, Kessler MR (2014) Reduction of epoxidized vegetable oils: a novel method to prepare bio-based polyols for polyurethanes. Macromol Rapid Commun 35:1068–1074. https://doi.org/10.1002/marc.201400039

Zhang C, Xia Y, Chen R et al (2013) Soy-castor oil based polyols prepared using a solvent-free and catalyst-free method and polyurethanes therefrom. Green Chem 15:1477–1484. https://doi.org/10.1039/C3GC40531A

Effect of Biodegradable Adhesive on Bleached Agave Americana Fibre Reinforced Epoxy Composites

Nadendla Srinivasababu

Abstract The environment can be best conserved through the effective utilization of the naturally available renewable-based materials like plant fibres whose cells were surrounded by the rigid cell wall. The mechanical performance of the composites can be enhanced through different methods of fibre treatment like physical and chemical means. So, in this work, the author had taken efforts in treating the agave Americana fibres with sodium hypochlorite and this bleaching process makes the fibres impurities free and better adhesion with the matrix. A biodegradable adhesive i.e. white dextrin was coated over the reinforcements i.e. agave Americana, bleached agave Americana manually. Ambient dried, uncoated, coated fibres were dried in an oven for the removal of moisture in the reinforcements completely. These fibres were reinforced into aerospace epoxy resin for the fabrication of the composites as per ASTM D 790-03 and ASTM D 6110-08 standards for flexural and impact specimens respectively. Load-elongation/deflection curves describe the role of white dextrin on the flexural performance of the composites. Based on the properties determined, comparisons made from the flexural and impact tests the impact of different types of reinforcements was understood. Based on the experimental results, it was noticed that there is a clear role of white dextrin in the enhancement of the flexural strength, modulus. Further, bleaching chemical treatment of the agave Americana fibres enhanced the bonding of the white dextrin to them and more performance under three-point bending was achieved. However, the impact performance of the uncoated composites was superior to other types experimentally studied in this work.

Keywords Agave American fibre · Sodium hypochlorite · Bleaching · White dextrin · Biodegradable adhesive · Flexural properties · Charpy impact strength

N. Srinivasababu (✉)
Fibrous Composites Research Lab, Department of Mechanical Engineering, Vignan's Lara Institute of Technology & Science, Vadlamudi 522213, Andhra Pradesh, India
e-mail: drnsbabu_vlits@vignan.ac.in; sbabunva@gmail.com; cnjlms22@yahoo.co.in

M. Jawaid et al. (eds.), *Eco-Friendly Adhesives for Wood and Natural Fiber Composites*, Composites Science and Technology,
https://doi.org/10.1007/978-981-33-4749-6_13

1 Introduction

In earlier days several hundreds of natural fibres were collected and examined for fabricating clothes. One of the important classes of the natural fibres i.e. vegetable fibres originated from nature (Gordon Cook 1984). Renewability and biodegradability, low density are the important factors perceived by the researchers in utilizing them in different sectors like automotive, construction. Fibre type, variability with its chemical composition, damage during processing and testing equipment/procedure adopted are some of the key influencing factors which caused variability in the reported results (Hughes 2012).

Commercial full potential applications of natural fibres (Jochen et al. 2000) were drawn after modification of their surface or development of new matrix system(s). Some of the investigators (Luner and Sandell 1969) have regenerated cellulose films from viscose and cellulose acetate for testing cellulose. The wettability of the films/samples was influenced by the cellulose sources and their method of preparation, physical state.

Adhesion and compatibility between sisal fibres and matrix (Viviana Cyras et al. 2004) were enhanced with alkaline and acetylation treatments. They observed that elastic modulus, yield stress and elongation at break were increased and with alkali, acetylation treatment respectively. Rough surface topography was generated after sisal fibres were treated with 10% NaOH along with the increase in aspect ratio by the reduction in fibre diameter (Kuruvilla et al. 1996). The hydrophilic property of the fibre surface (Stefan et al. 2008) prevents the adhesion to hydrophobic synthetic polymers.

Moisture absorption (Ramadevi et al. 2012) of natural fibres was high due to the presence of hydroxyl and polar groups which retard the compatibility with matrices. Dewaxed abaca fibres (ethanaol:benzene, 1:2) are subjected to alkaline and acetylation treatment. Moisture absorption studies of them indicate that 67% resistance was noticed when compared with untreated fibre. Cellulose content (Sreekala et al. 1997) plays a predominant role in the mechanical performance of the composites along with the micro-fibrillar angle, crystallinity index. Alkali treatment on fibres results in the enhancement of amorphous cellulose at the expense of amorphous cellulose through an irreversible mercerization effect. Several numbers of micropores of 0.2 μm were observed on oil-palm EFB after silane treatment.

Sodium bicarbonate ($NaHCO_3$) 10 wt% solution was used to treat sisal fibres (Fiore et al. 2016). Carbonic acid and hydroxide ion formation during chemical reactions of the solution was slightly alkaline. The presence of OH groups in fibres corresponds to weak acids/alcoholic hydroxyls which was similar to mercerization treatment. Removal of hemicellulose, lignin results in good bonding between fibre and matrix and was observed after the chemical treatment of 120 h.

Researchers have done physical treatments like corona, plasma, electron beam irradiation on natural fibres for changing their surface properties without altering the chemical composition and enhances the bonding between reinforcement and matrix. High alpha, hardwood fibres were subjected to corona treatment at different current (mA) (Belgacem et al. 1994). Modification of cellulose fibre alone increases 24% yield stress when compared with untreated fibre composites.

The surface of the sisal fibres was modified by argon, air—plasma treatments (Yuan et al. 2002). An in interfacial shear strength was increased by about 20% under an air-plasma treatment for 30 s. For improving the compatibility between the sisal fibres and polypropylene matrix, fibres were subjected to argon, air plasma treatments using modified plasma cleaner PDC-32 to modify their surface (Yuan et al. 2004). They concluded that most of the air plasma treated fibre polymer composites had shown superior performance than argon plasma treated fibre polymer composites. Tensile, flexural strength of air plasma treated polypropylene composites was increased by 21%, 19% respectively. Different compositions of polypropylene and natural fibres—kenaf, hemp, flax, sisal were subjected to the doses ranges from 0 to 20 kGy by electron beam irradiation (Kim et al. 2007) at a speed of 300 mm/s and electron energy of 0.5 meV was employed. The tensile strength of the composites was increased with an increase in the dose of the beam and reaches to the minimum at a dose of 10 kGy and then the value was increased and reaches a maximum at 15 kGy due increase in the degree of adherence of natural fibre with the polypropylene matrix.

The researchers have struggled in enhancing the adherence of the natural fibres with various matrices through chemical, physical treatments which ultimately produce good mechanical properties. But there is a research gap exits in the utilization of adhesive(s) on the natural fibres and make them compatible with the resins/polymers. So, in this work, the author has introduced a biodegradable adhesive namely white dextrin and attempted to coat it on agave Americana fibres. A predetermined, measured quantity of dextrin and water was mixed and a solution was prepared. This adhesive solution was coated on the agave Americana natural fibres manually through sequential steps of wetting, soaking and pressing under mild pressure in a tub. Agave Americana fibres were also subjected to bleaching treatment and adhesive coating one after another. All these uncoated, adhesive coated, chemically treated and chemically treated cum adhesive coated fibres were air, oven-dried. The detailed procedures of each case is readable from sections to come in this chapter. All the dried reinforcements were utilized in the fabrication of the epoxy composites and tested under flexural and impact loadings.

2 Materials and Methods

2.1 Fibre, Matrix and Chemicals

Agave Americana (Bezazi et al. 2014) is weather-resistant when compared with the group of agave plants. Leaves of the plant are about 2.3 m long and 250 m wide and were grey-blue, 130 mm thick with sharp spines (more than 25 mm long) on the edges. Tough and rigid leaves yield fibres and were used to produce ropes, matting

Fig. 1 Agave Americana fibre after oven drying, cutting

cloth, and embroidery of leather. Fibres were extracted by mechanical, chemical, or water (running/sea/distilled) immersion methods. Agave Americana fibres extracted through earth buried procedure have 0 mm gauge length, 310 ± 93 μm diameter had shown the tensile strength, modulus of 132 ± 69 MPa, 1.81 ± 0.55 GPa respectively during a tensile test done at a crosshead speed of 1 mm/min. So, based on the good fibre properties reported in the literature, the author in this work has considered agave Americana fibre as the reinforcement and was purchased from the local market. The fibres used in this work were properly arranged as shown in Fig. 1.

Aerospace-grade epoxy resin LY 5052 and Aradur 5052 hardener was employed as matrix received from Sree Industrial Composite Products, Hyderabad, Telangana, India. The resin can be processed easily for making composites by hand lay-up method. Some of the important properties of these materials are low viscosity and easy impregnation of reinforcements, long pot life, good mechanical properties as given in the datasheet of Huntsman epoxy resin/hardener.

In 1774 Swedish chemist Carl Wilhelm Scheele (Ponzano 2007) discovered a gas namely 'dephlogisticated salt spirit' and thereafter French chemist Claude-Louis Bertholet in 1787 obtained bleaching solutions after dissolving the gas in water. An Englishman Davy renamed this gas 'Chlorine'. Gaseous chlorine was dissolved in water to produce a bleaching solution on an industrial scale. Sodium hypochlorite is the salt of 'hypochlorous acid' and the hypochlorite ion is the active bleaching agent. The reactions of NaOCl are given below (Khan and Ahmad 1996).

$$NaOCl + H_2O \rightarrow NaOH + HOCl$$

$$HOCl \rightarrow H^+ + OCl$$

$$HOCl + H^+ + Cl^- \rightarrow Cl_2 + H_2O$$

The Paper industry uses NaOCl treatment as one of the very common bleaching process aimed to enhance the brightness of the pulp and removes impurities from the natural fibres. Sodium hypochlorite was procured from Deluxe Scientific & Chemical Company, Vijayawada, Andhra Pradesh, India and used in this work for bleaching of the agave Americana fibres.

2.2 Dextrin Adhesive

Starch was heated in presence of little amounts of moisture and acid to obtain different products which were called 'dextrin' in generic (BeMiller 2003, pg. 1773). By using any starch dextrin can be made and are classified as white dextrin, yellow/canary dextrin and British gum. Dextrins are soluble in cold water and produce less viscous solutions when compared with their parent raw material i.e. starch and after drying they form a film. The combination of depolymerization/hydrolysis and transglycosylation i.e. molecular rearrangement results in the production of dextrins. Dextrins were normally used as adhesives for paper products. Heating a dried acidified starch result in the formation of white dextrin and they were used in food in little amount.

In the presence of a high concentration catalyst (Melissa and Conner 1994), at low temperatures generally, 120–130 °C, 3–7 h of roasting time white dextrins are produced due to the primary reaction i.e. hydrolysis of starch molecules. Small and highly branched dextrins are formed due to very small repolymerizaton occurs and forms white/buff powder with the degree of polymerization about 20. The solubility of white dextrin's in the water ranges from 1 to 95%.

In this experimentation white dextrin was purchased from Stardex Pvt. Ltd., Hyderabad, Telangana, India and used as a biodegradable adhesive coating over the agave Americana fibres and the procedure as detailed in the subsequent section.

2.3 Reinforcement—Adhesive Coating, Bleaching

Agave Americana fibres in as-is condition were kept in an oven, heated at 70 °C up to 2 h for removal of moisture and labeled as '**Type 1**' reinforcement. A predetermined amount of white dextrin was added slowly to the premeasured quantity of water in a tub. Using a steel stirring rod, the dextrin and water were thoroughly mixed and ensured that there was no agglomeration visually. Type 1 reinforcement was placed in a tub contains white dextrin solution of 0.2 wt/vol%. The fibres were kept longitudinally and tilted several times till they were completely wet with the dextrin solution up to 6 days 18 h 30 min. Then the fibres were squeezed manually and kept in an oven at an average temperature of 120 °C up to 4 h. The moisture content of the fibres was checked through touch for every 45 min during they were in the oven. Because the fibres overheating leads to a twist of fibres which creates the difficulty during composites fabrication. Dried, adhesive coated agave Americana fibres were named as '**Type 2**' reinforcement.

To remove the impurities from the fibres bleaching treatment was performed on agave Americana fibres. Sodium hypochlorite (NaClO) of research-grade was poured into the water as per calculations and a solution of 0.1% was prepared. Then the fibres were completely dipped in the solution and allowed to soak in it up to 3 days. However, care was taken for complete wet of the fibres by tilting them many times in a solution. After bleaching the fibres were washed with a huge quantity of water till they were free from the chemical solution and was called '**Type 3**' reinforcement. Here the fibres were divided into two bunches and the first bunch was allowed to dry in the ambient atmosphere up to 6 h and then kept in an oven up to 4 h. The second bunch of agave Americana fibres was kept in a white dextrin solution of 0.2 wt/vol% in concentration. These fibres were allowed to soak up to 1 day 20 h 55 min and ensured their wetness completely through moving them in a dextrin solution manually. Again the second bunch fibres were squeezed and kept in a heating oven at an average temperature of 140 °C up to 3 h. Dried fibres were taken out from the oven and labeled as '**Type 4**' reinforcement.

2.4 Preparation and Testing of Composites

Type 1, 2, 3 and 4 agave Americana fibres were taken at different weight contents and were straightened by mild application of the load. Then the mold releasing wax purchased from Carbon Black Composites, Waliv Phata, Vasai East, Maharashtra, India was applied twice. Then the fibres were kept in a prepared mold and the resin and hardener mixture (100:50) was poured on them. Air bubbles were removed by rolling action on the fibres that exist in the mold. Then a polyester sheet was placed over the mold and an additional matrix present in the mold was spilled out after about 200 N load was placed on the mold. The mold was allowed to cure up to 24 h and the specimens were taken out. Using belt grinding machines sides of the specimens were finished and straight edges were obtained.

Though care was taken in the preparation of the composites there was a little variation in the thickness, width of the composites along the length of the specimens. So, based on the measurements taken at three different locations of the specimen i.e. (i) at a distance from one end; (ii) centre; (iii) at a distance from the other end as shown in Fig. 2a, b. The dimensions, average width and thickness of type 1 to 4 composites considered for flexural and impact test was given in Tables 1 and 2 respectively.

Fabricated composites specimens of type 1 to 4 were kept in a three-point-bend test attachment and tested at a crosshead speed of 2.3 mm/min according to ASTM D 790-03 standard. In this test, a support span of 80 mm distance was maintained with the radius of the supporter 5 mm. The test results were noted for further calculation of the flexural properties of the agave Americana fibre reinforced epoxy composites.

(a)
(b)

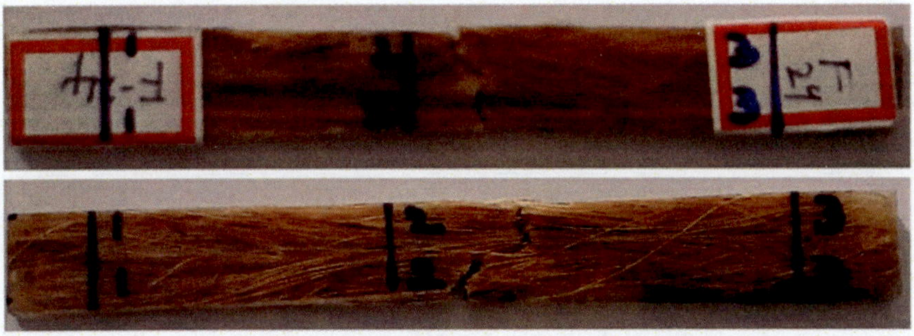

Fig. 2 Measurement of specimen dimensions at different sections, **a** flexural; **b** impact

Notch was cut up to a depth of 2.5 mm as per ASTM D 6110-08 on Type 1 to 4 agave Americana fibre reinforced epoxy composites using a notch cutter. Then the composite specimens were tested for knowing their impact energy performance on the Computerized Izod/Charpy impact tester. Care was taken to place the composite specimen on the simply supported beam. The pendulum was released according to the procedure for the Charpy impact test. The value of impact energy noted from the test was used for the calculation of the impact strength of the agave Americana epoxy composites.

3 Results and Discussion

Plain epoxy, type 1, 2, 3 and 4 reinforced epoxy composites were tested under a 3-point bending load and their flexural strength, modulus was calculated. Composites were also examined for determining their impact strength. Each mechanical property was elaborately discussed at each weight fraction of the composites hereafter.

3.1 Flexural Performance of the Composites

Plain epoxy and type 1, 2 3 and 4 epoxy composites were carefully analyzed for the load–deflection data to understand the impact of white dextrin coating on the reinforcements. Agave Americana fibre reinforced epoxy composites were flexural tested and load–deflection/elongation curves of a sample in each composition was drawn along with the no reinforcement as reference specimen i.e. plain epoxy. Plain epoxy specimen followed linear manner and reaches to the maximum load (Fig. 3) and fails from that load with no further displacement. At 5.87 wt% of type 1 reinforced epoxy composite's outer surface was broken at the maximum load of 178.49 N and then the load decreased drastically to 111.8 N (Fig. 4a) with a very minor increase in the

Table 1 Dimensions of the flexural test composite specimens considered for explaining load-elongation/deflection curves

Type of composites		Width at section 1-1 (mm)	Width at section 2-2 (mm)	Width at section 3-3 (mm)	Average width (mm)	Thickness at section 1-1 (mm)	Thickness at section 2-2 (mm)	Thickness at section 3-3 (mm)	Average thickness (mm)
Plain		14.547	14.011	13.041	13.866	4.537	4.510	4.520	4.522
Type 1	Composition 1	14.502	14.527	14.531	14.520	4.533	5.046	5.013	4.864
	Composition 2	14.522	14.501	14.509	14.510	5.034	5.014	5.044	5.031
	Composition 3	14.517	13.751	15.004	14.854	5.009	5.022	5.046	5.026
Type 2	Composition 1	14.925	13.751	14.872	14.516	4.659	4.652	4.730	4.680
	Composition 2	16.474	15.074	15.431	15.659	4.674	4.638	4.757	4.690
	Composition 3	15.175	14.723	14.442	14.780	4.826	4.833	4.878	4.846
Type 3	Composition 1	14.516	14.512	14.536	14.521	5.004	5.009	5.026	5.013
	Composition 2	15.517	15.010	15.413	15.313	5.045	5.250	5.053	5.116
	Composition 3	14.450	14.039	14.039	14.176	5.137	5.250	5.352	5.246
Type 4	Composition 1	15.005	15.040	15.021	15.002	4.549	5.070	5.021	4.880
	Composition 2	13.534	13.503	13.036	13.358	5.014	5.013	4.531	4.853
	Composition 3	14.002	14.512	14.035	14.183	4.535	4.535	5.008	4.693

Table 2 Dimensions of the random impact test composite specimens taken from each composition of the composites

Type of composites		Width at section 1-1 (mm)	Width at section 2-2 (mm)	Width at section 3-3 (mm)	Average width (mm)	Thickness at section 1-1 (mm)	Thickness at section 2-2 (mm)	Thickness at section 3-3 (mm)	Average thickness (mm)
Plain		14.547	14.505	14.031	14.361	4.547	4.543	4.547	4.546
Type 1	Composition 1	14.045	14.033	14.028	14.035	4.524	4.541	4.544	4.536
	Composition 2	14.931	15.137	15.274	15.114	4.838	5.142	5.035	5.005
	Composition 3	14.016	14.043	13.544	13.868	4.538	4.543	5.011	4.697
Type 2	Composition 1	14.374	15.046	15.336	14.919	4.946	5.033	5.019	4.999
	Composition 2	14.037	15.477	15.431	14.982	5.026	4.937	5.026	4.996
	Composition 3	15.515	14.502	14.532	14.850	5.139	5.127	5.029	5.098
Type 3	Composition 1	14.517	14.547	14.512	14.525	4.525	4.539	5.002	4.689
	Composition 2	14.533	15.526	15.005	15.021	4.531	4.535	5.006	4.691
	Composition 3	13.001	13.037	13.044	13.027	5.015	5.012	5.031	5.019
Type 4	Composition 1	13.020	13.047	13.017	13.028	5.005	4.549	4.549	4.701
	Composition 2	13.519	13.544	14.006	13.690	5.011	4.533	5.001	4.848
	Composition 3	13.511	13.036	12.528	13.025	5.010	5.003	5.009	5.007

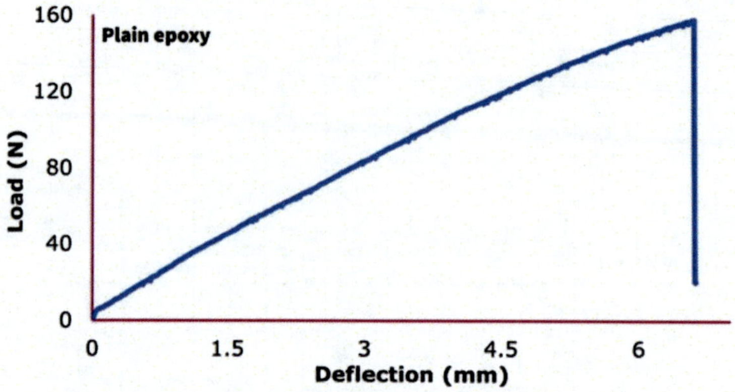

Fig. 3 Load–deflection curve of plain epoxy specimen

elongation from 4.79 to 4.83 mm. This lowest weight fraction composite had shown a gradual decrease in load with an increase in deflection during the test.

With an increase in elongation, the load was linearly increased in the case of type 1 epoxy composite at 13.9 wt% as observed in Fig. 4b. But after failure was initiated the composite's load-bearing ability was decreased rapidly with minor loading i.e. displacement changed from 6.43 to 7.42 mm. But at maximum content of the fibre the load–deflection curve was linear up to 9.966 mm deflection and load were more or less uniform up to 15.118 mm, the constant curve was visualized from Fig. 4c and a further increase in deflection lead to the sudden drop in the load till 17.394 mm deflection. Thence, the load was decreased gradually with an increase in elongation.

With the beginning of type 2 reinforcement linear (from 0 to 6.978 mm elongation) and non-linear (from 7.037 to 11.32 mm elongation) and constant curve (from 11.33 to 12.72 mm elongation) was shown in Fig. 5a and was failed with no reasonable indication of elongation. Though the reinforcement was about more than the doubled the load was increased linearly up to 5.925 mm deflection, Fig. 5b. Type 2 composites peak load was decreased slightly up to 10.721 mm deflection and then the specimen failed with no reasonable elongation. At maximum type 2 reinforcement, the composites had shown a gradual decrease in load (after the peak load, Fig. 5c) from 11.423 to 18.014 mm elongation before the complete failure. Type 3 reinforcement of 5.76% resulted in a linear increase in load with the deflection and the peak load of 193.2 N was shown at 5.25 mm, Fig. 6a. From the deflection of 5.483 mm, the curve was non-linear till maximum deflection was reached in the test. Figure 6b, shows that the specimen had shown a sudden decrease in the load after 9.838 mm deflection. The maximum reinforcement of type 3 composite had shown a clear decrease in the load from 7.869 to 10.883 mm deflection as noticed from Fig. 6c.

With minimum type 4 reinforcement, the composites had shown the sudden failure after 12.487 mm deflection, Fig. 7a. Beyond the peak deflection of 7.937 mm (Fig. 7b), the load was decreased gradually and reaches to the minimum before the

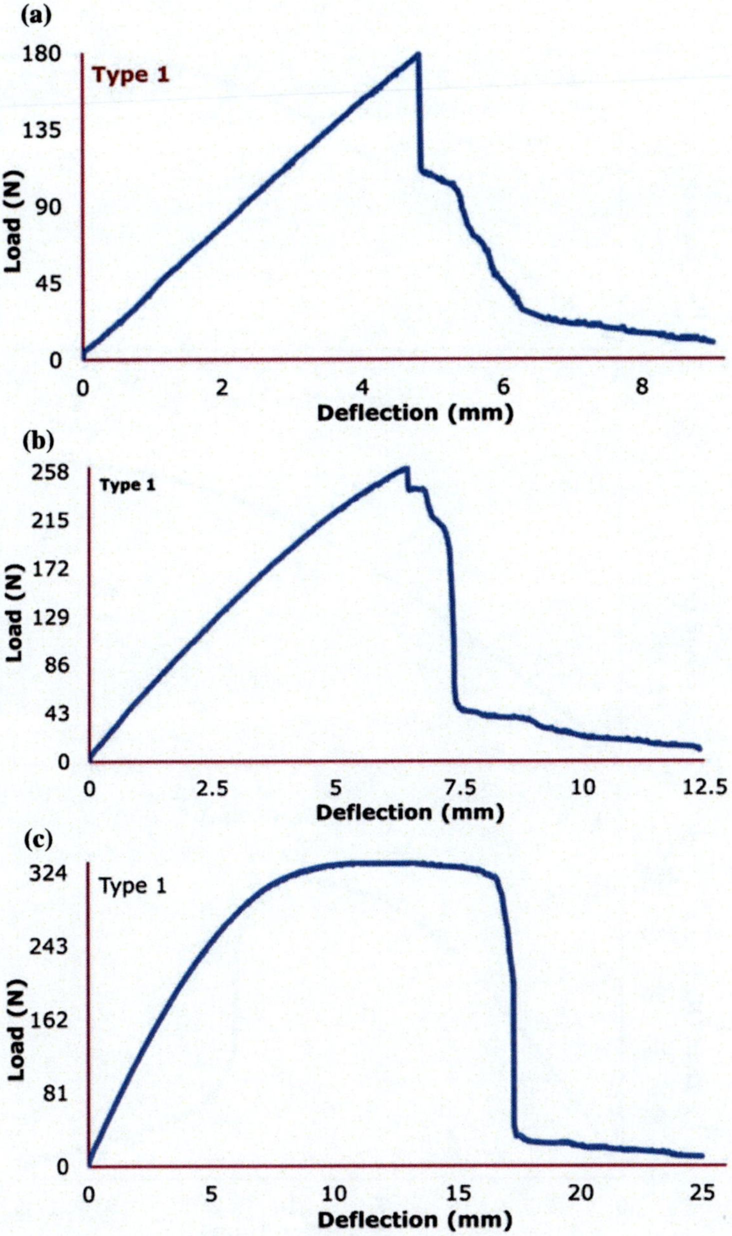

Fig. 4 Load–deflection curve of type 1 **a** composition 1; **b** composition 2; **c** composition 3 epoxy composite specimen

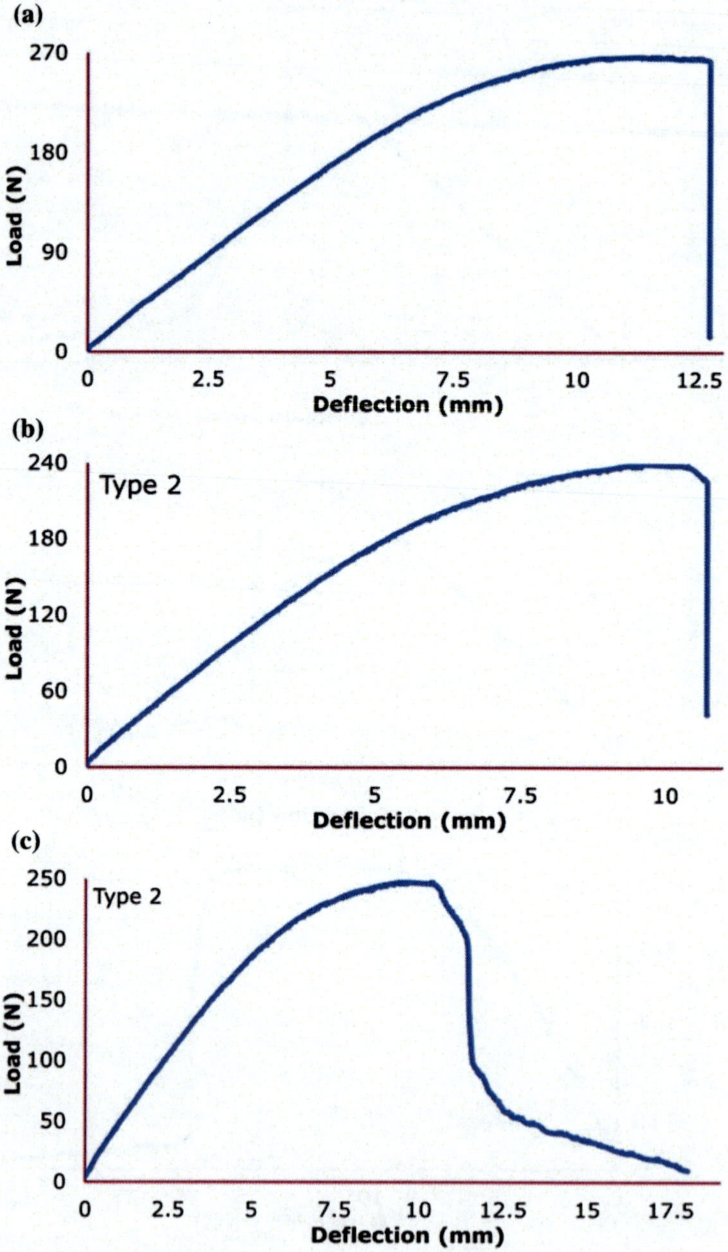

Fig. 5 Load–deflection curve of type 2 **a** composition 1; **b** composition 2; **c** composition 3 epoxy composite specimen

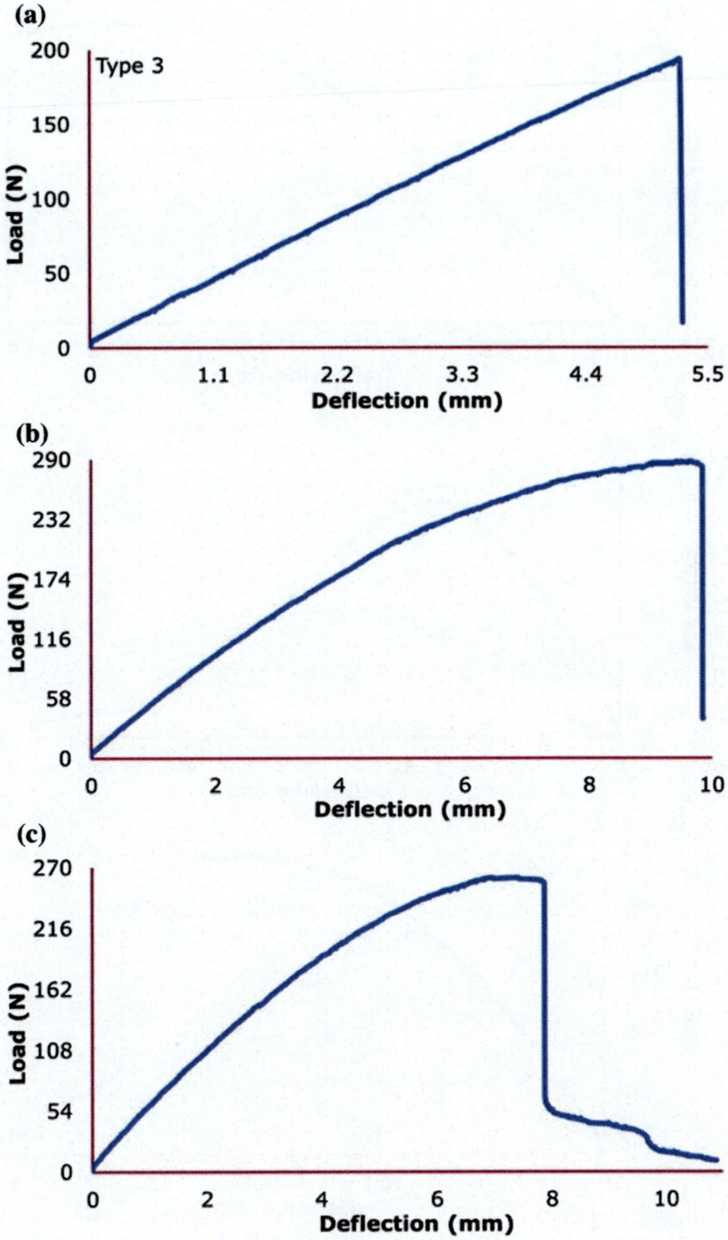

Fig. 6 Load–deflection curve of type 3 **a** composition 1; **b** composition 2; **c** composition 3 epoxy composite specimen

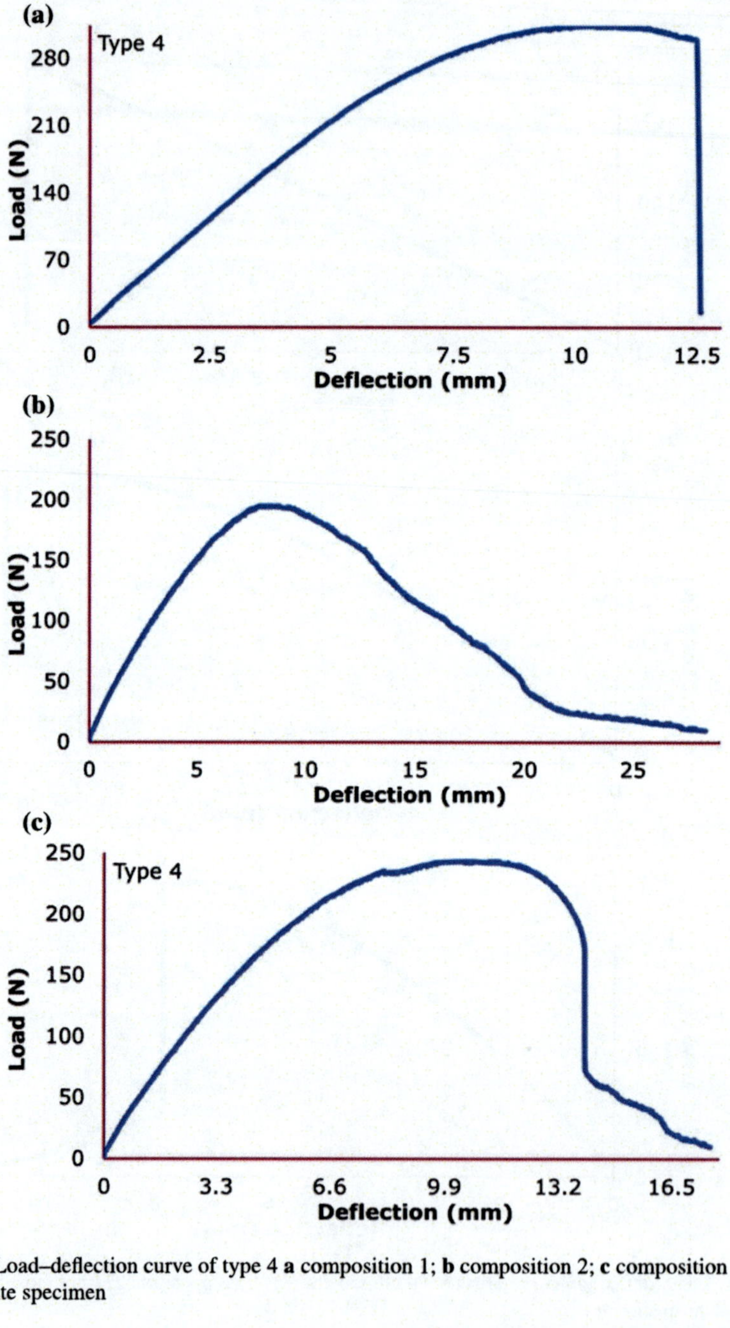

Fig. 7 Load–deflection curve of type 4 **a** composition 1; **b** composition 2; **c** composition 3 epoxy composite specimen

next type 4 level reinforced composite specimen test was stopped and the expected phenomenon. From 14.040 mm to 17.623 mm the type 4 composite specimen had exhibited a gradual decrease in load with the increased deflection, Fig. 7c.

From the load–deflection data analysis of type 1, composite specimens load at middle and maximum reinforcement was 1.46, 1.87 times more than the lowest reinforcement composites. With dextrin coating on the reinforcement, type 2 reinforced composite specimen load was 1.49 times more than the uncoated composites at the lowest weight fraction. But, within the type 2 composites, the second level of reinforcement resulted in a decrease in load and thereafter the load was increased at the maximum fibre content. This load was lower than that of the peak load obtained at 6.46% weight fraction of type 2 reinforcement. White dextrin coating increased elongation of the composite specimen when type 1, 2 reinforcements were compared at the lowest reinforcement content. Even then the dextrin coating of maximum level reinforcement i.e. type 2 had made the composites exhibit an indication of the load decrease with increased deflection.

The load bear by type 3 and 4 reinforced composites was increased and then decreased, and decreased and then increased respectively. Second, the lowest level of type 3, 4 reinforced composites had exhibited a maximum load of 288.3, 313.8 N respectively. Bleaching treatment on the agave Americana fibre i.e. type 4 created maximum impact on the elongation of the composites when compared with uncoated reinforced composites. White dextrin-coated reinforcement i.e. type 4 epoxy composites elongation was 2.39, 2.87 and 1.62 times more when compared with type 3 epoxy composites at lowest, second level and maximum reinforcements respectively.

With an increase in type 1 reinforcement, the flexural strength was increased up to the maximum content of the fibre. Whereas type 2 reinforcement of 6.46% epoxy composites flexural strength was 61.47% more than type 1 epoxy composites. This result indicates that the white dextrin may act as an interface between the reinforcements has improved the composite strength under bending load. With an increase in reinforcement to the next level of each type 1, 2 epoxy composites had shown more or less similar flexural strength, Fig. 8.

But at maximum reinforcement level type 1 epoxy composites flexural strength was 25.18% more than type 2 epoxy composites. At this junction of the type 2 reinforcement in the composites, the white dextrin coating may be insufficient or irregularly coated and resulted in inferior performance when compared with type 1 composites at the maximum fibre content. The flexural modulus of type 1 epoxy composites was increased from 2.518 to 3.847 GPa with an increase in fibre content from 0 to 20.78%.

At the lowest reinforcement content, type 2 epoxy composites flexural modulus was slightly more than that of the type 1 epoxy composites, Fig. 9. The flexural modulus of the type 2 composites at the second and third level of reinforcement was less by little amount when compared with type 1 composites. White dextrin had not much influenced the flexural modulus of the type 2 composites.

The flexural strength of type 3 epoxy composites was increased and then decreased with an increase in fibre content and the maximum flexural strength of 88.46 MPa was exhibited at 14.88% of the fibre as observed from Fig. 10. Bleaching has made

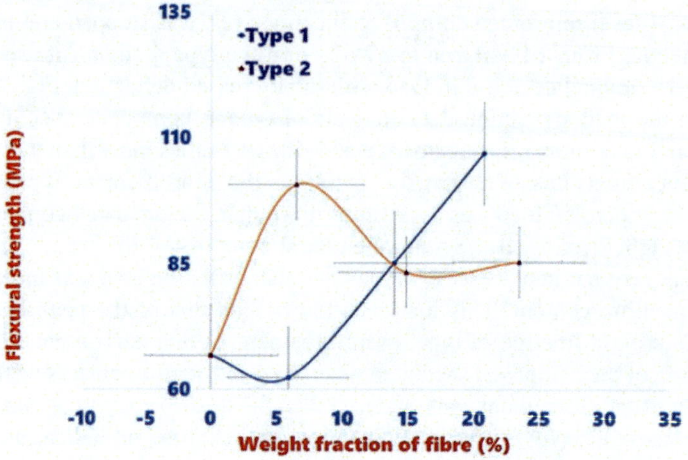

Fig. 8 Flexural strength of type 1, 2 epoxy composites with error bars

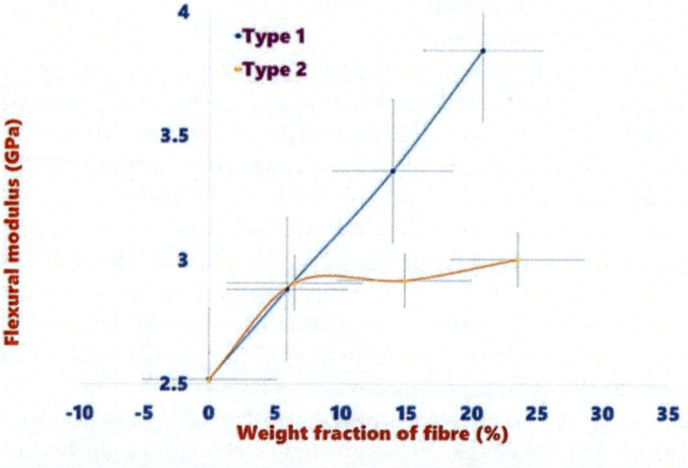

Fig. 9 Flexural modulus of type 1, 2 epoxy composites with error bars

the white dextrin to influence much on the flexural strength of the type 4 epoxy composites. At 5.58% of type 4 reinforced epoxy composites, flexural strength was 65.68% more when compared with the type 3 epoxy composites. The flexural strength of the type 4 composites was decreased and then increased with an increase in fibre content. Though the bleaching treatment has not much influenced the second level of reinforcement in terms of flexural strength at 22.67% maximum fibre content of type 4 reinforcement the flexural strength of the composites was 12.04% more when compared with type 3 epoxy composites.

Bleaching treatment on agave Americana fibres had enhanced the bond between the reinforcement and epoxy matrix. The flexural modulus of the composites was

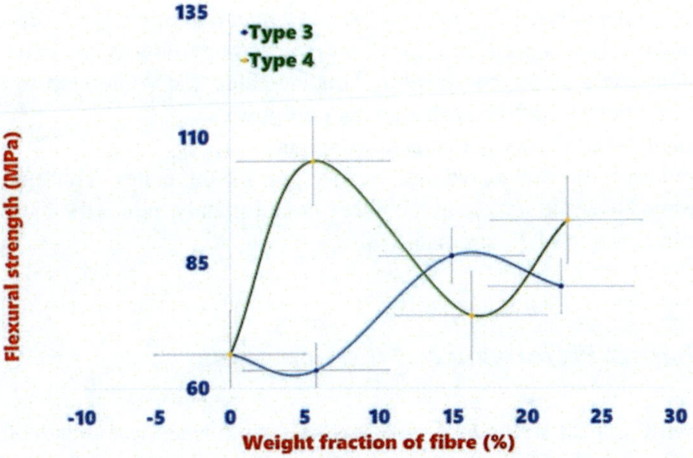

Fig. 10 Flexural strength of type 3, 4 epoxy composites with error bars

increased from 2.518 to 3.238 GPa when the type 3 reinforcement increased from 0 to 22.25 weight percentage as taken from Fig. 11. White dextrin coating on bleached fibres made the type 4 epoxy composites exhibit 28.63% more flexural modulus when compared with type 3 epoxy composites at the lowest weight fraction of the reinforcement. Flexural modulus of type 3, 4 composites at maximum weight content of

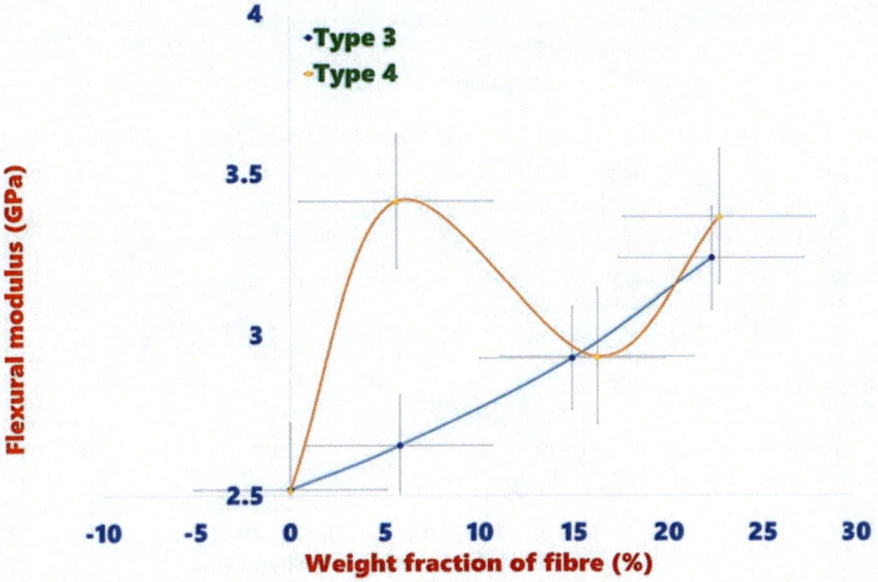

Fig. 11 Flexural modulus of type 3, 4 epoxy composites with error bars

reinforcement exhibited 7.75%, 12.01% more than that of the type 2 epoxy composites. However, type 4 epoxy composites flexural modulus was any extent more than that of the type 3 epoxy composites. This indicates that white dextrin improved in the flexural modulus of the type 4 epoxy composites.

Belched type 3 epoxy composites flexural strength was 24.98% less when compared with the raw agave Americana type 1 composites. This may due to the holes formed over the surface of the fibres and ultimately reduce their strength during loading in flexure (Al-Kaabi et al. 2005).

3.2 Impact Performance of the Composites

The Charpy impact strength of type 1 epoxy composites was drastically increased from 7.42 to 65.55 kJ/m^2 when the fibre content increased from 0 to 22.78%. White dextrin coating created an impact on the composites at 14.92 wt%. Here the impact strength of type 2 composites was 41.19% more when compared with type 1 composites. Except for this result, the remaining contents of the type 1 composites impact performance was better and more than that of the type 2 epoxy composites as recognized from Fig. 12.

The Charpy impact strength was increased with type 3, 4 fibre content in the epoxy matrix and was shown in Fig. 13. But the impact strength of type 3 epoxy composites was more than that of the type 4 epoxy composites. White dextrin coating

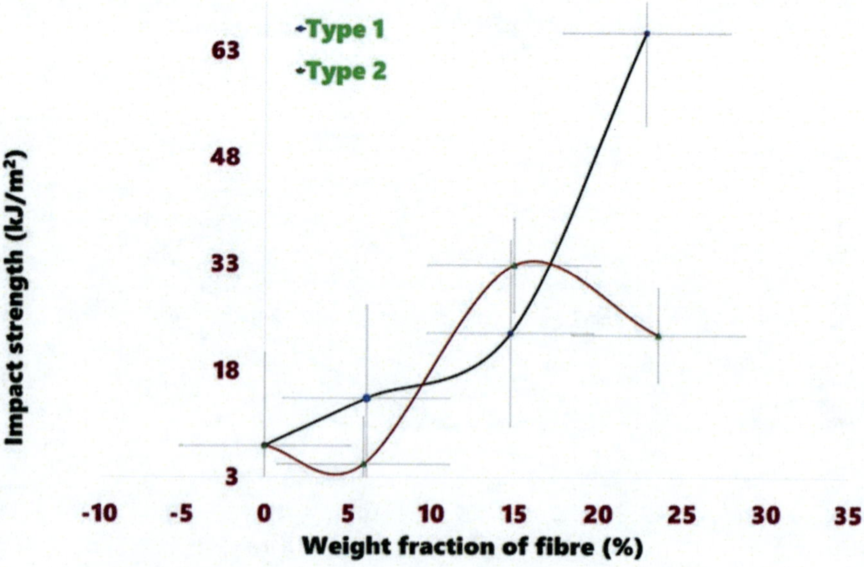

Fig. 12 Impact strength of type 1, 2 epoxy composites with error bars

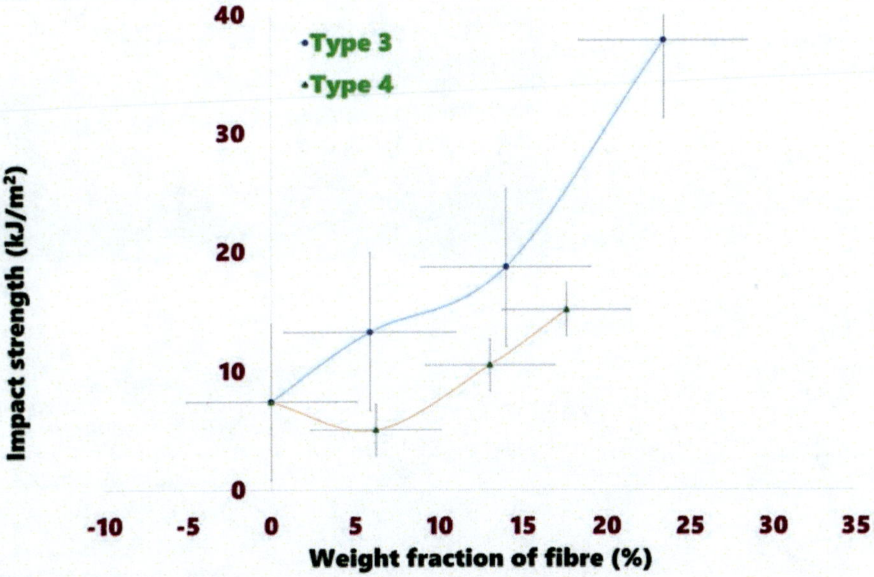

Fig. 13 Impact strength of type 3, 4 epoxy composites with error bars

on the reinforcement has not made any impact on the impact performance of the epoxy composites in a positive trend direction. Instead, it has decreased the impact performance of the composites.

The Charpy impact strength was decreased in the decreasing order of type 1, 2, 3 and 4 epoxy composites. White dextrin adhesive coating, bleaching, bleached-adhesive coating resulted in good enhancement of bonding between the reinforcements and matrix. This condition requires relatively low impact energy which may be varied depending on the bond strength between reinforcement and matrix (Sreekumar et al. 2009).

3.3 Flexural and Impact Specimens' Failure Surfaces

The tested specimen was randomly picked and physically seen and measured for its failure in thickness direction during flexural loading conditions using a steel rule. Plain epoxy showed brittle failure surface, Fig. 14. Type 1, composition 1 to 3 specimen failed and varied along with thickness based on the reinforcement, Fig. 15a–c. No fibre pull-out was observed in type 3, composition 1 (Fig. 16a) whereas composition 2 (Fig. 16b) little fibres were pulled out of the matrix. Type 2, composition 3 specimen failed up to a certain extent in thickness and fibres came out at fractured surfaces, Fig. 16c.

Fig. 14 Flexural fractured specimen of plain epoxy

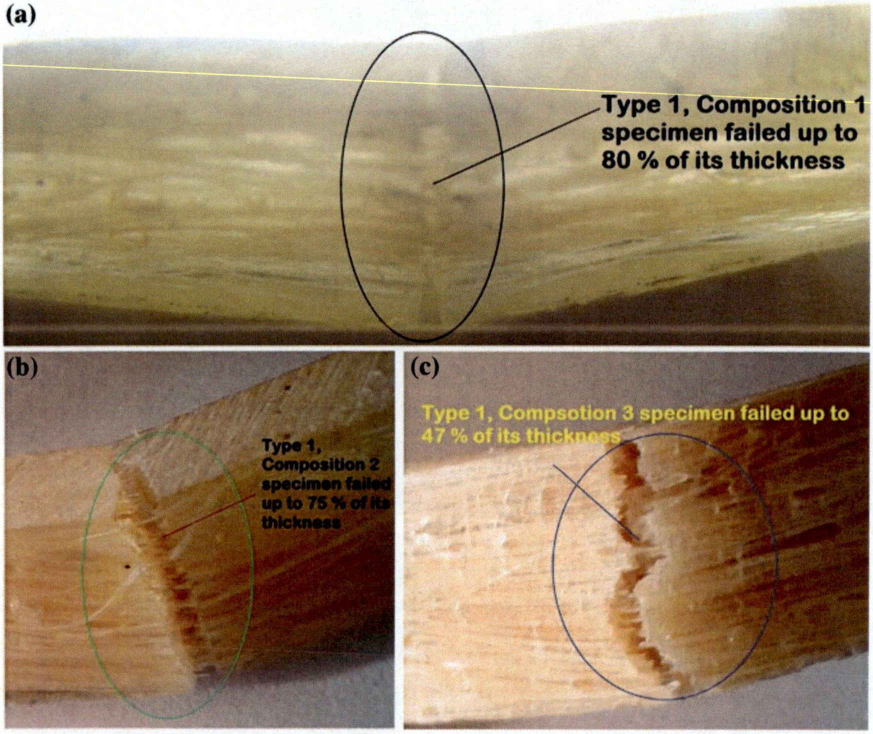

Fig. 15 Flexural fractured type 1 **a** composition 1; **b** composition 2; **c** composition 3 epoxy composite specimen

Fibres were not at all came out from the matrix and clear failure surface see in type 3, composition 1 specimen, Fig. 17a. Whereas composition 2 and 3 specimen (Fig. 17b, c) failed with little pull out of the fibres due to good interlocking with the matrix created by bleaching treatment on reinforcements. A similar trend was seen in type 4, composition 1 to 3 specimen, Fig. 18a–c.

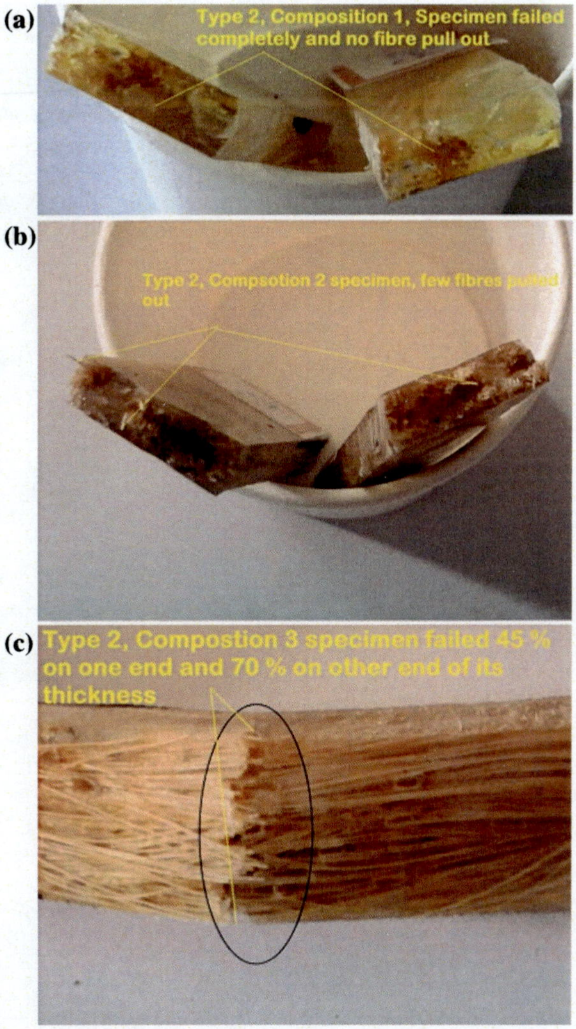

Fig. 16 Flexural fractured type 2 **a** composition 1; **b** composition 2; **c** composition 3 epoxy composite specimen

All the impact specimens (except type one composition) failed with 'complete break type' as per ASTM D6110-08, Fig. 19. Type 1, composition 1 specimen has shown 'hinge break' and was shown in Fig. 20a–c.

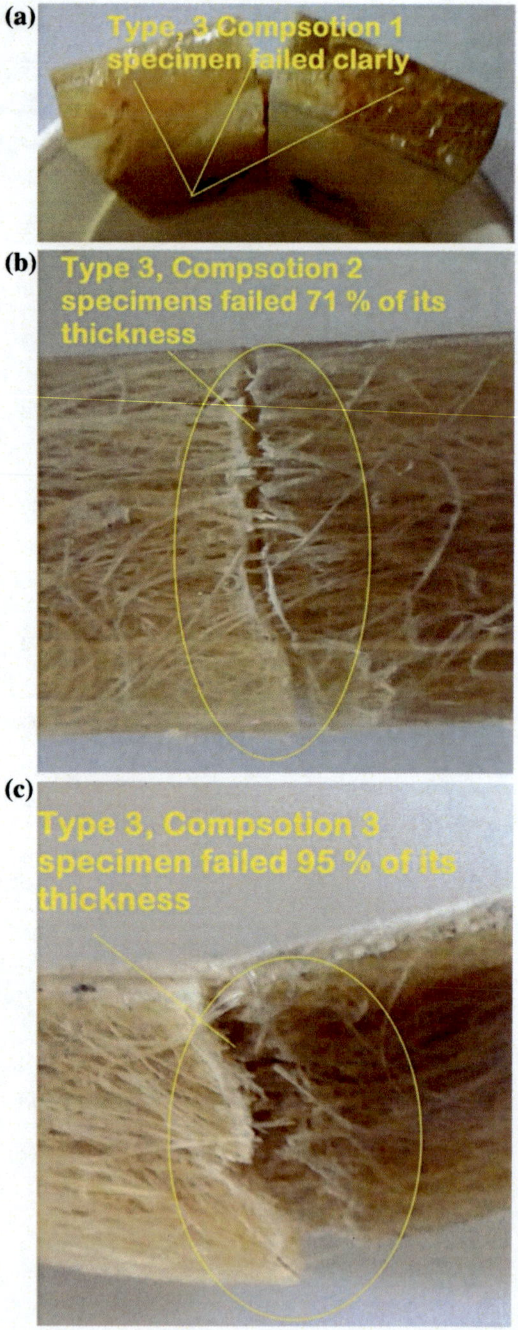

Fig. 17 Flexural fractured type 3 **a** composition 1; **b** composition 2; **c** composition 3 epoxy composite specimen

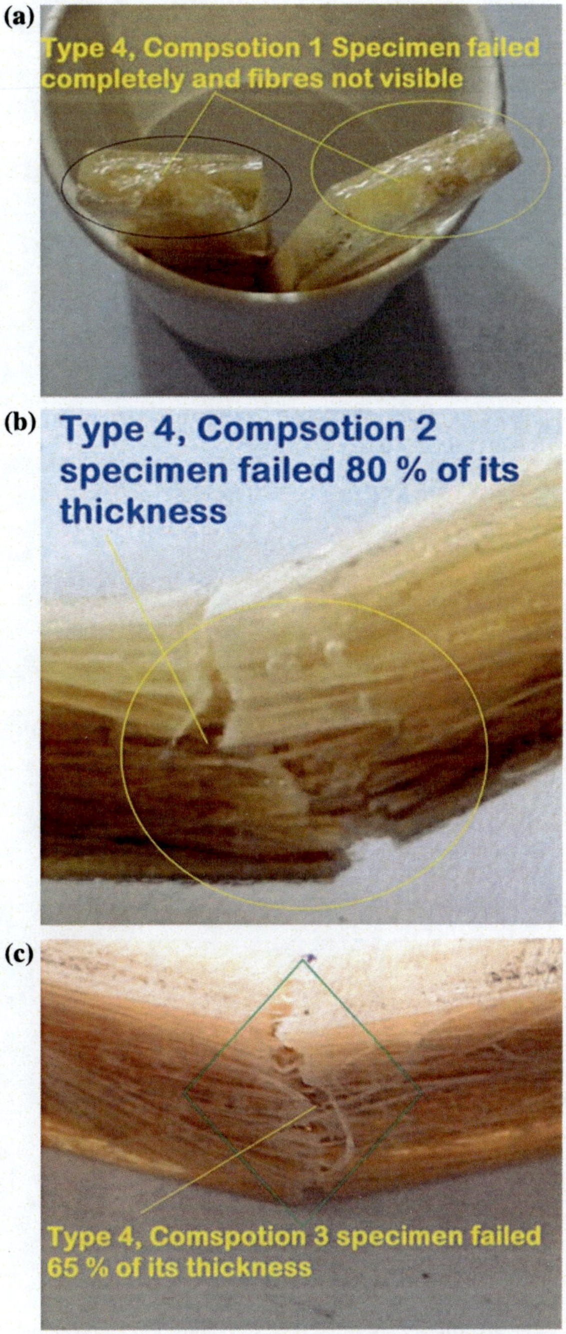

Fig. 18 Flexural fractured type 4 **a** composition 1; **b** composition 2; **c** composition 3 epoxy composite specimen

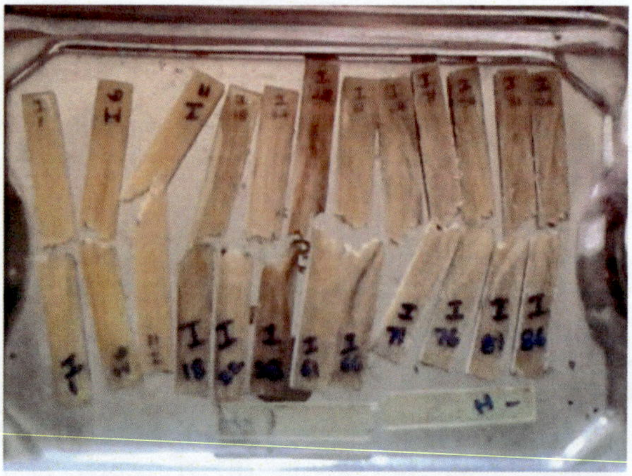

Fig. 19 Impact fractured type 1, 2, 3 and 4 and composition 1, 2, and 3 epoxy composite specimens

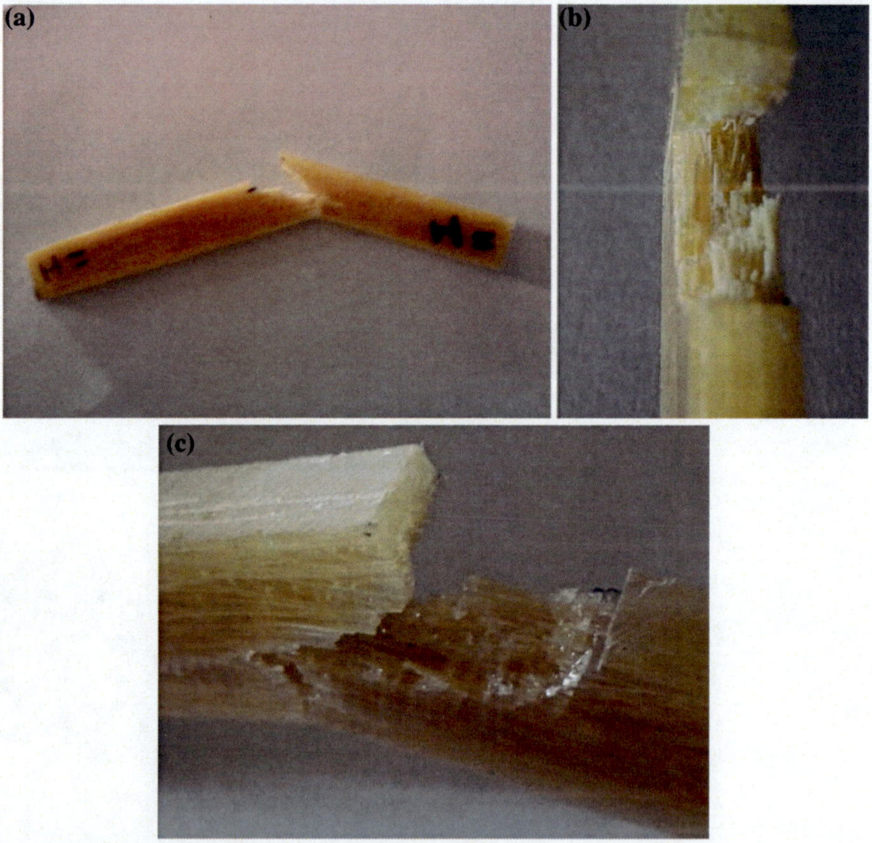

Fig. 20 Impact fractured type 1 composition 1 **a** specimen; **b** fibres pull out from one end; **c** no fibre pull out at other end

4 Conclusions

White dextrin coating was done on the agave Americana fibres manually with ease and composite specimens of type 1, 2, 3 and 4 were fabricated by wet lay-up successfully. White dextrin coated type 2 epoxy composites enhanced the flexural strength at the lowest weight fraction of the reinforcement in comparison to type 1 epoxy composites. There was a consistent increase in the flexural modulus of type 2 epoxy composites. The flexural modulus of bleached type 3 epoxy composites was slightly more when compared with type 2 epoxy composites where their reinforcement was dextrin coated. Dextrin based type 4 epoxy composites showed better flexural modulus than type 3 and 2 epoxy composites. Bleach treatment of the agave Americana fibres had resulted in better adhesion of the white dextrin and was result in more flexural modulus. Bleach treatment results in good bonding and the impact failure was 'clear break' type as per the standard. Nevertheless, there was an impact of white dextrin on the reinforcements used in this work and thereby the performance of the composites wasvaried in flexural and impact loading.

References

Al-Kaabi K, Al-Khanbashi A, Hammami A (2005) Date palm fibers as polymeric matrix reinforcement: DPF/polyester composite properties. Polym Compos 26:604–613

Belgacem MN, Bataille P, Sapieha S (1994) Effect of corona modification on the mechanical properties of polypropylene/cellulose composites. J Appl Polym Sci 53:379–385

BeMiller JN (2003) Dextrins. In: Caballero B (ed) Encyclopaedia of food sciences and nutrition. Academic Press, pp 1773–1775

Bezazi A, Belaadi A, Bourchak M, Scarpa F, Boba K (2014) Novel extraction techniques, chemical and mechanical characterisation of Agave americana L. natural fibres. Compos B Eng 66:194–203

Fiore V, Scalici T, Nicoletti F, Vitale G, Prestipino M, Valenza A (2016) A new eco-friendly chemical treatment of natural fibres: effect of sodium bicarbonate on properties of sisal fibre and its epoxy composites. Compos B Eng 85:150–160

Gordon Cook J (1984) Handbook of textile fibres: volume 1 natural Fibres. Woodhead Publishing, xi, xvii

Hughes M (2012) Defects in natural fibres: their origin, characteristics and implications for natural fibre-reinforced composites. J Mater Sci 47:599–609

Jochen G, Voytek Gutowski S, Andrzej Bledzki K (2000) About the surface characteristics of natural fibres. Macromol Mater Eng 283:132–139

Khan F, Ahmad SR (1996) Chemical modification and spectroscopic analysis of jute fibre. Polym Degrad Stab 52:335–340

Kim SW, Ohb S, Lee K (2007) Variation of mechanical and thermal properties of the thermoplastics reinforced with natural fibers by electron beam processing. Radiat Phys Chem 76:1711–1714

Kuruvilla J, Sabu T, Pavithran C (1996) Effect of chemical treatment on the tensile properties of short sisal fibre-reinforced polyethylene composites. Polymer 37:5139–5149

Luner P, Sandell M (1969) The wetting of cellulose and wood hemicelluloses. J Polym Sci: Part C 28:115–142

Melissa GDB, Conner AH (1994) Carbohydrate polymers as adhesives. In: Pizzi A, Mittal KL (eds) Handbook of adhesive technology. Marcel Dekker, pp 299–307

Ponzano GP (2007) Sodium hypochlorite: history, properties, electrochemical production. In: Ronco C, Mishkin GJ (eds) Disinfection by sodium hypochlorite: dialysis applications, vol 154. Karger, pp 7–8

Ramadevi P, Sampathkumar D, Srinivasa CV (2012) Effect of alkali treatment on water absorption of single cellulosic abaca fiber. Biresources 7:3515–3524

Sreekala MS, Kumaran MG, Sabu T (1997) Oil palm fibers: morphology, chemical composition, surface modification, and mechanical properties. J Appl Polym Sci 66:821–835

Sreekumar PA, Thomas SP, Saiter JM, Joseph K, Unnikrishnan G, Thomas S (2009) Effect of fiber surface modification on the mechanical and water absorption characteristics of sisal/polyester composites fabricated by resin transfer molding. Compos Part A Appl Sci Manuf 40:1777–1784

Stefan R, ElS A, Leif S, Basavaraju B (2008) Improvement of fibre-matrix-adhesion of natural fibres by chemical treatment. Macromol Symp 262:170–181

Viviana Cyras P, Vallo C, Jose Kenny M, Vazquez A (2004) Effect of chemical treatment on the mechanical properties of starch-based blends reinforced with sisal fibre. J Compos Mater 38:1387–1399

Yuan X, Jayaraman K, Bhattacharyya D (2002) Plasma treatment of sisal fibres and its effects on tensile strength and interfacial bonding. J Adhes Sci Technol 16:703–727

Yuan X, Jayaraman K, Bhattacharyya D (2004) Mechanical properties of plasma-treated sisal fibre-reinforced polypropylene composites. J Adhes Sci Technol 18:1027–1045